Jacobi-Davidson Type Methods
for Computing Rovibronic Energy Levels
of Triatomic Molecules

Zur Erlangung des akademischen Grades eines

Doktors der Naturwissenschaften

am Fachbereich Mathematik und Naturwissenschaften der
Bergischen Universität Wuppertal
genehmigte

Dissertation

von

Dipl.-Math. Peter Langer

aus Wuppertal

Tag der mündlichen Prüfung: 11. April 2008
Referent: Prof. Dr. Andreas Frommer
Koreferent: Prof. Dr. Bruno Lang

Bibliografische Information der Deutschen Nationalbibliothek

Die Deutsche Nationalbibliothek verzeichnet diese Publikation in der
Deutschen Nationalbibliografie; detaillierte bibliografische Daten sind
im Internet über http://dnb.d-nb.de abrufbar.

ISBN 978-3-8325-2251-3

Logos Verlag Berlin GmbH
Comeniushof, Gubener Str. 47,
10243 Berlin
Tel.: +49 030 42 85 10 90
Fax: +49 030 42 85 10 92
INTERNET: http://www.logos-verlag.de

Abstract

The task of solving the stationary Schrödinger equation is a longstanding and enormous challenge in many important areas of natural sciences. As explicit symbolic solutions of the operator eigenvalue problem are only attainable in very rare cases, one mostly has to resort to numerical techniques, especially to methods for giant Hermitian eigenvalue problems. In this thesis we are concerned with the specific case of triatomic molecules that exhibit the so called Double-Renner effect. To begin with, we explain the origin and the theoretical background of the abstract Schrödinger problem and we discuss viable techniques for the transition to suitable finite dimensional Hermitian matrices that approximate the original Hamiltonian in a reasonable fashion, and thus, make it accessible for a numerical treatment. However, due to tremendous storage requirements and computing times, that may soon extend to a couple of weeks, the use of conventional so-called direct solvers (QR method, RRR algorithm) is either not feasible or not advisable. Therefore, our main focus for the treatment of the matrix eigenvalue problem is on Jacobi-Davidson type methods that belong to the alternative class of iterative projection algorithms. Our aim is to show that these methods may be successfully applied in our context, in the sense that they are more efficient in terms of computing time than direct eigensolvers on the one hand, and the fellow algorithms of the iterative projection method class (Lanczos, Davidson, Olsen) on the other hand. To do so, we have to construct and to identify suitable preconditioners for the arising shift-and-invert systems that take advantage of the inherent information of the specific problem. Besides, efficient and problem-adjusted routines for matrix-vector multiplication are decisive for the success of our approach. Our ideas are illustrated and confirmed by extensive numerical experiments and results.

◇

Die Aufgabe, Lösungen der zeitunabhängigen Schrödinger-Gleichung zu bestimmen, ist eine schon seit langem bestehende enorme Herausforderung in vielen wichtigen naturwissenschaftlichen Themenfeldern. Da symbolische Lösungen des Operator-Eigenwertproblems nur in den seltensten Fällen explizit angegeben werden können, ist man zumeist gezwungen, auf numerische Techniken – insbesonders auf Verfahren für gigantische hermitesche Eigenwertprobleme – zurückzugreifen. In der vorliegenden Arbeit befassen wir uns mit dem speziellen Fall drei-atomiger Moleküle, die den sogenannten "doppelten Renner Effekt" aufweisen. Wir erläutern zunächst den Ursprung sowie den theoretischen Hintergrund des abstrakten Schrödinger Problems und diskutieren gangbare

Techniken für den Übergang zu endlich dimensionalen hermiteschen Matrizen, die den ursprünglichen Hamilton-Operator hinreichend gut approximieren und somit einer numerischen Behandlung zugänglich machen. Auf Grund des enormen Speicherbedarfs und der immensen Rechenzeiten, die sich schnell auf bis zu mehrere Wochen erstrecken können, ist die Anwendung konventioneller, so genannter direkter Löser (QR Verfahren, RRR Algorithmus) entweder nicht möglich oder nicht ratsam. Unser Hauptaugenmerk bei der Behandlung des Matrix-Eigenwertproblems liegt daher auf Varianten des Jacobi-Davidson Verfahrens, die zur alternativen Klasse der iterativen Projektionsmethoden zählen. Unser Ziel ist es zu zeigen, dass diese Verfahren in unserem Kontext erfolgreich angewendet werden können, d.h. dass sie – was die Rechenzeit anbelangt – zumeist effizienter sind als direkte Eigenlöser einerseits und die übrigen Verfahren aus der Klasse der iterativen Projektionsmethoden (Lanczos, Davidson, Olsen) andererseits. Um dies bewerkstelligen zu können, müssen wir geeignete Präkonditionierer für die auftretenden Shift-and-Invert Systeme identifizieren und konstruieren, welche die Problemspezifischen Informationen ausnutzen. Außerdem sind dem Problem angepasste Routinen für die Matrix-Vektor Multiplikation entscheidend für den Erfolg unseres Ansatzes. Wir erläutern unsere Ideen an Hand umfangreicher numerischer Experimente und Resultate.

Contents

II. Quantum Chemistry   119

5. Eigenvalue Problems in Theoretical Spectroscopy   121

6. The Double Renner Effect for Triatomic Molecules   159

1. Introduction

A fundamental principle of quantum mechanics (see [109] for instance) implies that physical systems at atomic and sub-atomic level almost always (cf. Example 5.34) only have discrete energy states which are characterized by the eigenvalues E_m of the *Hamilton operator* $\widehat{H}$ in the famous *Schrödinger equation*

$$\widehat{H}\psi_m = E_m\psi_m \qquad (1.1)$$

Knowing its solutions E_m enables one to explain the properties of atoms and molecules, the systems that we are primarily interested in, and in principle, it is possible to predict physical and chemical phenomena by purely computational means, without the help of any additional experiments (*ab initio calculation*). This explains the central importance of the Schrödinger equation and the fact that it has various applications in physics, chemistry and astronomy. A transition between two allowed states E_i and E_j arises when the energy difference

$$\Delta E = E_i - E_j = h \cdot \nu_{ij} \qquad (1.2)$$

is absorbed or emitted in the form of radiation with the frequency ν_{ij}. The resulting characteristic absorption and emission spectra are "finger prints" that allow for conclusions with respect to the material, e.g. in astronomical observations.

The field of *Theoretical Spectroscopy*, a branch of quantum chemistry, is concerned with the computational prediction of such spectra by solving (1.1), and in this thesis we will specifically examine the numerical computation of *rovibronic energy levels* for triatomic molecules that exhibit the so called *Double Renner effect* (see Section 6.1 for an explanation as well as for the etymology of *rovibronic*). More precisely, we will consider

- the **MgCN** molecule (as a representative of **ABC** type molecules, i.e. molecules with three different nuclei)

- the **HOO** molecule (as a representative of **ABB** type molecules, i.e. molecules with two identical nuclei)

Our considerations are building upon the PhD thesis by ODAKA [86] where a full theoretical account is given. Before we go into the details, let us have a closer look at the general problem for arbitrary N-atomic molecules and let us briefly outline the state-of-the-art with respect to its solution.

1

1.1. The General Problem for N-atomic Molecules

Unfortunately, the seeming simplicity of (1.1) by no means reflects the actual computational complexity of the problem and, about 80 years after SCHRÖDINGER formulated his fundamental equation [105, 106, 107, 108], its solution still poses an enormous challenge. Even modern computers often have trouble to cope with the tremendous storage requirements and computing times of several weeks are not unusal. The reasons for these difficulties are manifold: First of all, apart from very few exceptions, such as for the hydrogen atom, it is in general not possible to state explicit solutions in terms of closed analytic expressions (cf. discussion in Example 5.34, Section 5.3). For more complicated systems, one is forced to rely on a couple of compromises and approximations in order to make the problem tractable. This leads to simplified instances of the Schrödinger equation that may be solved numerically, e.g. using a *variational approach*: The Hamilton operator $\widehat{H}$ is projected onto a finite dimensional subspace of $L^2(\mathbb{R}^n)$ by means of a *Rayleigh-Ritz projection* such that one arrives at an ordinary Hermitian matrix eigenvalue problem

$$\mathbf{H}\widetilde{\mathbf{c}} = \widetilde{E}_m \widetilde{\mathbf{c}} \tag{1.3}$$

$\mathbf{H}$ is commonly referred to as an FBR (="Finite Basis Representation"), because it is the representation matrix of $\widehat{H}$ with respect to the basis of a finite dimensional subspace of $L^2(\mathbb{R}^n)$. For reasons that will be explained in Section 5.8, it is natural to employ a product basis with $3N - 6$ factor bases for N-atomic molecules which explains that one also speaks of a *product basis problem*. By MacDonald's theorem (a well-known variational principle in quantum chemistry, see Theorem 5.45 and Section 5.7) the *Ritz values* $\widetilde{E}_m$ are known to be upper bounds to the true eigenvalues E_m of $\widehat{H}$, and the larger one chooses the size n of the Hamiltonian matrix $\mathbf{H}$ the better these approximations $\widetilde{E}_m$ will be. The straightforward application of this variational approach, however, is impeded by some obstacles:

- Due to the inherent product basis structure already simple systems may lead to huge problem dimensions n and small changes in the sizes of the involved factor bases can make the size n "explode", as the following example nicely illustrates: A molecule with $N = 5$ nuclei has a matrix representation with respect to a product basis made up of 9 factor bases which, for the sake of simplicity, we assume to have equal sizes n_i. The overall problem size for the choice $n_i = 3$ is $n = 3^9 = 19,386$ and corresponds to 2.8 GB memory required for storing the corresponding FBR (provided that the matrix entries are represented by double precision variables). However, the computed approximations may be still too crude, and to obtain tighter upper bounds, one has to increase the basis sizes. The choices $n_i = 4$ and $n_i = 5$ result in total dimensions of $n = 4^9 = 262,144$ and $n = 5^9 = 1,953,125$, respectively. Clearly, the corresponding storage requirements (512 G and 28422 G) are now far beyond the possibilites of many modern computers.

- the computation of the matrix elements $\mathbf{H}_{ij}$ can be rather time consuming (due to the underlying algorithms for numerical integration)

A measure to cope with these difficulties may be a further reduction of the problem size by means of an additional subsequent Rayleigh-Ritz projection (called *contraction scheme*)

$$\mathbf{V}^*\mathbf{H}\mathbf{V} = \widetilde{\mathbf{H}} \tag{1.4}$$

Consequently, the resulting eigenvalue problem

$$\widetilde{\mathbf{H}}\widetilde{\mathbf{c}} = \widetilde{\widetilde{E}}_m\widetilde{\mathbf{c}} \tag{1.5}$$

is called a *contracted basis problem*. Obviously, the contraction defined by (1.4) is lossy, because the approximations $\widetilde{\widetilde{E}}_m$, in turn, are upper bounds to the solutions of the FBR problem $\widetilde{E}_m$ and the original problem E_m, i.e.

$$\widetilde{\widetilde{E}}_m \geq \widetilde{E}_m \geq E_m \tag{1.6}$$

However, if one succeeds in finding a suitable contraction matrix $\mathbf{V}$ and in making a reasonable trade-off between the problem size and the desired accuracy, this approach may lead to satisfactory results. A general description of contraction schemes along with a discussion of their pros and cons will be given in Section 5.8.

We have seen that the Schrödinger equation (1.1) can be reduced to ordinary finite-dimensional Hermitian eigenproblems (1.3) and (1.5) in order to obtain approximate solutions. Hence, apart from the construction of the matrix elements, the key problem lies in computing eigensystems and it is hardly surprising that efficient numerical algorithms are crucial for the success of a computational approach.

1.2. General Solution Strategies

As for virtually any numerical task, there is no philosopher's stone and there exists a rather wide variety of different approaches for computing eigensystems. Essentially, one can distinguish between

- direct methods

- iterative methods

If the problem under consideration is small, then *direct methods*, which are well-known from standard textbooks on numerical analysis (see [121, 122], [49] or [91], for instance), are preferred because they are black boxes and because they are superior in performance. There exists a couple of well-established algorithms which are available in form of reliable software, most notably the LAPACK library [2, 7]. The user only needs to supply the problem size n, the part of the spectrum he (or she) is interested in and has to store the matrix explicitly in memory. Whether or not a problem can be considered "small", in turn, is a matter of the available computer architecture, and thus the answer will change from case to case and from year to year. PARLETT [91] reports that in 1978 a matrix with the size $n = 12000$ was regarded "big" and that the computation of 30 eigenpairs was highly expensive. By contrast, about 30 years later, in the year 2007, it is not uncommon that computers have available several gigabytes of memory and the smallest matrix arising in our numerical experiments (see Chapter 7) has the dimension $n = 11952 \approx 12000$. Clearly, the storage requirements, $8 \cdot 12000^2$ Bytes $= 1.06$ GigaBytes, are no problem for the SUN^{TM} Fire workstation (with 32 GB memory), on which we carried out our numerical experiments. Using an efficient state-of-the-art direct solver the complete eigensystem can be computed in less than one hour. This example nicely illustrates the amazing technical and scientific progress that has been made over the last decades. Nevertheless, the principal limitations of direct methods still exist: Starting from a certain size n, explicit storage of the matrix is no more possible and, rather soon, the unfavorable time complexity $\mathcal{O}(n^3)$ of direct solvers becomes a severe drawback. Therefore, *iterative methods* (more precisely: *iterative projection methods*) may be an interesting alternative, because they rely on a different concept: Information of the matrix $\mathbf{H}$ is only accessed implicitly by means of matrix-vector products, and in fortunate cases, e.g. if $\mathbf{H}$ has a regular sparsity pattern, it is only necessary to store a few non-zero elements h_{ij}. The idea is to successively build up a small subspace $\mathcal{K} \subset \mathbb{R}^n$ by means of suitable expansions of $\mathcal{K}$ and to project the input matrix $\mathbf{H}$ onto $\mathcal{K}$ which, again, is accomplished by means of Rayleigh-Ritz projections. The resulting *interaction matrices* are small and their eigensystems can be easily computed using direct methods. A detailed description of this general concept and its ingredients will be given in Section 3.3. On the other hand, the convergence behavior of iterative projection methods may not be easy to predict, and in many cases quite a lot of knowledge on the properties and the structure of $\mathbf{H}$ is required, which makes these methods more difficult to use and less attractive for non-expert users. Furthermore, the methods are only suitable for computing small partial eigensystems, which fortunately meets our demands, as one is typically only interested in the lower part of the spectrum. An excellent state-of-the-art survey of iterative projection methods may be found in [8], further interesting material is collected in [100] and [125]. The historical evolution of eigenvalue methods is summarized in [48].

Of course both, direct and iterative projection methods, may be applied to the Hermitian eigenvalue problems (1.3) and (1.5) in our context, and to facilitate the discussion, CARRINGTON ET AL. [19] have introduced the following terminology for the four possible combinations:

- *direct-product approach*, i.e. application of a direct solver to (1.3)

- *direct-contracted approach*, i.e. application of a direct solver to (1.5)

- *iterative-product approach*,
  i.e. application of an iterative projection method to (1.3)

- *iterative-contracted approach*,
  i.e. application of an iterative projection method to (1.5)

1.3. Objective of the Thesis

Let us now come back to our problem, the computation of energy levels for triatomic molecules with the *Double Renner effect*. Apart from the theoretical analysis of the *Double Renner effect*, ODAKA describes in her thesis [86] the FORTRAN 90 [85] software DR, which she developped to determine energy levels of arbitrary triatomic molecules. Specifically, the numerical results for the **MgCN** and the **HOO** molecule (the examples she examines in her thesis) have been computed with the aid of her software. After the user has supplied some input data (see input files in Appendix A.5 for full detail), such as

- the type of the molecule (**ABC** or **ABB** type)

- the masses of the involved nuclei

- a potential energy surface (see Sections 5.5 and 6.1)

- the sizes of the factor bases

- parameters/thresholds for numerical integration

the software proceeds in three steps:

1. it computes the non-zero blocks of the FBR $\mathbf{H}$ using numerical integration schemes

2. it determines a block diagonal contraction matrix $\mathbf{V}$ and computes the non-zero blocks of the contracted Hamiltonian matrix $\widetilde{\mathbf{H}}$

3. it determines the energy levels of interest (specified by some upper bound) by computing the eigenvalues of $\widetilde{\mathbf{H}}$ with the expert driver xSYEVX of the LAPACK library [2].

Hence, with respect to our terminology, the program DR pursues a *direct-contracted approach*. Of course, it is natural to ask about the use of the other algorithmic combinations listed above and, secondly, what eigensolver one should choose as an ingredient in each case. An attempt to answer the question for the *direct-contracted* case is made in [87], where the authors optimize the numerical computation of energy levels in the DR program. They propose to replace the LAPACK [2] standard approach by a combination of the two-stage tridiagonalization introduced in [12] and [13] and the recent RRR algorithm for symmetric tridiagonal eigenproblems [32]. In principle, the authors' recommendation is not only valid for the rather specific problem of triatomic molecules, but also carries over to all eigenproblems where a relatively small fraction of the total spectrum is sought-after (e.g. 6%-10% of the smallest eigenvalues). In other words, product-basis problems may be addressed in the same fashion, provided that the computation is not obviated by memory restrictions.

Why should one consider product basis problems at all? First of all, the approximations obtained are more accurate in the sense of (1.6), i.e. one obtains tighter upper bounds to the exact solutions of (1.1). Even for high quality contraction schemes it may be a delicate matter to figure out whether a computed approximation (especially the *Ritz vector*) is trustworthy, as the Examples 7.6 and 7.7 in Section 7.3 impressively reveal. Secondly, one should keep in mind that the construction and application of a contraction can be rather time consuming as well. Then an *iterative-product calculation* may be even more effective if one is only interested in a small part of the spectrum. Last but not least, for triatomic molecules the situation is not as unfavorable as the above example for molecules with $N = 5$ nuclei suggests. The total problem dimension increases less rapidly and, exploiting the sparsity of the Hamiltonian matrix **H**, one can make use of an iterative method which allows to tackle rather large problems. At any rate, considering product basis problems gives one more flexibility and an additional means of verification. So far, especially the *Lanczos method* has been studied rather intensively as a component in the framework for iterative approaches, e.g. by CARRINGTON and his co-workers (see [18, 19], [92], [104] and [129, 130] for instance). In this context, one should also mention *Davidson's method* [29, 30], which was designed for eigenvalue computations arising in quantum chemistry, but is only applicable under rather specific conditions (i.e. the matrix must be strongly diagonal dominant, see Section 3.3.2.3 for more details). Both, *Davidson's* and the *Lanczos method* (which we will briefly outline in Section 3.3), are often only a last resort in practical situations (when direct solvers are no more applicable), because they are slow to converge and because they are no "black boxes". This is the starting point and motivation for the investigations in this thesis, and similar to direct solvers (the RRR method developped by DHILLON [32]) we can benefit from the progress in the field of iterative projection methods that has been made during the 90s of the last century:

The *Jacobi-Davidson method* was proposed by SLEIJPEN and VAN DER VORST [114] as an improvement over *Davidson's method*, and it is an attempt to remove the conceptual weaknesses of the hitherto methods. It attracted a great deal of attention since its introduction in 1996 and it re-aroused the interest of numerical analysts in iterative projection methods, because it turned out to be superior in a couple of important and interesting applications (see [44], [45], e.g.) and is based on an interesting idea. Basically, it can be regarded as an inexact Rayleigh-Quotient process where in every step of the iteration a shift-and-invert system (called *Jacobi-Davidson correction equation*)

$$(I - uu^*)(A - \theta I)(I - uu^*)t = -(A - \theta I)u = -r \qquad (1.7)$$

with respect to an approximate eigenpair (θ, u) is solved approximately (with increasing accuracy), e.g. by means of *Krylov subspace methods*. The approximate solution t is used as a new direction for the search space $\mathcal{K}$. A motivation, a detailed introduction, and a theoretical discussion of the algorithm along with its variants will be given in Chapter 4. In spite of its success, the JD method, which has been investigated for more than one decade, is still far from well-understood. The algorithm is rather sophisticated and its success seems to be highly depending upon the structure of the problem. This opens up a wide and vital field of research, and the promising results in related fields of applications are additional motivation for our considerations. A near-complete state-of-the-art survey of material (sorted by topic, people, references and software) on Jacobi-Davidson type methods may be found in the recently established web portal [54]. The main concern of this thesis is to analyze to what extent the *Jacobi-Davidson method* may successfully be applied to the Hermitian eigenproblems (1.3) and (1.5) arising in the context of triatomic molecules with the *Double Renner effect*. To do so, we have to answer a couple of questions:

- How can one take advantage of the sparsity of the Hamiltonian matrix $\mathbf{H}$? We will discuss this issue in Chapter 7, where we derive an optimized storage scheme, which exploits the regular sparsity pattern of $\mathbf{H}$ (only non zero elements are stored) and results in an efficient procedure for matrix-vector multiplications.

- How can the *Jacobi-Davidson correction equation* (1.7) be solved efficiently? Since we will be using iterative Krylov solvers for this purpose, it is a key issue to find suitable *preconditioners K* as a decisive ingredient for the success of these methods. More precisely, these are matrices that approximate the coefficient matrix A well (i.e. $K \approx A$ in some sense) and that are used to transform the equation $Ax = b$ into the equivalent linear system

$$K^{-1}Ax = K^{-1}b \qquad (1.8)$$

  Applying the Krylov method to the preconditioned operator $B := K^{-1}A$ leads to a faster convergence of the solver, provided that K is chosen appropriately and that the additional costs for constructing and applying K are outweighed by an overall reduction of computing time. A general description of the concept may be found in Section 4.2.3.5. For the *Jacobi-Davidson method* things are more involved because the projections $(I - uu^*)$ in the correction equation have to be taken into account and because the coefficient matrix changes in every step of the iteration due to the shift parameter θ. We will derive and present preconditioned variants of the JD method, that take into acount these special features, in Section 4.3.6. The identification and construction of suitable preconditioners is discussed in Section 7.5.1.

- To assess the performance of the *Jacobi-Davidson method*, we have to compare it to other iterative projection methods (*Lanczos, Davidson*). The numerical results of our experiments in Chapter 7 reveal that the JD method is superior in many cases, once a suitable preconditioner has been identified. Besides, we have to analyze under what circumstances the *Jacobi-Davidson method* can outperform direct solvers. To this end, we will compare it with the performance of the optimized direct solver which is suggested and employed in [87] and which is specifically designed for our problem, the computation of rovibronic energy levels of triatomic molecules (a description of the underlying ideas will be given in Section 3.2.1.2).

- Is it feasible to compute eigenpairs at arbitrary regions of the spectrum (especially interior eigenvalues)? This question is of general interest for iterative projection methods, and not only important for our application. The further one moves into the interior of the spectrum, the less satisfactory the convergence of iterative projection methods in general becomes. The reason for these problems results from a superposition of two general numerical difficulties:

  - it is hard to obtain suitable approximations for internal eigenpairs by means of an ordinary Rayleigh-Ritz procedure. We will analyze the general weaknesses of the approach and present alternative *extraction methods* in Section 3.3.1.

  - looking for interior eigenvalues inevitably leads to indefinite linear systems of the type $(A - \theta I)x = b$. Solving them by means of Krylov solvers often leads to disappointing results, because it is difficult to find suitable preconditioners. This is also true of our situation, but we will outline some possible remedies in Section 7.5.2.

- Finally, as an important by-product, in Section 7.7 we shall see how the Jacobi-Davidson approaches may be parallelized on shared memory architectures. However, it will turn out that the essential restriction is the lack of preconditioners with inherent parallelism for the computation of interior eigenvalues.

The main focus of our considerations will be on product-basis problems, as they are more complicated to handle and as they exhibit more structure that may be exploited in devising preconditioners. For the contracted case things are getting considerably simpler (one only needs to exploit the block sparsity of $\tilde{\mathbf{H}}$ and the preconditioners simplify as well), and one proceeds analogously. Therefore, the corresponding description in Section 7.6 can be kept brief.

For the most part, the results of the JD-related numerical experiments in this thesis were obtained by means of our software package JACDAV (written in the C programming language [67]) which we specifically developped for our purposes. Basically, the implementation follows the ideas and techniques used in SLEIJPEN's MATLAB® JDQR package [112] and GEUS' JDBSYM package [46]. However, it incorporates important additional features (option for harmonic and refined extraction, modified subspace expansion) which are important in our context and which have not been covered by the state-of-the-art software so far. For more details, we refer to the corresponding description in the Appendix A.4. For very few exceptions, it was more appropriate to employ the MATLAB® environment (see the product website [5] and the introduction in [94] for further information), especially when it comes to producing sparsity plots and when the computing times of the experiments have no particular relevance (see Result 7.24 in Section 7.5.2, for instance, where we employed SLEJPEN's JDQR MATLAB® software [112] and the ILUPACK MATLAB® driver [15] as a "plug in" for the preconditioner to be supplied by the user). The test results for the IRL method (Alg. 3.16, see Section 3.3.2.2) in Section 7.5.3 were produced using the FORTRAN 77 [85] software ARPACK [75]. For our direct-product and direct-contracted experiments in Sections 7.4.1 and

7.4.2, we developped simple driver routines (written in the C programming language [67]) that connect to the required LAPACK [2] and SBR routines.

1.4. Structure and Organization of the thesis

The agenda shows that we do not only have to discuss general aspects from numerical linear algebra (direct methods and iterative projection methods), but also the theoretical background (quantum chemistry, theoretical spectroscopy) of the eigenvalue problems we are dealing with. For this reason, it is sensible to subdivide the thesis into two introductory parts, which are largely independent of each other, and a concluding part that brings together numerical analysis and quantum chemistry:

- Part I: Chapters 2 - 4
  We recall some basics on Hermitian eigenproblems as well as some important technical prerequisites (orthogonalization, orthogonal bases of Krylov spaces, singular value decompositions and Kronecker products) in Chapter 2 before we begin with the discussion of numerical algorithms in Chapter 3. The introductory remarks already show that we are not only concerned with the JD method, but also with the general dichotomy between direct and iterative solvers. Since we will be using an optimized direct solver in our experiments (cf. [87]), it is appropriate to outline the underlying framework and to give a survey of its components, including a brief state-of-the-art review of the most important direct solvers for tridiagonal Hermitian matrices and a description of tridiagonalization strategies. Furthermore, we briefly go into single-vector iteration methods (power method, inverse iteration and Rayleigh quotient iteration) as they form the basis for the iterative projection methods that we will describe in Section 3.3. In Chapter 4 we finally come to the *Jacobi-Davidson method* and its variants.

- Part II: Chapters 5 - 6
  Chapter 5 gives a survey of what theoretical spectroscopy is about and describes for the general case of N atomic molecules how Hermitian eigenvalue problems are derived. Furthermore, the dichotomy between product basis and contracted basis problems is illustrated, and the resulting four general solution strategies are discussed. In Chapter 6 we come to our concrete problem, i.e. triatomic molecules that exhibit the *Double Renner effect*. We describe the construction and structure of the Hamiltonian matrices in full detail.

- Part III: Chapters 7 - 8
  In Chapter 7 we eventually combine the insights of the introductory parts, and we show in detail how to apply the JD method in our context (efficient matrix-vector multiplication, preconditioners, product basis vs. contracted basis) including extensive numerical results. Finally, Chapter 8 summarizes the essential results and gives an outlook for future research.

1.5. Acknowledgements

This thesis is the result of my research during my employment as a scientific assistant at the University of Wuppertal from October 2002 until February 2007, and I would like to express my gratitude to a couple of people who contributed to the success of my work, professionally, as well as on a human level. First of all, thanks go to all members of the Scientific Computing group for the pleasant and uncomplicated working atmosphere during all these years. Especially, I would like to point out my dear colleagues DR. THOMAS BEELITZ, DR. KARSTEN BLANKENAGEL, DR. ELTON BOJAXHIU, KARSTEN KAHL and DR. JEAN-HONORÉ TAPAMO for their amicable helpfulness as well as for their steady and sincere interest in my work. It would have been impossible to complete this thesis without the support of the people from the Theoretical Spectroscopy group, PROF. DR. PER JENSEN, DR. VLADLEN MELNIKOV and, especially, DR. TINA ERICA ODAKA. Their kindness, their willingness for discussions and their patience in answering all my questions over and over again were decisive to understand the theoretical background and to be able to use the *Double Renner software* DR. I am also grateful for helpful discussions with PROF. DR. BRUNO LANG, as an expert on direct eigensolvers, and with PROF. DR. MICHIEL HOCHSTENBACH, as an expert on iterative projection methods. Besides, I appreciate the kind support of DR. GERARD SLEIJPEN and PROF. DR. MATTHIAS BOLLHÖFER who did not hesitate to provide me with the latest versions of their software, the JDQR MATLAB® package and the ILUPACK library. Last but not least, I am indebted to my advisor, PROF. DR. ANDREAS FROMMER, for accepting me as a member in his group, for proposing this interesting topic and for his valuable comments and corrections.

It was a nice co-incidence that many friends of mine were working on their doctoral theses during the same time as I was. Their research has made them experts in such interesting and exciting scientific branches as computer science, economics, historical science and human medicine. I greatly benefited from the lively exchange on our projects in innumerous inspiring discussions which broadened my intellectual horizon and encouraged me to continue my work in difficult stages. Therefore, I cordially thank DR. JÖRN GROTHE, DR. GERO LÜCKEMEYER, DR. CHRISTIAN MÜLLER, DR. PASCAL NEVRIES, DR. KLAUS SEGBERS, MIKOŁAJ SIEMASZKO and DR. BENJAMIN ZEMLIN for their friendship and their catching scientific enthusiasm.

Finally, I could not have succeeded without the constant love and support of my family – my parents and my brother. They always encouraged me to go my way, cheered me up and gave me distraction whenever it was necessary. I am grateful to my parents that they conveyed me the value of a good academic education.

Wuppertal, April 2009

Part I.

Numerical Linear Algebra

2. Preliminaries

In this chapter we briefly review the most important definitions and theoretical facts on the numerics of eigenvalue problems with special emphasis on the Hermitian case. Furthermore, some useful tools of trade including special matrix decompositions and orthogonalization techniques are recalled.

2.1. Eigensystems of General Matrices

Definition 2.1 (Eigenpairs, eigenvalues, eigenvectors, eigensystems)
Let $A \in \mathbb{C}^{n \times n}$ an arbitrary square matrix. The pair $(\lambda, x) \in \mathbb{C} \times \mathbb{C}^n$ is called an eigenpair of A if

1. $x \neq 0$

2. $Ax = \lambda x$

The scalar λ is called an eigenvalue and the vector x is called an eigenvector of A. Furthermore, we call the set of all eigenpairs an eigensystem and a subset a partial eigensystem. The task of finding one or more eigenpairs of a given matrix A is referred to as an eigenproblem.

Definition 2.2 (Spectrum and spectral radius)
The set of all $\lambda \in \mathbb{C}$ that are eigenvalues of $A \in \mathbb{C}^n$ is called the spectrum of A and is denoted by $\sigma(A)$. The spectral radius of A is the nonnegative real number

$$\kappa(A) := \max\{\, |\lambda| \,:\, \lambda \in \sigma(A) \,\} \tag{2.1}$$

This is just the radius of the smallest disc centered at the origin in the complex plane that includes all the eigenvalues of A.

Theorem 2.3 (Algebraic characterization of eigenvalues)
For $A \in \mathbb{C}^{n \times n}$ we call $p_A(\mu) = \det(A - \mu I)$ the characteristic polynomial of A. The following conditions are equivalent

1. λ is an eigenvalue of A

2. $p_A(\lambda) = 0$

Proof: well-known, see [71] e.g. □

Definition 2.4 (Similarity, Similarity transformation)
*Two matrices $A, B \in \mathbb{C}^{n \times n}$ are said to be similar, iff there is a non-singular $T \in \mathbb{C}^{n \times n}$,
such that*

$$T^{-1}AT = B \tag{2.2}$$

One also says that B results from a similarity transformation of A.

Proposition 2.5 (Eigenvalues of similar matrices)
If $A, B \in \mathbb{C}^{n \times n}$ are similar, then $p_A(\mu) = p_B(\mu)$ and A and B have the same eigenvalues.

Proof: well-known, see [71] e.g.   □

Theorem 2.6 (Continuous dependence on the matrix coefficients)
*The eigenvalues $\lambda_1, \ldots, \lambda_n$ of $A \in \mathbb{C}^{n \times n}$ depend continuously on the matrix coefficients
a_{ij}. More precisely, for any given $\epsilon > 0$ there is a $\delta = \delta(\epsilon) > 0$ such that for $B \in \mathbb{C}^{n \times n}$
with $\|B - A\| \leq \delta$ we find an ordering of B's eigenvalues $\mu_1, \ldots, \mu_n$ with the property*

$$|\mu_j - \lambda_j| \leq \epsilon \qquad (j = 1, \ldots, n)$$

Proof: It is clear that the coefficients of the characteristic polynomial $p_A(\mu) = \det(A - \mu I)$ depend continuously on the matrix coefficients a_{ij}. Thus, it remains to show that the zeros $\lambda_1, \ldots, \lambda_n$ of a polynomial

$$q(z) = z^n + b_{n-1}z^{n-1} + \ldots + b_1 z + b_0$$

depend continuously on the coefficents b_i of the polynomial as well. The proof uses auxiliary results from the field of complex analysis (Rouché's theorem, cf. [6]) and is discussed in detail e.g. in [131].   □

Remark 2.7
*Notice, that by contrast eigenvectors are not necessarily continuous functions of the
matrix coefficients, as can be recognized from the following example which is due to
PARLETT [91]: Consider the symmetric matrix*

$$A = \begin{bmatrix} \nu + \delta & \gamma \\ \gamma & \nu - \delta \end{bmatrix}, \qquad \delta, \gamma, \nu \in \mathbb{R} \tag{2.3}$$

whose eigenvalues λ_1 and λ_2 are computed according to Theorem 2.3 as

$$\lambda_1 = \nu - \sigma, \qquad \lambda_2 = \nu + \sigma \tag{2.4}$$

where

$$\sigma = \sqrt{\delta^2 + \gamma^2} \tag{2.5}$$

Depending on the sign of δ the eigenvector x_1 related to λ_1 is

$$x_1 = \begin{bmatrix} 1 \\ \pm\gamma/(\sigma + |\delta|) \end{bmatrix} \tag{2.6}$$

Consider now the following particular cases

- $\delta = 0, \gamma = 0$. *Then*

$$x_1(\delta, \gamma) = \begin{bmatrix} 1 \\ 0 \end{bmatrix} \tag{2.7}$$

- $\delta \neq 0, \gamma \neq 0$. *Then*

$$x_1(\delta, \gamma) \rightarrow \begin{bmatrix} 1 \\ 0 \end{bmatrix} \quad (\gamma \rightarrow 0) \tag{2.8}$$

- $\delta = 0, \gamma \neq 0$. *Then*

$$x_1(\delta, \gamma) \rightarrow \begin{bmatrix} 1 \\ \pm 1 \end{bmatrix} \quad (\gamma \rightarrow 0, \gamma \neq 0) \tag{2.9}$$

This shows, that the discontinuity arises when A has multiple eigenvalues, i.e. in the neighborhood of $(\delta, \gamma) = (0, 0)$ □

Theorem 2.8 (Geršgorin)
Let $A = (a_{ij}) \in \mathbb{C}^{n \times n}$. For $i = 1, \ldots, n$ we define the Geršgorin [1] discs

$$G_i := \{z \in \mathbb{C} \; : \; |z - a_{ii}| \leq r_i\} \qquad \text{where} \qquad r_i := \sum_{\substack{j=1 \\ j \neq i}} |a_{ij}| \tag{2.10}$$

Then the following statements hold:

1. *Any eigenvalue λ of A is located in one of the Geršgorin discs G_i.*

2. *Suppose that there are m Geršgorin discs G_i whose union S is disjoint from all other discs. Then S contains exactly m eigenvalues (counted with their algebraic multiplicities).*

Proof: well known, see [56] and [100], e.g.   □

2.2. Eigensystems of Hermitian Matrices

Much more can be said, when dealing with Hermitian matrices. They exhibit very nice and useful theoretical properties, and their eigensystems can be characterized in more detail than it is the case for arbitrary square matrices. Fortunately, the matrices arising in our considerations and computations later on are symmetric, i.e. real-Hermitian, such that we will be able to take advantage of what is discussed in the following.

---

[1]see also remarks on transliteration of Russian names in Appendix A.2

2.2.1. Basic Properties and Definitions

First of all, it is appropriate to distinguish some important classes of matrices:

Definition 2.9 (Adjoint matrix)
For $A \in \mathbb{C}^{n \times m}$ its adjoint matrix A^* is defined as

$$A^* := \overline{A}^T \tag{2.11}$$

where the conjugate-complex matrix $\overline{A}$ is defined component wise, i.e.

$$\overline{A} := [\overline{a_{ij}}] \tag{2.12}$$

Definition 2.10 (Hermitian, symmetric, orthogonal, unitary)
   1. $A \in \mathbb{C}^{n \times n}$ is called Hermitian, iff $A = A^*$

   2. For the special case that $A \in \mathbb{R}^{n \times n}$ is Hermitian, it is called symmetric, i.e. it holds $A = A^T$.

   3. $U \in \mathbb{C}^{n \times n}$ is called unitary, iff $U^*U = \overline{U}^T U = I$

   4. For the special case that $U \in \mathbb{R}^{n \times n}$ is unitary, it is called orthogonal, i.e. it holds $U^T U = I$.

   5. For the sake of simplicity, one often uses the neutral term orthonormal matrix when referring to both orthogonal and unitary matrices. Sometimes also the appelation self-adjoint matrix may be found in the literature as a comprehensive term for Hermitian and symmetrix matrices. Notice, however, that these terms need not necessarily coincide in the context of unbounded linear operators acting on arbitrary Hilbert spaces (see Def. 5.9 and the related discussion in Chapter 5).

The definition reveals that everything what is stated for complex-Hermitian matrices in the following, is especially also valid for real-symmetric matrices and all one has to do is replace the terms "Hermitian" by "symmetric" resp. "unitary" by "orthogonal" in the corresponding theorems.

Proposition 2.11 (General properties of eigenpairs of Hermitian matrices)
Let $A \in \mathbb{C}^{n \times n}$ be a Hermitian matrix.

   1. All eigenvalues λ_i $(i = 1, \ldots, n)$ of A are real

   2. For two eigenpairs (λ_i, u_i) and (λ_j, u_j) with $\lambda_i \neq \lambda_j$ we have $u_i^* u_j = 0$

Proof: well known, see [71] e.g.   □

Theorem 2.12 (Eigendecomposition of Hermitian matrices)
Any Hermitian matrix $A \in \mathbb{C}^{n \times n}$ is unitarily similar to a real diagonal matrix, i.e. there is a unitary matrix $U \in \mathbb{C}^{n \times n} = [u_1, \ldots, u_n]$ whose columns u_i are eigenvectors of A such that

$$U^* A U = \Lambda = \mathrm{diag}(\lambda_1, \ldots, \lambda_n) \tag{2.13}$$

Unless otherwise stated we assume the eigenvalues $\lambda_i \in \mathbb{R}$ to be ordered by ascending magnitude

$$\lambda_1 \leq \lambda_2 \leq \ldots \lambda_n \tag{2.14}$$

Proof: well-known, see [71] e.g.   □

2.2.2. Variational Characterisations

The *Rayleigh quotient* defined and discussed below is an important tool to obtain an optimal eigenvalue approximation related to a given eigenvector approximation.

Definition 2.13 (Rayleigh quotient)
For a Hermitian matrix $A \in \mathbb{C}^{n \times n}$ and an arbitrary non-zero vector $u \in \mathbb{C}^n$ we call the ratio

$$\rho(u) \equiv \rho(u, A) \equiv \frac{u^* A u}{u^* u} \tag{2.15}$$

the Rayleigh quotient.
If u is an eigenvector of A the Rayleigh quotient obviously reproduces the associated eigenvalue λ, i.e. $\rho(u) = \lambda$.

Proposition 2.14 (Minimal residual property of the Rayleigh quotient)
For each $u \in \mathbb{C}^n$ the residual

$$r(u, \mu) = A u - \mu u \tag{2.16}$$

is minimized in the $\|\cdot\|_2$-norm if we choose μ to be the Rayleigh quotient $\rho(u)$, i.e.

$$\| A u - \rho(u) u \|_2 \leq \| A u - \mu u \|_2 \quad \forall \mu \in \mathbb{C} \tag{2.17}$$

Furthermore, the residual $r(u, \rho(u))$ is perpendicular to u:

$$r(u, \rho(u)) = A u - \rho(u) u \perp u \tag{2.18}$$

Proof: Without loss of generality we may assume that u is of norm unity. Let now $[u, U]$ be unitary and set

$$\begin{bmatrix} u^* \\ U^* \end{bmatrix} A[u, U] = \begin{bmatrix} \rho(u) & h^* \\ g & B \end{bmatrix} \tag{2.19}$$

Then

$$\begin{bmatrix} u^* \\ U^* \end{bmatrix} r(u, \mu) = \begin{bmatrix} \rho(u) & h^* \\ g & B \end{bmatrix} \cdot \begin{bmatrix} u^* \\ U^* \end{bmatrix} u - \mu \cdot \begin{bmatrix} u^* \\ U^* \end{bmatrix} u = \begin{bmatrix} \rho(u) - \mu \\ g \end{bmatrix} \qquad (2.20)$$

As $\| \cdot \|_2$ is unitarily invariant,

$$\|r(u, \mu)\|_2^2 = |\rho(u) - \mu|^2 + \|g\|_2^2 \qquad (2.21)$$

Clearly, the above expression takes its minimum for $\mu = \rho(u)$. (2.18) follows by direct verification. $\qquad \square$

Proposition 2.15 (Rayleigh-Ritz)
Under the assumptions of Definition 2.13 the following properties hold: For non-zero $x \in \mathbb{C}^n$ $\rho(x)$ ranges over the interval $\mathcal{I} := [\lambda_1, \lambda_n]$, i.e.

1. $\lambda_1 x^* x \leq x^* A x \leq \lambda_n x^* x$ for all $x \in \mathbb{C}^n$

2. $\lambda_{max} = \lambda_n = \max\limits_{x \neq 0} \frac{x^* A x}{x^* x} = \max\limits_{x^* x = 1} x^* A x$

3. $\lambda_{min} = \lambda_1 = \min\limits_{x \neq 0} \frac{x^* A x}{x^* x} = \min\limits_{x^* x = 1} x^* A x$

Proof: According to Theorem 2.12 there exists a unitary matrix $U \in \mathbb{C}^{n \times n}$ such that $A = U^* \Lambda U$ with $\Lambda = \text{diag}(\lambda_1, \lambda_2, \ldots, \lambda_n)$. For any $x \in \mathbb{C}^n$ we have

$$x^* A x = x^* U \Lambda U^* x = (U^* x)^* \Lambda (U^* x) = \sum_{i=1}^n \lambda_i |(U^* x)_i|^2$$

As each term in the above sum is nonnegative, it follows

$$\lambda_{min} \sum_{i=1}^n |(U^* x)_i|^2 \leq x^* A x = \sum_{i=1}^n \lambda_i |(U^* x)_i|^2 \leq \lambda_{max} \sum_{i=1}^n |(U^* x)_i|^2$$

Since U is unitary,

$$\sum_{i=1}^n |(U^* x)_i|^2 = \sum_{i=1}^n |x_i|^2 = x^* x$$

and, hence, we have shown that

$$\lambda_1 x^* x = \lambda_{min} x^* x \leq x^* A x \leq \lambda_{max} x^* x = \lambda_n x^* x$$

These inequalities are sharp, for if x is an eigenvector of A corresponding to the eigenvalue λ_1, the *Rayleigh quotient* $\rho(x) = \frac{x^* A x}{x^* x}$ reproduces λ_1. The same argument applies for λ_n. $\qquad \square$

Remark 2.16
Proposition 2.15 shows that the Rayleigh quotient $\rho(x)$ of an arbitrary vector $x \in \mathbb{C}^n$ with respect to a Hermitian $A \in \mathbb{C}^{n \times n}$ can take any value in the interval $\mathcal{I} = [\lambda_1, \lambda_n]$. As a trivial consequence, $\rho(x)$ can be written as a convex combination of the extremal eigenvalues λ_1 and λ_n. We can say even more, i.e. the expansion of x by A's eigenvectors

$$x = \sum_{i=1}^{n} \xi_i u_i \tag{2.22}$$

uniquely determines a convex combination of all eigenvalues λ_i for the Rayleigh quotient

$$\rho(x) = \frac{(Ax, x)}{(x, x)} = \frac{\sum_{k=1}^{n} \lambda_k |\xi_k|^2}{\sum_{k=1}^{n} |\xi_k|^2} = \sum_{k=1}^{n} \beta_k \lambda_k, \tag{2.23}$$

where obviously

$$0 \leq \beta_i = \frac{|\xi_i|^2}{\sum_{i=1}^{n} |\xi_k|^2} \leq 1, \quad \text{and} \quad \sum_{i=1}^{n} \beta_i = 1 \tag{2.24}$$

See [100] for a more general version of this relation.   □

Theorem 2.17 (Courant-Fischer, Min-Max principle)
Let $A \in \mathbb{C}^{n \times n}$ be a Hermitian matrix with eigenvalues $\lambda_1 \leq \lambda_2 \leq \ldots \leq \lambda_n$, let k be a given integer with $1 \leq k \leq n$ and let $\mathcal{C}_j$ resp. $\mathcal{S}_j$ denote subspaces of $\mathbb{C}^n$ with dimension j. Then λ_k can be characterized by either of the two following relations:

$$\lambda_k = \min_{\mathcal{S}_{n-k}} \max_{u \perp \mathcal{S}_{n-k}} \frac{u^* A u}{u^* u} \tag{2.25}$$

$$\lambda_k = \max_{\mathcal{C}_{k-1}} \min_{v \perp \mathcal{C}_{k-1}} \frac{v^* A v}{v^* v} \tag{2.26}$$

Proof: Since

$$\dim \mathcal{S}_{n-k}^{\perp} + \dim \mathcal{C}_{k-1}^{\perp} = k + (n - k + 1) = n + 1 > n$$

it follows that the subspaces $\mathcal{S}_{n-k}^{\perp}$ and $\mathcal{C}_{k-1}^{\perp}$ must have a non-empty intersection, i.e. $\mathcal{I} := \mathcal{S}_{n-k}^{\perp} \cap \mathcal{C}_{k-1}^{\perp} \neq \emptyset$. Let $w \in \mathcal{I}$, $w \neq 0$ any nonzero vector in both subspaces, so

$$\min_{v \perp \mathcal{C}_{k-1}} \rho(v) \leq \rho(w) \leq \max_{u \perp \mathcal{S}_{n-k}} \rho(u)$$

These inequalities hold for all choices of $\mathcal{S}_{n-k}$ and $\mathcal{C}_{k-1}$, thus we can choose those subspaces that maximize the left-hand side and minimize the right-hand side of the inequality:

$$\max_{\mathcal{C}_{k-1}} \min_{v \perp \mathcal{C}_{k-1}} \rho(v) \leq \min_{\mathcal{S}_{n-k}} \max_{u \perp \mathcal{S}_{n-k}} \rho(u)$$

To show equality, we use $\mathcal{Z}_{n-k} = \mathrm{span}\{z_1, \ldots, z_{n-k}\}$ for $\mathcal{S}_{n-k}$ and $\mathcal{Z}_{k-1}$ for $\mathcal{C}_{k-1}$ analogously (z_k denotes the eigenvector associated with λ_k), and the chain of inequalities extends to

$$\min_{x \perp \mathcal{Z}_{k-1}} \rho(x) \leq \max_{\mathcal{C}_{k-1}} \min_{v \perp \mathcal{C}_{k-1}} \rho(v) \leq \min_{\mathcal{S}_{n-k}} \max_{u \perp \mathcal{S}_{n-k}} \rho(u) \leq \max_{w \in \mathcal{Z}_k} \rho(w)$$

and from the properties of a *Rayleigh quotient* we finally obtain

$$\lambda_k \leq \max_{\mathcal{C}_{k-1}} \min_{v \perp \mathcal{C}_{k-1}} \rho(v) \leq \min_{\mathcal{S}_{n-k}} \max_{u \perp \mathcal{S}_{n-k}} \rho(u) \leq \lambda_k$$

$\square$

The following two theorems are subsumed as so-called *interlacing theorems* resp. *interleaving theorems* in the literature and will be of importance later on, when we consider matrices that originate from deleting rows and columns of a larger matrix or from orthogonal projections on subspaces of smaller dimension.

Lemma 2.18 (Inclusion principle)
Let $A \in \mathbb{C}^{n \times n}$ be a Hermitian matrix, let r be an integer with $1 \leq r \leq n$, and let A_r denote any $r \times r$ principal submatrix of A (obtained by deleting $n - r$ rows and the corresponding columns from A). For each integer k such that $1 \leq k \leq r$ we have

$$\lambda_k(A) \leq \lambda_k(A_r) \leq \lambda_{k+n-r}(A) \tag{2.27}$$

Proof: Suppose that $A_r \in \mathbb{C}^{n \times n}$ is formed by deleting rows $i_1, \ldots, i_{n-r}$ and the corresponding columns from A and let $\mathcal{E}_{n-r} = \mathrm{span}\{e_{i_1}, \ldots, e_{i_{n-r}}\}$ denote the subspace spanned by the related canonical unit vectors. Application of Theorem 2.17 (first part) yields:

$$\lambda_{k+n-r}(A) = \min_{\mathcal{S}_{r-k}} \max_{u \perp \mathcal{S}_{r-k}} \frac{u^* A u}{u^* u} \geq \min_{\mathcal{S}_{r-k}} \max_{\substack{u \perp \mathcal{S}_{r-k} \\ u \perp \mathcal{E}_{n-r}}} \frac{u^* A u}{u^* u} = \min_{\mathcal{S}_{r-k}} \max_{u \perp \mathcal{S}_{r-k}} \frac{u^* A_r u}{u^* u} = \lambda_k(A_r)$$

The lower estimate can be obtained with the same reasoning:

$$\lambda_k(A) = \max_{\mathcal{S}_{k-1}} \min_{u \perp \mathcal{S}_{k-1}} \frac{u^* A u}{u^* u} \leq \max_{\mathcal{S}_{k-1}} \min_{\substack{u \perp \mathcal{S}_{k-1} \\ u \perp \mathcal{E}_{n-r}}} \frac{u^* A u}{u^* u} = \max_{\mathcal{S}_{k-1}} \max_{u \perp \mathcal{S}_{k-1}} \frac{u^* A_r u}{u^* u} = \lambda_k(A_r)$$

$\square$

Corollary 2.19 (Poincaré separation theorem)
Let $A \in \mathbb{C}^{n \times n}$ be a Hermitian matrix, let r be a given integer with $1 \leq r \leq n$, and let $U_r = [u_1, \ldots, u_r] \in \mathbb{C}^{n \times r}$ be an orthonormal matrix, i.e. $U_r^* U_r = I_r$. Let $B_r = U_r^* A U_r$. If the eigenvalues of A and B_r are arranged in increasing order, we have

$$\lambda_k(A) \leq \lambda_k(B_r) \leq \lambda_{k+n-r}(A) \tag{2.28}$$

Proof: If $r < n$, supplement U_r by $n - r$ additional vectors $u_{r+1}, \ldots, u_n$ to obtain a unitary matrix $U \in \mathbb{C}^{n \times n}$. $U^* A U$ has the same eigenvalues as A and the matrix B_r in the statement of the corollary is a principal submatrix of $U^* A U$ obtained by deleting the last $n - r$ rows and columns. The assertion now follows directly from Lemma 2.18.
□

2.2.3. Perturbation Analysis and Error Bounds

In the following, we collect and review some useful results on how perturbations in the matrix coeffcents may affect the eigenvalues. Furthermore, essential error bounds for approximate eigenpairs (θ, u) are derived, which are of importance, when one has to assess the quality of approximations obtained in numerical algorithms for eigensystems we will be discussing later on.

Notation 2.20
Let $x, y \in \mathbb{C}^n$. Then we denote

- $|x| := [\, |x_i| \,]$,
  i.e. the vector $|x|$ is formed from the moduli of its components.

- $|x| \leq |y|$, *iff $|x_i| \leq |y_i|$ for all $i = 1, \ldots, n$,*
  i.e. the vectors $|x|$ and $|y|$ are compared component wise.

□

Definition 2.21 (Absolute and Monotone Norms)
A norm $\| \cdot \|$ is called monotone, *iff $\|x\| \leq \|y\|$ for all $x, y \in \mathbb{C}^n$ with $|x| \leq |y|$. It is called* absolute, *iff $\| \, |x| \, \| = \|x\|$ for all $x \in \mathbb{C}^n$*

Lemma 2.22
Let $\| \cdot \|$ be a norm resp. the associated matrix norm. Then the following statements are equivalent:

1. *$\| \cdot \|$ is a monotone norm, i.e. $\|x\| \leq \|y\|$ whenever $|x| \leq |y|$*

2. *For any diagonal matrix $D := \operatorname{diag}(d_1, \ldots, d_n) \in \mathbb{C}^{n \times n}$ it holds*

$$\|D\| = \max_{j=1,\ldots,n} |d_j|$$

3. *$\| \cdot \|$ is an absolute norm, i.e. $\| \, |x| \, \| = \|x\|$ for all $x \in \mathbb{C}^n$*

Proof: see [56] □

Proposition 2.23 (Bauer-Fike)
Let $A \in \mathbb{C}^{n \times n}$ be diagonalizable and $P \in \mathbb{C}^{n \times n}$ the non-singular matrix that transforms it into diagonal form, i.e. $P^{-1}AP = \text{diag}(\lambda_1, \ldots, \lambda_n) =: D$. Furthermore, let $A + E \in \mathbb{C}^{n \times n}$ be a perturbation of A and λ an eigenvalue of $A + E$. Then

$$\min_{j=1,\ldots,n} |\lambda - \lambda_j| \leq \|P^{-1}EP\| \leq \text{cond}(P)\|E\| \tag{2.29}$$

Here $\| \cdot \|$ denotes the matrix norm related to an absolute vector norm and $\text{cond}(P) := \|P\|\|P^{-1}\|$ the condition number of P with respect to this matrix norm.

Proof: Without loss of generality we can assume that $\lambda \neq \lambda_j$ (for $\lambda = \lambda_j$ the assertion of the theorem is trivial) and we denote by x the eigenvector related to λ. Because of $(A + E)x = \lambda x$ we have

$$Ex = (\lambda I - A)x = (\lambda I - PDP^{-1})x = P(\lambda I - D)P^{-1}x$$

and it follows

$$P^{-1}x = (\lambda I - D)^{-1}(P^{-1}EP)P^{-1}x$$

By Lemma 2.22 we obtain

$$\|P^{-1}x\| \leq \|(\lambda I - D)^{-1}\| \, \|P^{-1}EP\| \, \|P^{-1}x\| = \max_{j=1,\ldots,n} \frac{1}{\lambda - \lambda_j}\|P^{-1}EP\| \, \|P^{-1}x\|$$

and the assertion follows readily. $\qquad \square$

Corollary 2.24
Let $A \in \mathbb{C}^{n \times n}$ be Hermitian and $A + E$ a perturbation of A. If $\lambda_1, \ldots, \lambda_n$ are the eigenvalues of A and λ an eigenvalue of $A + E$ then

$$\min_{j=1,\ldots,n} |\lambda - \lambda_i| \leq \|E\|_2 \tag{2.30}$$

Proof: We choose the Euclidean norm $\| \cdot \|_2$ as the absolute vector norm in Proposition 2.23. Since A is Hermitian, it can be transformed to diagonal form by means of a unitary matrix P. Now the assertion directly follows, as $\text{cond}_2(P) = 1$. $\qquad \square$

Corollary 2.25
Let (θ, u) be an approximate eigenpair of $A \in \mathbb{C}^{n \times n}$ where $\|u\|_2 = 1$. Then we can establish the following bound, where $r = Au - \theta u$ and λ is the eigenvalue of A closest to θ:

$$|\theta - \lambda| \leq \|r\|_2 \tag{2.31}$$

Proof: Define $E := (\theta u - Au)u^*$. Then

$$(A + E)u = Au + (\theta u - Au)u^* u = \theta u$$

Thus, θ is an eigenvalue of $A + E$. Then

$$\|E\|_2 = \kappa(E^* E)^{\frac{1}{2}} = \|\theta u - Au\|_2 = \|r\|_2 \tag{2.32}$$

where κ is the spectral radius (see Definition 2.2). Corollary 2.24 immediately yields the assertion. $\square$

Proposition 2.26 (Error bounds for approximate eigenpairs)
Let $u \in \mathbb{C}^n$ be an approximate eigenvector of norm unity of $A \in \mathbb{C}^{n \times n}$. We obtain an approximate eigenpair (θ, u) by forming the associated Rayleigh quotient $\theta = u^* Au^*$. Let λ be the eigenvalue closest to θ and x the corresponding eigenvector. If we now define the so-called "gap" δ by

$$\delta = \min_i \{\, |\lambda_i - \theta|, \; \lambda_i \neq \lambda \,\} \tag{2.33}$$

then we can relate the Euclidean norm of the residual $r = Au - \theta u$ to the following error bounds

$$|\theta - \lambda| \;\leq\; \frac{\|r\|_2^2}{\delta} \tag{2.34}$$

$$\sin \theta(u, x) \;\leq\; \frac{\|r\|_2}{\delta} \tag{2.35}$$

Proof: The proof of (2.34) is elementary, but somewhat longish as it requires additional technical lemmas. We thus refer the reader to [91] and [100].
To verify (2.35) we decompose $u = x \cos\varphi + z \sin\varphi$ where $z \perp u$:

$$\begin{aligned}
(A - \theta I)u &= \cos\varphi(A - \theta I)x + \sin\varphi(A - \theta I)z \\
&= \cos\varphi(\lambda - \theta)x \;+ \sin\varphi(A - \theta I)z
\end{aligned}$$

The two vectors on the right hand side are orthogonal to each other, as

$$\langle x, (A - \theta I)z \rangle = \langle (A - \theta I)x, z \rangle = (\lambda - \theta)\langle x, z \rangle = 0$$

Thus,

$$\begin{aligned}
\|r\|_2^2 = \|(A - \theta I)u\|_2^2 &= \sin^2\varphi\|(A - \theta I)z\|_2^2 + \cos^2\varphi|\lambda - \theta|^2 \\
&\geq \sin^2\varphi\|(A - \theta I)z\|_2^2
\end{aligned}$$

Since $z \perp u$, $\|(A - \theta I)z\|_2$ is larger than the smallest eigenvalue of $A - \theta I$ restricted to the subspace $\langle x \rangle^T$, which is exactly the gap δ. $\square$

We see that the quality of an eigenvector approximation u may be evaluated by means of (2.35). However, the general difficulty is that in many cases no information on the distribution of the eigenvalues is available beforehand, such that nothing, or only little can be said about the gap δ. Therefore, a small norm of the residual $r = Au - \theta u$ does not automatically imply that u approximates the exact eigenvector x well, unless it is known that δ is sufficiently large.

As for the bounds on the approximate eigenvalue, we can always pick the better of the two bounds (2.34) and (2.31), provided that we possess information about the gap δ (e.g. a lower bound). This leads to the following

Corollary 2.27
Let (θ, u) be an approximate eigenpair of $A \in \mathbb{C}^{n \times n}$ where $\|u\|_2 = 1$, then

$$|\theta - \lambda| \leq \min \left\{ \|r\|_2, \frac{\|r\|_2^2}{\delta} \right\} \tag{2.36}$$

The following proposition is a useful tool to characterize the deviation of an approximate eigenvector from its exact counterpart:

Proposition 2.28 (Cosine and scalar product)
Let $u, v \in \mathbb{R}^n$. Then the angle $\phi = \angle(u, v) \in [0, \pi]$ is determined by

$$\cos \phi = \frac{\langle u, v \rangle}{\|u\|_2 \cdot \|v\|_2} \tag{2.37}$$

Proof: The assertion is a direct consequence of the well-known Law of Cosines from trigonometry. □

2.3. Technical Tools

2.3.1. Orthogonal and Unitary Matrices

Orthogonal and unitary matrices $Q \in \mathbb{K}^{n \times n}$ play a prominent role in numerical algorithms for eigenvalue problems as they share several useful and important properties:

1. the inverse Q^{-1} is readily available, as $Q^*Q = QQ^* = I_n$, i.e. $Q^* = Q^{-1}$ (cf. Def. 2.10)

2. the spectrum of $A \in \mathbb{K}^{n \times n}$ is invariant w.r.t. orthogonal resp. unitary similarity transformations Q^*AQ (cf. Prop. 2.5)

3. orthogonal transformations are numerically stable (cf. [91])

2.3.1.1. Householder Reflections

For any nonzero vector $v \in \mathbb{R}^n$ one can construct a related matrix $P_v \in \mathbb{R}^{n \times n}$, such that its application to a vector $x \in \mathbb{R}^n$ effects a reflection in the hyperplane $\text{span}\{v\}^T$ (i.e. the hyperplane that is perpendicualar to v). This is made precise by the following

Definition 2.29 (Householder matrix)
Let $v \in \mathbb{R}^n$ be nonzero. Then we call

$$P_v = I_n - \frac{2}{v^*v}vv^* \tag{2.38}$$

a Householder reflection (synonymous terms: Householder matrix, Householder transformation) and the vector v Householder vector. It is easy to verify that Householder matrices are both orthogonal and symmetric.

We are specifically interested in the *Householder matrix* $P_v \in \mathbb{R}^n$ that transforms a given vector $x \in \mathbb{R}^n$ to a scalar multiple of the first unit vector, i.e. $P_v x = \alpha \cdot e_1$. These matrices are beneficial for computing matrix decompositions to be discussed later on.

Corollary 2.30 (Transformation $x \to \pm \|x\|_2 \cdot e_1$)
Let $x \in \mathbb{C}^n$ be nonzero. Define

$$v := x \pm \|x\|_2 \cdot e_1 \quad \text{and} \quad \tau = \frac{2}{v^*v} \tag{2.39}$$

Then application of the Householder reflection

$$P = I_n - \frac{2}{v^*v}vv^* = I_n - \tau vv^* \tag{2.40}$$

on x yields

$$Px = \mp\|x\|_2 \cdot e_1 = \mp(\|x\|_2, 0, \ldots, 0)^* \tag{2.41}$$

Proof: Straightforward verification. □

Note that *Householder matrices* are never explicity computed, as for their application knowledge of the *Householder vector* v and the related parameter $\tau = 2/(v^*v)$ is sufficient. The following algorithm (cf. [49]) determines these parameters according to (2.39) such that cancellation is avoided, $v(1) = 1$ and $P_v x = +\|x\|_2 \cdot e_1$ (positive multiple of e_1):

Algorithm 2.1: Generation of a Householder matrix

---
1 **function** $[v, \tau] = \mathbf{house}(x)$
2 $n = \mathbf{length}(x)$
3 $\sigma = x(2:n)^* x(2:n)$
4 $v = \begin{bmatrix} 1 \\ x(2:n) \end{bmatrix}$
5 **if** $\sigma = 0$ **then**
6 $\quad \tau = 0$
7 **else**
8 $\quad \mu = \sqrt{x(1)^2 + \sigma}$
9 $\quad$ **if** $x(1) \leq 0$ **then**
10 $\quad\quad v(1) = x(1) - \mu$
11 $\quad$ **else**
12 $\quad\quad v(1) = -\sigma/(x(1) + \mu)$
13 $\quad$ **end if**
14 $\quad \tau = 2v(1)^2/(\sigma + v(1)^2)$
15 $\quad v = v/v(1)$
16 **end if**
17 **return** $[v, \tau]$

Remark 2.31 (Applying Householder transformations)
Given a Householder transformation defined by $P = I - \beta vv^*$, it is important not to explicitly perform a matrix-matrix multiplication. The pre-multiplication of P to a matrix A is realized by

$$PA = A - vw^* \qquad (2.42)$$

instead, where $w = \beta A v$, i.e. applying a Householder transform involves one matrix-vector product and one outer product update. □

Remark 2.32 (Generalization of Householder matrices to the complex case)
As is pointed out in [74], one has to be careful when trying to generalize Householder reflections to the application on complex Hermitian matrices. Unlike the real case, it is not always possible to find a Hermitian matrix H such that $Hx = \alpha \cdot e_1$ for an arbitrary non-zero vector $x \in \mathbb{C}^n$. In general, it is only possible to construct a unitary matrix P with the desired property. For further details on this issue and a discussion on software implementations, e.g. in the form of the LAPACK [2] routine CLARFG, see [74]. □

2.3.1.2. Givens and Jacobi Rotations

Householder reflections are very helpful when several successive elements of a matrix row or a matrix column are supposed to be annihilated. However, rather often the situation arises, that one wishes to zero single elements and then *Givens rotations* are the appropriate means:

Definition 2.33 (Givens rotation)
A Givens matrix $G(i, k, \theta) \in \mathbb{R}^{n \times n}$, where $i, k \in \{1, \ldots, n\}$ and $\theta \in \mathbb{R}$ is a rank-two correction to the identity defined by

$$
G(i, k, \theta) = \begin{bmatrix}
1 & \cdots & 0 & \cdots & 0 & \cdots & 0 \\
\vdots & \ddots & \vdots & & \vdots & & \vdots \\
0 & \cdots & c & \cdots & s & \cdots & 0 \\
\vdots & & \vdots & \ddots & \vdots & & \vdots \\
0 & \cdots & -s & \cdots & c & \cdots & 0 \\
\vdots & & \vdots & & \vdots & \ddots & \vdots \\
0 & \cdots & 0 & \cdots & 0 & \cdots & 1
\end{bmatrix}
\begin{matrix} \\ \\ i \\ \\ k \\ \\ \\ \end{matrix}
\qquad (2.43)
$$

where $c = \cos\theta$ and $s = \sin\theta$. Clearly, $G(i, k, \theta)$ is orthogonal.

Multiplication by $G(i, k, \theta)^*$ results in a counterclockwise rotation by the angle θ in the (i, k) plane. More precisely, if $x \in \mathbb{R}^n$ and $y = G(i, k, \theta)^* x$ then

$$
y_j = \begin{cases}
cx_i - sx_k & j = i \\
sx_i + cx_k & j = k \\
x_j & j \neq i, k
\end{cases}
\qquad (2.44)
$$

More specifically, we are interested in *Givens rotations* that annihilate y_j for $j = i$. The following procedure, which avoids overflow (see [49]), determines the parameters c and s, such that

$$
\begin{bmatrix} c & s \\ -s & c \end{bmatrix}^* \begin{bmatrix} a \\ b \end{bmatrix} = \begin{bmatrix} r \\ 0 \end{bmatrix}
\qquad (2.45)
$$

Algorithm 2.2: Generation of a Givens rotation

1 **function** $[c, s] = \textbf{givens}(a, b)$
2 **if** $b = 0$ **then**
3 | $c = 1, s = 0$
4 **else**
5   **if** $|b| > |a|$ **then**
6 | $\tau = -a/b; s = 1/\sqrt{1 + \tau^2}; c = s\tau$
7   **else**
8 | $\tau = -b/a; c = 1/\sqrt{1 + \tau^2}; s = c\tau$
9   **end if**
10 **end if**
11 **return** $[c, s]$

Analogous to *Householder transforms*, the application of *Givens rotations* is never accomplished by means of explicit multiplication of the corresponding matrix, as only 2 columns resp. rows are affected. The update $A \rightarrow G(i, k, \theta)^* A$ defined by

$$A([i, k], :) = \begin{bmatrix} c & s \\ -s & c \end{bmatrix} A([i, k], :) \tag{2.46}$$

should be better realized by the following procedure

Algorithm 2.3: Application of a Givens rotation $G(i, k, \theta)$

1 **for** $j = 1, \ldots, n$ **do**
2     $\tau_1 = A(i, j)$
3     $\tau_2 = A(k, j)$
4     $A(1, j) = c\tau_1 - s\tau_2$
5     $A(2, j) = s\tau_1 - c\tau_2$
6 **end for**

Remark 2.34 (Givens rotations vs. Jacobi rotations)
In Section 3.2.3, where we describe Jacobi's method for computing eigensystems, we will also encounter so-called Jacobi rotations which are employed to annihilate particular elements of a given matrix by means of similarity transformations (cf. Def. 2.4). By contrast, we speak of Givens *rotations when referring to simple one-sided matrix-matrix and matrix-vector multiplications.* □

2.3.2. QR Factorisation and Orthonormalisation of Vector Sets

The construction of orthonormal vectors sets is an important ingredient for a multitude of algorithms in the field of numerical linear algebra, especially for the iterative projection methods described in this thesis. The following definition gives an algebraic top-level formulation for the relation of a vector set given by the columns of the matrix A and the columns of Q originating from the orthogonalization of A:

Definition 2.35

Let $A \in \mathbb{C}^{n \times m}$, where $n \geq m$. If there exist a unitary matrix $Q \in \mathbb{C}^{n \times n}$, i.e. $Q^*Q = I$ and a matrix

$$R = \left[\frac{R_1}{0} \right] \in \mathbb{C}^{n \times m} \tag{2.47}$$

where $R_1 \in \mathbb{C}^{m \times m}$ is upper triangular, such that $A = QR$, then the factorization $A = QR$ is called a QR decomposition of A.

The simplest way to orthonormalize the column vectors of a given matrix $V \in \mathbb{C}^{n \times m}$ is well-known under the name *Gram-Schmidt procedure*:

Algorithm 2.4: Classical Gram-Schmidt orthonormalisation (CGS)

    Input: $V = [v_1, \ldots, v_m] \in \mathbb{C}^{n \times m}$

1   compute $r_{11} = \|v_1\|_2$

2   **if** $r_{11} = 0$ **then**

3 | **stop**

4   **else**

5 | $q_1 = v_1/r_{11}$

6   **end if**

7   **for** $j = 2, \ldots, m$ **do**

8 | **for** $i = 1, \ldots, j-1$ **do**

9 | | $r_{ij} = (v_j, q_i)$

10 | **end for**

11 | $\hat{q} = v_j - \sum_{i=1}^{j-1} r_{ij} q_i$

12 | $r_{jj} = \|\hat{q}\|_2$

13 | **if** $r_{jj} = 0$ **then**

14 | | **stop**

15 | **else**

16 | | $q_j = \hat{q}/r_{jj}$

17 | **end if**

18   **end for**

    Output: $V^*V = I_m$

It is easy to verify that the above algorithm does not break down, if and only if the vectors $v_1, \ldots, v_m$ are linearly independent. From Lines 9–16 of Alg. 2.4 one can see that the following relation holds

$$v_j = \sum_{i=1}^{j} r_{ij} q_i \tag{2.48}$$

Letting $Q_1 = [q_1, \ldots, q_m]$ and R_1 be the upper $m \times m$ triangular matrix whose nonzero elements are the r_{ij} defined in Alg. 2.4, then (2.48) can be re-written as $V = Q_1 R_1$. If we supplement Q_1 by $Q_2 = [q_{m+1}, \ldots, q_n]$ such that $Q = [Q_1, Q_2] \in \mathbb{C}^{n \times n}$ is orthonormal and form R from R_1 according to (2.47), we obtain a QR decomposition of V in the sense of Definition 2.35, i.e. $V = QR$. These considerations prove the following

Theorem 2.36 (Existence of a QR decomposition)
Let $A \in \mathbb{C}^{n \times m}$ and $\text{rank}(A) = m$, i.e. $m \leq n$. Then there exists a QR factorization of A, where R (resp. R_1) has non-vanishing diagonal entries r_{ii}, $i = 1, \ldots, m$

It is known that the classical Gram-Schmidt algorithm is numerically unstable. Depending on the structure of A and the desired quality one should resort to better alternatives, e.g. a Modified Gram-Schmidt procedure, which is obtained by a suitable reorganization of Alg. 2.4, or QR decompositions obtained by means of *Householder reflections* or *Givens rotations*. The latter is advantageous for the application to Hessenberg matrices. For a more detailed discussion see [49].

The iterative projection algorithms for eigenvalue computations we will be dealing with later on construct orthogonal vector sets successively, rather than in one batch for a given vector set, as suggested in Alg. 2.4. We thus require specialized variants which orthogonalize a given vector t against a matrix $Q \in \mathbb{C}^{n \times m}$ whose columns are already orthonormal. It has proven advantageous (see [14]) to make use of a scheme that iteratively applies the classical Gram-Schmidt procedure:

Algorithm 2.5: Iterative classical Gram-Schmidt orthonormalisation (ICGS)

---

1   **function** $u = \mathbf{orth}(Q, t)$
2   $u = t$
3   $\alpha = 0.5$, $it_{max} = 3$, $it = 1$
4   $r_0 = \|u\|_2$
5   **loop**
6      $u = u - Q(Q^*u)$
7      $r_1 = \|u\|_2$
8   **if** $r_1 > \alpha r_0$ ***or*** $it \geq it_{max}$ **then**
9   **exit loop**
10   **end if**
11     $it = it + 1$, $r_0 = r_1$
12 **end loop**
13 **if** $r_1 \leq \alpha r_0$ **then**
14   **error**('loss of orthogonality')
15 **end if**
16 $u = 1/r_1 \cdot u$
17 **return** u
    Output: $\text{span}\{Q, t\} = \text{span}\{Q, u\}$, $[Q, u]^*[Q, u] = I$

The orthogonalization can be done using matrix-vector operations (BLAS 2 [1] in a computer code) and, in general, it is sufficient to perform two passes of the loop (Lines 5-9 in Alg. 2.5), where the quality of orthogonality is assessed by means of the parameter α.

2.3.3. Orthogonal Bases of Krylov Spaces

Krylov subspaces, i.e. subspaces of the form

$$\mathcal{K}_m(A, v) = \{v, Av, A^2v, \ldots, A^{m-1}v\} \tag{2.49}$$

have a prominent role to play in both iterative methods for eigenvalue problems and linear systems. As one can recognize from the definition (2.49), they are associated with a vector $v \neq 0$ and a matrix $A \in \mathbb{C}^{n \times n}$ and obviously dim $\mathcal{K}_m(A, v) \leq m$. We are especially interested in constructing orthogonal bases for $\mathcal{K}_m(A, v)$ and depending on whether A is Hermitian or not we can state different algorithms for this purpose.

2.3.3.1. Arnoldi's Procedure

Let us first consider the general case that the matrix $A \in \mathbb{C}^{n \times n}$ is non-Hermitian. Then Algorithm 2.6 which is referred to as *Arnoldi's procedure* multiplies at each step the previous Arnoldi vector v_j by A and orthonormalizes the resulting vector w_j against all previous v_i's by means of a classical Gram-Schmidt procedure. It can be shown by a simple induction argument that the Arnoldi procedure generates a basis of $\mathcal{K}_m(A, v_1)$ whose vectors are assembled as columns in $V \in \mathbb{C}^{n \times m}$.

Algorithm 2.6: Arnoldi

1 choose a vector v_1 such that $\|v_1\|_2 = 1$
2 **for** $j = 1, 2, \ldots, m$ **do**
3   compute $h_{ij} = (Av_j, v_i)$ for $i = 1, 2, \ldots, j$
4   compute $w_j := Av_j - \sum_{i=1}^{j} h_{ij} v_i$
5      $h_{j+1,j} = \|w_j\|_2$
6   **if** $h_{j+1,j} = 0$ **then**
7 | stop
8   **end if**
9      $v_{j+1} = w_j / h_{j+1,j}$
10 **end for**

The following proposition is very useful for theoretical purposes as it provides a compact matrix formulation for the relation between A and V:

Proposition 2.37 (Arnoldi relation)
Denote by V_m the $n \times m$ matrix with column vectors $v_1, \ldots, v_m$; by $\bar{H}_m$ the $(m+1) \times m$ Hessenberg matrix whose nonzero entries h_{ij} are defined by Algorithm 2.6; and by H_m the matrix obtained from $\bar{H}_m$ by deleting its last row. Then the following relations hold:

$$\begin{aligned} AV_m &= V_m H_m + w_m e_m^* & (2.50) \\ &= V_{m+1} \bar{H}_m & (2.51) \\ V_m^* A V_m &= H_m & (2.52) \end{aligned}$$

Proof: From Lines 4, 5 and 9 of Algorithm 2.6 one easily recognizes that

$$Av_j = \sum_{i=1}^{j+1} h_{ij} v_i \qquad j = 1, 2, \ldots, m \tag{2.53}$$

which proves (2.50) and (2.51).
(2.52) is readily obtained by multiplying both sides of (2.50) by V_m^* and exploiting the orthonormality of V_m. $\qquad\qquad\qquad\qquad\qquad\qquad\qquad\qquad\qquad\qquad\qquad\qquad$ □

2.3.3.2. Lanczos Procedure

For the case that $A \in \mathbb{C}^{n \times n}$ is Hermitian things are getting considerably simpler and this is explained in the following theorem:

Theorem 2.38
Assume that Arnoldi's procedure (Alg. 2.6) is applied to a Hermitian matrix $A \in \mathbb{C}^{n \times n}$. Then the coefficients h_{ij} generated by the algorithm are such that

$$h_{ij} = 0 \qquad \text{for } 1 \leq i < j - 1 \tag{2.54}$$

$$h_{j,j+1} = h_{j+1,j}, \qquad j = 1, 2, \ldots, m \tag{2.55}$$

It is important to note that all coefficients h_{ij} are real, even if the input matrix A has complex entries. In other words, the matrix H_m obtained from the Arnoldi process is tridiagonal and real-symmetric.

Proof: To see that H_m is Hermitian, apply the Arnoldi relation (2.52) and exploit that A is Hermitian:

$$H_m = V_m^* A V_m = V_m^* A^* V_m = (V_m^* A V_m)^* = H_m^*$$

Since H_m is not only Hermitian, but also upper Hessenberg, we see that H_m is tridiagonal, i.e. we have (2.54). The diagonal entries h_{ii} of a Hermitian matrix H_m are known to be real because of $h_{ii} = \overline{h_{ii}}$. The fact that the entries $h_{j,j+1}$ are real, too, follows from their construction as a norm of a vector in Line 5 of the Arnoldi procedure (Alg. 2.6). $\qquad$ □

It is common to use the following

Notation 2.39

$$\alpha_j \equiv h_{jj} \quad \text{and} \quad \beta_j \equiv h_{j-1,j}$$

The resulting H_m matrix is denoted by

$$T_m = \begin{bmatrix} \alpha_1 & \beta_2 & & & \\ \beta_2 & \alpha_2 & \beta_3 & & \\ & & \ddots & & \\ & & \beta_{m-1} & \alpha_{m-1} & \beta_m \\ & & & \beta_m & \alpha_m \end{bmatrix} \tag{2.56}$$

□

The Gram-Schmidt orthogonalization in Alg. 2.6 obviously simplifies to a simple three-term recurrence and along with the notation 2.39 we can formulate the following simplification of Alg. 2.6 for the Hermitian case which is known as *Lanczos algorithm*:

Algorithm 2.7: Lanczos

---

1   choose a vector v_1 such that $\|v_1\|_2 = 1$. Set $\beta_1 \equiv 0$, $v_0 \equiv 0$
2   **for** $j = 1, 2, \ldots, m$ **do**
3   compute $w_j = Av_j - \beta_j v_{j-1}$
4   compute $\alpha_j = (w_j, v_j)$
5      $w_j = w_j - \alpha_j v_j$
6      $\beta_{j+1} = \|w_j\|_2.$
7   **if** $\beta_{j+1} = 0$ **then**
8   stop
9   **end if**
10      $v_{j+1} = w_j/\beta_{j+1}$
11 **end for**

Proposition 2.37 directly carries over to the following analogon for the Lanczos procedure:

Proposition 2.40 (Lanczos relation, Lanczos factorization)
Denote by V_m the $n \times m$ matrix with column vectors $v_1, \ldots, v_m$; by $\bar{T}_m$ the $(m+1) \times m$ triangular matrix whose nonzero entries α_i and β_i are defined by Algorithm 2.7; and by T_m the matrix obtained from $\bar{T}_m$ by deleting its last row. Then the following relations hold:

$$\begin{aligned} AV_m &= V_m T_m + w_m e_m^* & (2.57) \\ &= V_{m+1} \bar{T}_m & (2.58) \\ V_m^* AV_m &= T_m & (2.59) \end{aligned}$$

Remark 2.41
In exact arithmetic the above algorithm guarantees the orthonormality of the Lanczos vectors assembled in V and for computational purposes it would be sufficient to store the two most recent vectors v_{j-1} and v_j in order to compute v_{j+1}. However, in practice, due to inevitable rounding errors, the Lanczos vectors v_j rather soon lose their global

orthogonality. Consequently, one may be forced to take measures in order to recover the orthogonality and for this purpose several strategies have been developped, e.g. partial or selective reorthogonalization (for more details on this issue see [91]).   □

2.3.4. Singular Value Decomposition (SVD)

Singular value decompositions are a generalization of the concept of eigendecompositions to rectangular matrices $A \in \mathbb{C}^{n \times p}$ and will turn out useful later on as a device to solve least squares problems (see Section 3.3.1.3). The following theorem gives a precise definition and guarantees the existence of such decompositions:

Theorem 2.42 (Singular value decomposition (SVD))
Let $A \in \mathbb{C}^{n \times p}$, where $n \geq p$. Then there exist orthonormal matrices

$$U = [u_1, \ldots, u_n] \in \mathbb{C}^{n \times n} \quad \text{and} \quad V = [v_1, \ldots, v_p] \in \mathbb{C}^{p \times p} \tag{2.60}$$

such that

$$U^* A V = \begin{bmatrix} \Sigma \\ 0 \end{bmatrix} \in \mathbb{R}^{n \times p} \tag{2.61}$$

where

$$\Sigma = diag(\sigma_1, \ldots, \sigma_p) \in \mathbb{R}^{p \times p} \tag{2.62}$$

and

$$\sigma_1 \geq \sigma_2 \geq \ldots \geq \sigma_p \geq 0 \tag{2.63}$$

i.e. the scalars σ_i $(i = 1, \ldots, p)$ are always non-negative real numbers.

Proof: see [49]   □

Definition 2.43
We call

- σ_i $(i = 1, \ldots, p)$ singular values *of A*
- u_i $(i = 1, \ldots, n)$ left singular vectors *of A*
- v_i $(i = 1, \ldots, p)$ right singular vectors *of A*
- σ_{min} the smallest singular value *of A and v_{min} the related singular vector*
- σ_{max} the largest singular value *of A and v_{max} the related singular vector*

Corollary 2.44
Under the preliminaries of Theorem 2.42 it holds

$$\left. \begin{array}{rcl} A v_i & = & \sigma_i u_i \\ A^* u_i & = & \sigma_i v_i \end{array} \right\} \quad i = 1, \ldots, p \tag{2.64}$$

Proof: straightforward verification □

The following corollary highlights the relation between singular value decompositions and eigendecompositions:

Corollary 2.45
Let $A \in \mathbb{C}^{n \times p}$, where $n \geq p$ and $U^* A V = \begin{bmatrix} \Sigma \\ 0 \end{bmatrix}$ be an SVD of A. Then

$$V^* A^* A V = \Sigma^2 = \mathrm{diag}(\sigma_1^2, \ldots, \sigma_m^2) \in \mathbb{R}^{p \times p} \tag{2.65}$$

This shows that the squares of the singular values of A are the eigenvalues of the cross-product matrix $A^* A$. The right singular vectors of A are the eigenvectors of $A^* A$.

Proof: simple verification □

Corollary 2.45 allows us to derive an analogon of the variational characterization of eigenvalues stated in Proposition 2.15 for the case of singular values:

Corollary 2.46
Let $A \in \mathbb{C}^{n \times p}$, where $n \geq p$ and $A = U^* A V$ its SVD according to Theorem 2.42. Then

$$\min_{\substack{y \in \mathbb{C}^m \\ \|y\|_2 = 1}} \|A y\|_2 = \sigma_{min} \quad \text{and} \quad \|A v_{min}\|_2 = \sigma_{min} \tag{2.66}$$

$$\max_{\substack{y \in \mathbb{C}^m \\ \|y\|_2 = 1}} \|A y\|_2 = \sigma_{max} \quad \text{and} \quad \|A v_{max}\|_2 = \sigma_{max} \tag{2.67}$$

Proof: Application of Proposition 2.15 to $B = A^* A$ results in

$$\sigma_{min}(A)^2 = \lambda_{min}(B) = \min_{y \neq 0} \frac{y^* A^* A y}{y^* y} = \min_{\|y\|_2 = 1} y^* A^* A y = \min_{\|y\|_2 = 1} \|A y\|_2^2 \tag{2.68}$$

Furthermore, because of (2.64)

$$\|A v_{min}\|_2 = \|\sigma_{min} u_{min}\|_2 = \sigma_{min} \|u_{min}\|_2 = \sigma_{min} \tag{2.69}$$

which proves the second part of (2.66). The assertion (2.67) follows analogously. □

2.3.5. Kronecker Products

Kronecker products of will turn out useful later on as matrix representations of abstract operator tensor products (see Sections 5.8 and 6.4.3). A theoretical introduction with respect to the use of Kronecker products in multilinear algebra is given e.g. in [79].

Definition 2.47 (Kronecker product)
If $B \in \mathbb{C}^{n \times m}$ and $C \in \mathbb{C}^{p \times q}$, then their Kronecker product is given by

$$A = B \otimes C = \begin{bmatrix} b_{11}C & b_{12}C & \cdots & b_{1n}C \\ b_{21}C & b_{22}C & \cdots & b_{2n}C \\ \vdots & \vdots & \ddots & \vdots \\ b_{n1}C & b_{m2}C & \cdots & b_{nm}C \end{bmatrix} \in \mathbb{C}^{(np) \times (mq)} \tag{2.70}$$

Some useful properties of Kronecker products are summarized in the following lemma:

Lemma 2.48 (Kronecker product properties)

1. The Kronecker product is associative, i.e. it holds

$$(A \otimes B) \otimes C = A \otimes (B \otimes C) \tag{2.71}$$

2. If $I_r \in \mathbb{C}^{r \times r}$ denotes the identity matrix of order r, then

$$I_p \otimes (I_q \otimes A) = I_{pq} \otimes A \tag{2.72}$$

3. The Kronecker product is distributive, i.e. it holds

$$\begin{aligned} A \otimes (B + C) &= A \otimes B + A \otimes C \tag{2.73} \\ (A + B) \otimes C &= A \otimes C + B \otimes C \tag{2.74} \end{aligned}$$

   provided that the arising ordinary matrix multiplications are defined.

4. Notice, that the Kronecker product is in general not commutative.

5. If the ordinary matrix multiplications AC and BD are defined, then

$$(A \otimes B)(C \otimes D) = (AC) \otimes (BD) \tag{2.75}$$

   This property is often referred to as *mixed-product property* of the Kronecker product.

6. If A and B are nonsingular square matrices, then

$$(A \otimes B)^{-1} = A^{-1} \otimes B^{-1} \tag{2.76}$$

Proof: straightforward verification by application of Def. 2.47, see [49] and [126] $\square$

3. Methods for Computing Partial Eigensystems of Hermitian Matrices

In this chapter we give a general survey of different classes of eigensolvers and formulate abstract generic algorithms to outline the similarities of the algorithms within a class. The *Jacobi-Davidson method*, which is in the center of our interest, will be treated in detail in the following chapter taking advantage of the preliminary work on iterative projection methods in this chapter. Since the dichotomy between direct and iterative methods will be of importance later on, and since the state-of-the-art situation is not well-reflected in the standard textbooks, we will also give brief descriptions of the most important direct methods along with useful references.

3.1. Iterative Single Vector Methods

Iterative single vector methods are the simplest algorithms for eigenvalue computations. They are easy to understand and straightforward to implement in a computer code, but they are in general only capable of computing one approximate eigenpair. As they are often related to more sophisticated methods to be discussed later on, they are notwithstanding of general interest, and a brief discussion of these methods shall be given below. There are also *iterative multiple vector methods* (for details see [49], [91] or [100]), which are generalizations of *iterative single vector methods* designed for the computation of several eigenpairs. Their convergence, however, is often rather slow, and for this reason they are seldomly used. In practice, one prefers to use the more efficient and modern *iterative projection methods* instead which will be discussed later on.

3.1.1. Power Method

Let us without loss of generality assume that the eigenvalues $\lambda_i \in \mathbb{R}$ of a Hermitian matrix $A \in \mathbb{C}^{n \times n}$ be ordered by ascending magnitude of their moduli

$$|\lambda_1| \leq |\lambda_2| \leq \ldots < |\lambda_n|$$

i.e. λ_n is the dominant eigenvalue of A and q_n the associated eigenvector. If one chooses an appropriate starting vector z_0 (i.e. z_0 is not orthogonal to q_n), then the sequence of vectors

$$q^{(i)} = \frac{A^i z_0}{\|A^i z_0\|_2}$$

Algorithm 3.1: Power method for Hermitian matrices

---

1 **function** (λ, u)=power$(A, z^{(0)})$

2 $k = 0$

3 **repeat**

4     $\quad q^{(k)} = z^{(k)}/\|z^{(k)}\|_2$

5     $\quad z^{(k+1)} = Aq^{(k)}$

6     $\quad \lambda^{(k+1)} = [q^{(k)}]^* z^{(k+1)}$

7     $\quad r = z^{(k+1)} - \lambda^{(k+1)} q^{(k)}$

8     $\quad k = k + 1$

9 **until** $(\|r\|_2 \leq \epsilon_M |\lambda|)$

10 **return** (λ_k, q_k)

---

converges to the sought-after eigenvector q_n, i.e. the vectors $q^{(i)}$ become increasingly parallel to q_n.

Essentially, the convergence of this *power method* is linear and depends on the ratio $|\lambda_{n-1}/\lambda_n|$. We cite the following theorem from [49], which renders this more precise for the Hermitian case:

Theorem 3.1 (Convergence of the power method for Hermitian matrices)
Suppose $A \in \mathbb{C}^{n \times n}$ is Hermitian and that

$$Q^* AQ = \mathrm{diag}(\lambda_1, \ldots, \lambda_n)$$

where $Q = [q_1, \ldots, q_n]$ is unitary and $|\lambda_1| \leq |\lambda_2| \leq |\lambda_{n-1}| < |\lambda_n|$. Let the vectors $q^{(k)}$ be specified by Algorithm 3.1 and define the "error-angle" $\theta_k \in [0, \pi/2]$ by

$$\cos(\theta_k) = |q_n^* q^{(k)}|$$

If $\cos(\theta_0) \neq 0$, then

$$|\sin(\theta_k)| \;\leq\; \tan(\theta_0) \left| \frac{\lambda_{n-1}}{\lambda_n} \right|^k$$

$$|\lambda^{(k)} - \lambda_n| \;\leq\; |\lambda_1 - \lambda_n| \, \tan^2(\theta_0) \left| \frac{\lambda_{n-1}}{\lambda_n} \right|^{2k}$$

Proof: see [49]   □

The power method (Alg. 3.1) is slow to converge, especially if λ_n and λ_{n-1} are not well-separated from each other, and it is only appropriate for finding the dominant eigenpair (λ_n, q_n). Nonetheless, it is of theoretical interest, as it is related to other algorithms, such as the *inverse iteration* (Alg. 3.2) to be discussed in the following, the *QR method* (Alg. 3.7) or the *Lanczos method* (Alg. 3.15).

3.1.2. Inverse iteration (INVIT)

The disadvantages of the power method can be eliminated by a simple, but fundamental idea. If an approximation $\sigma \neq \lambda_k$ to an eigenvalue λ_k of A is available, such that

$$|\lambda_k - \sigma| < |\lambda_i - \sigma| \qquad \forall i \neq k$$

then obviously $(\lambda_k - \sigma)^{-1}$ becomes the single dominant eigenvalue of $B := (A - \sigma I)^{-1}$. In order to obtain an approximation to A's eigenvector q_k, it is now straightforward to apply the power method to B. The resulting algorithm is called *inverse iteration* or *inverse power method* and, apart from a starting vector $z^{(0)} \neq 0$, it requires an approximation σ to the sought-after eigenvalue λ_k:

Algorithm 3.2: Inverse iteration for Hermitian matrices (INVIT)

1 **function** $(\lambda, u) = $invit$(A, \sigma, z^{(0)})$
2 $k = 0$
3 **repeat**
4 $\quad$ $x_k = z_k / \|z_k\|_2$
5 $\quad$ solve $(A - \sigma I)z^{(k+1)} = x^{(k)}$ for $z^{(k+1)}$
6 $\quad$ $\theta^{(k+1)} = [x^{(k)}]^* z^{(k+1)}$
7 $\quad$ $r = z^{(k+1)} - \theta^{(k+1)} x^{(k)}$
8 $\quad$ $k = k + 1$
9 **until** $(\|r\|_2 \leq \epsilon)$
10 **return** $(\sigma + 1/\theta^{(k)}, x^{(k)})$

Remark 3.2

- *The parameter σ is usually referred to as a* shift, *and the idea behind the inverse iteration is therefore called shift-and-invert-approach. Note that this a general strategy, which also extends to other methods, and which may be used whenever the convergence of a method is slow and unsatisfactory.*

- *As indicated in Line 5 of Algorithm 3.2, one does not explicitly compute the inverse of $(A - \sigma I)$, but solves the arising linear systems in each step of the iteration. Hence, a suitable decomposition (e.g. LU factorization) of $(A - \sigma I)$ is required, which may be a drawback when such a factorization is expensive, e.g. when A is large and dense.*

- *Unlike the power method, the inverse iteration is of great use in practice. Applying Theorem 3.1 to $(A - \sigma I)^{-1}$ shows that the convergence rate of inverse iteration is determined by the factor*

$$\kappa = \left| \frac{\lambda_k - \sigma}{\lambda_j - \sigma} \right|$$

where $\lambda_j - \sigma$ is the second smallest eigenvalue of $(A - \sigma I)$ in modulus. Thus, if σ is a very good approximation to an eigenvalue λ_k, rapid convergence to the eigenpair (λ_k, q_k) can be expected.

- At first glance, it appears to be dangerous to choose the shift σ too close to an eigenvalue λ_k, since then $(A - \sigma I)$ is nearly singular and solving the associated ill-conditioned linear systems seems almost inevitably to lead to erroneous results. However, PARLETT ([91], Chapter 4.3, "Advantages of an Ill-Conditioned System") exposes in a detailed analysis that these fears are unjustified, as only the "direction" of an eigenvector is of interest, and as the error introduced by solving the nearly singular systems is almost entirely in the direction of the sought-after eigenvector q_k.

- Careful implementations of inverse iteration are available, e.g. in LAPACK [2, 7], in the form of the routines xSTEIN, and these are employed to obtain eigenvector approximations of high quality if only eigenvalue approximations are available (e.g. obtained by means of bisection, see Paragraph 3.2.2.2).

$\square$

3.1.3. Rayleigh Quotient Iteration (RQI)

A natural extension of the *inverse iteration* is to replace the constant shift σ in Alg. 3.2 by the *Rayleigh quotient* related to the current iterate $x^{(k)}$, which is known to be an optimal eigenvalue approximation (cf. Prop. 2.14). Consequently, the shift now varies in each iteration step, and the resulting *Rayleigh quotient iteration* scheme reads as given in Alg. 3.3:

Algorithm 3.3: Rayleigh quotient iteration for Hermitian matrices (RQI)

---
1 **function** (λ, u)=RQI(A, x_0)
2 $k = 0$, $\mu_0 = x_0^* A x_0$
3 **repeat**
4   solve $(A - \mu_k I) z^{(k+1)} = x_k$ for $z^{(k+1)}$
5     $x^{(k+1)} = z^{(k+1)} / \|z^{(k+1)}\|_2$
6     $\mu^{(k+1)} = [x^{(k+1)}]^* A x^{(k+1)}$
7     $r = A x^{(k+1)} - \mu^{(k+1)} x^{(k+1)}$
8     $k = k + 1$
9 **until** $(\|r\|_2 \le \epsilon)$
10 **return** $(\mu^{(k)}, x^{(k)})$

Remark 3.3
- The RQI converges cubically in most cases. We refer to the book by PARLETT [91] for a detailed convergence analysis and an in-depth discussion of the topic.

- Despite its rapid convergence the RQI is seldomly used in practice, as the shift μ now varies, and thus, in every iteration step a new factorization of $(A - \mu I)$ is required, which is often too expensive, and which is almost never compensated for by possible savings in the number of iteration steps.

- Nevertheless, the RQI is of importance in the following, as it is closely related to the Jacobi-Davidson method (Alg. 4.2) which is in the focus of our interest.

□

3.2. Direct Methods

The term *direct method* for an eigenproblem is common in the literature, but somewhat misleading, as looking for eigenvalues λ of $A \in \mathbb{C}^{n \times n}$ is equivalent to determining the roots of its characteristic polynomial $p_A(\mu)$ (Theorem 2.3). For $n > 4$ a well-known result from Galois theory [72] states, that there is in general no explicit formula for the roots of $p_A(\mu)$, which implies that any numerical method for computing eigenvalues λ must necessarily have an iterative component. A more suitable explanation for *direct* lies in the fact that methods of this class explicitly access and manipulate the matrix coefficients a_{ij}, and that they "immediately" yield the sought-after eigensystems (at least for small and medium-sized problems, say $n < 2000$). Most often, these methods rely on the fact that the application of similarity transformations on A does not affect its eigenvalues (Proposition 2.5), and thus, they attempt to obtain a matrix of simpler structure whose eigenvalues are cheaper to compute. For this reason, one often also encounters the terms *transformation method* or *direct transformation method*. If one is not only interested in the eigenvalues of A alone, but also in all or some of their corresponding eigenvectors, it is necessary to store the transformation matrix U in order to recover the eigenvectors of A. To avoid computational overhead, it has proven advantageous to transform the Hermitian matrix $A \in \mathbb{C}^{n \times n}$ to tridiagonal form

$$T = U_{AT}^* A U_{AT} = \begin{bmatrix} d_1 & e_1 & & & & \\ e_1 & d_2 & e_2 & & & \\ & e_2 & d_3 & e_3 & & \\ & & \ddots & \ddots & \ddots & \\ & & & e_{n-2} & d_{n-1} & e_{n-1} \\ & & & & e_{n-1} & d_n \end{bmatrix} \in \mathbb{R}^{n \times n} \qquad (3.1)$$

by means of a suitable unitary U_{AT} before carrying out the actual eigenvalue computation on T. It is important to note that the coefficients of T can always be achieved to *real*, even if A is complex-Hermitian, which is analogous to the situation of the Lanczos procedure (Alg. 2.7) which generates tridiagonal matrices T_m that are guaranteed to be real-symmetric. This approach is formalized in the generic three-step algorithm, Alg. 3.4, when one is interested in the (partial) eigensystem of A with the index set for $k \leq n$ eigenvectors of A given by $\mathcal{I} = \{i_1, \dots, i_k\}$:

Algorithm 3.4: Generic direct algorithm for dense Hermitian matrices

  function $[\Lambda, \widehat{Q}] = eigsys(A, \mathcal{I})$

1 reduce A to tridiagonal form T by means of a suitable unitary transformation U_{AT}:

$$T = U_{AT}^* \cdot A \cdot U_{AT}$$

2 compute eigendecomposition of T by an algorithm of choice:

$$\Lambda = U_T^* \cdot T \cdot U_T$$

$$\Lambda = \text{diag}(\lambda_1, \ldots, \lambda_n)$$

3 form $\widehat{U}_T = U_T[\,[i_1 : i_k]\,, :]$ and back-transform $\widehat{U}_T$ to obtain the eigenvectors of A:

$$\widehat{Q} = U_{AT} \cdot \widehat{U}_T$$

Algorithm 3.4 provides the framework for the LAPACK [2] driver routines for symmetric and Hermitian eigenproblems (xSYEV, xSYEVX, xSYEVR and xSYEVD, cf. [7]).

The steps of of Algorithm 3.4 involve the following costs:

1. Tridiagonalization: $\mathcal{O}(\frac{4}{3}n^3)$ flops

2. Computation of the desired eigenpairs of T: $\mathcal{O}(n^2)$ / $\mathcal{O}(k \cdot n)$ flops (depending on the choice of algorithm, see below)

3. Back-transformation: $\mathcal{O}(2kn^2)$ flops

In what follows, we can only give a brief state-of-the-art survey and a rather rough outline of how reduction of a general symmetric matrix to tridiagonal form and the solution of the resulting eigenproblem can be accomplished. For a thorough and comprehensive exposition of the matter we refer to [49], [91], [122] and the references in the following paragraphs.

3.2.1. Reduction to Tridiagonal Form

As can be seen from the introductory part, the costs for the tridiagonalization of A become dominant for $k \ll n$. Since this situation is typical of the eigenproblems we are dealing with in this thesis, it is rewarding to have a closer look at the technical details. In the following, we will restrict ourselves to the case that the matrix to be tridiagonalized is real-symmetric. However, paying attention to Remark 2.32 there is no major difficulty in devising analogous algorithms for complex-Hermitian input matrices, implementations of which are to be found e.g. in the form of the LAPACK [2] routines xCHTRD (see [7]).

3.2.1.1. Standard Approach

The common approach to obtain (3.1) is to successively apply $n - 2$ *Householder transformations* H_k on A. To see how a transformation H_k is computed, let us suppose that we have already determined $k - 1$ *Householder matrices* $H_1, \ldots, H_{k-1}$ such that $A_{k-1} = (H_1 \cdots H_{k-1})^* A(H_1 \cdots H_{k-1})$ and

$$A_{k-1} = \begin{bmatrix} T_{k-1} & a & 0 \\ a^* & \delta & b^* \\ 0 & b & R_{n-k} \end{bmatrix}$$

where the leading principal submatrix $T_{k-1} \in \mathbb{R}^{(k-1)\times(k-1)}$ is tridiagonal, $a \in \mathbb{R}^{k-1}$, $b \in \mathbb{R}^{n-k}$, $\delta \in \mathbb{R}$ a scalar and $R_{n-k} \in \mathbb{R}^{(n-k)\times(n-k)}$ the remainder matrix. Let now $P_k \in \mathbb{R}^{(n-k)\times(n-k)}$ be the *Householder matrix*, such that $\tilde{b} = P_k \cdot b = \|b\|_2 \cdot e_1 \in \mathbb{R}^{n-k}$, then the sought-after transformation H_k is made up by

$$H_k = \begin{bmatrix} I_{n-k} & 0 \\ 0 & P_k \end{bmatrix}$$

and one obtains

$$A_k = H_k A_{k-1} H_k = \begin{bmatrix} T_{k-1} & a & 0 \\ a^* & \delta & \tilde{b}^* \\ 0 & \tilde{b} & \tilde{R}_{n-k} \end{bmatrix}$$

A_k has the desired property, i.e. the leading $k \times k$ principal submatrix is now tridiagonal. Clearly, if $U_{AT} = H_1 \cdots H_{n-2}$, then $T = U_{AT}^* A U_{AT}$ is tridiagonal. This is formalized in the following algorithm which makes use of Alg. 2.1 to determine y and τ:

Algorithm 3.5: Householder tridiagonalization

---

1 **for** $k = 1, \ldots, n - 2$ **do**
2      $[y, \tau] = \mathbf{house}(A(k + 1 : n, k))$
3      $z = \tau A(k + 1 : n, k + 1 : n)y$
4      $v = z - (\tau z^* y/2)y$
5      $A(k + 1, k) = \|A(k + 1 : n, k)\|_2,\ A(k, k + 1) = A(k + 1, k)$
6      $A(k + 1 : n, k + 1 : n) = A(k + 1 : n, k + 1 : n) - yv^* - vy^*$
7 **end for**

However, for an implementation in a computer code it is desirable to have a high fraction of matrix-matrix operations (level 3 BLAS [1]), because the available computer architecture and memory hierarchy can efficiently be exploited, unecessary memory traffic is avoided and considerable gain in performance can be achieved. To this end, a blocked version of Algorithm 3.5 has been proposed in [35], the main idea being to aggregate p *Householder transforms* and to perform the matrix updates block-wise in $(n-2)/p$ steps:

Algorithm 3.6: Blocked Householder tridiagonalization

1 $N = (n-2)/p$
2 $Y = [], V = []$
3 **for** $k = 1, N$ **do**
4      $s = (k-1)p + 1$
5   **for** $j = s, s + p - 1$ **do**
6          $a_j = a_j - YV(j,:)^* - VY(j,:)^*$;   /* update column j of A */
7          $[y_{j+1}, \tau_{j+1}] = \mathbf{house}(A(j+1:n, j))$
8          $z = \tau_j A y_j - \tau_j Y(V^* y_j) - \tau_j V(Y^* y_j)$
9          $v_j = z - (\frac{\tau_j}{2} z^* y_j) y_j$
10         $Y = [Y, y_j], V = [V, v_j]$
11   **end for**
12     $A = A - Y \cdot V^* - V \cdot Y^*$
13 **end for**

This is essentially the algorithm used in the LAPACK [2] routine xSYTRD to reduce an arbitrary real symmetric matrix to tridiagonal form. However, about 50% of the operations are still matrix-vector operations, such that the advantage of Algorithm 3.6 over 3.5 may not be that great.

3.2.1.2. Two-stage Approach

BISCHOF, LANG and SUN [12, 13] have proposed a strategy which is more appropriate for our situation, where the number of sought-after eigenpairs k is small as compared to the problem size n. They suggest to do the tridiagonalization in two steps:

1. Reduce the full symmetric matrix A to a banded matrix B with semi-bandwith $b > 0$ by means of a suitable unitary transformation U_{AB}:

$$B = U_{AB}^* A U_{AB} \tag{3.2}$$

2. Tridiagonalize B by means of a unitary transformation U_{BT}

$$T = U_{BT}^* B U_{BT} \tag{3.3}$$

The costs for the intermediate bandwidth reduction are almost as high as the costs for the complete tridiagonalization in Algorithms 3.5 and 3.6. However, almost all of the involved operations now can be realized using matrix-matrix operations (level 3 BLAS [1]). The subsequent tridiagonalization of the banded matrix only takes additional $\mathcal{O}(6bn)$ flops, but none of these can be done by means of level 3 BLAS [1] operations. Although slightly more expensive in terms of flops, this two-step approach can lead to a significant speedup, provided that the machine in use has fast processors and a comparatively slow memory. It is important to note, however, that this advantage only exists if one is interested in the tridiagonalization of A alone. If one also opts for eigenvectors (after computing m relevant eigenpairs of the tridiagonal T), one has to take into account the back-transformation, which now consists of 2 steps, *each of them* involving $\mathcal{O}(2mn^2)$ flops. Clearly, this is twice as expensive as the corresponding back-transformation in the standard approach, and consequently, the two-stage approach can only be competitive if the number of sought-after eigenpairs m is a small fraction of n. Luckily, this meets our situation, as we are typically only interested in about 10% of the lowest eigenvalues. Numerical experiments in [12] and especially in [87] demonstrate that two-step reduction in combination with an efficient tridiagonal eigensolver (*RRR method*, see below) applied to eigenproblems that are also examined in this thesis, is superior to the corresponding combination of standard tridiagonalization and tridiagonal solver.

3.2.2. Methods for the Symmetric Tridiagonal Eigenproblem

We have already seen that it is sufficient to consider algorithms for computing eigensystems of real-symmetric tridiagonal matrices T, as complex entries in T can be avoided by an appropriate choice of the transformation matrix. There exists a number of well-established methods for computing eigensystems of tridiagonals, most notably the *QR method*, which was introduced in the beginning of the 60s of the last century, and for which efficient and reliable software is available. In the meantime, a lot of progress has been made in the theoretical understanding of symmetric tridiagonal eigenproblems, and this has consequently lead to methods with even superior properties, e.g. the recent *RRR method*. However, rather surprisingly, about ten years after its introduction, it is still hard to find textbooks on numerical linear algebra that provide a comprehensive survey of methods for the symmetric tridiagonal eigenproblem and almost always the RRR algorithm is not even mentioned, although it is going to be the standard method in the forthcoming LAPACK [2] release. Thus, for non-experts the state-of-the-art situation is somewhat obscure, and for this reason, it is appropriate to give a brief survey of existing methods, their software availability and their pros and cons.

3.2.2.1. QR Algorithm

For a long time, the *QR method* which is due to FRANCIS [38, 39] and KUBLANOVSKAÂ[1] [69] has been the standard approach for computing eigenpairs of small and medium-sized matrices. Basically, it relies on the following simple iteration scheme:

---

[1]see also remarks on transliteration of Russian names in Appendix A.2

Algorithm 3.7: QR iteration (explicitly shifted)

---

1 **for** $k = 0, 1 \ldots$ **do**
2 | choose a suitable shift $\sigma_k \in \mathbb{R}$
3 | $T_k - \sigma_k I = Q_k R_k$ (QR decomposition)
4 | $T_{k+1} = R_k Q_k + \sigma_k I$
5 **end for**

Obviously, Algorithm 3.7 generates a sequence of unitarily similar matrices

$$T_{k+1} = Q_k^* T_k Q^k = (Q_0 \cdot Q_1 \cdots Q_k)^* T_0 (Q_0 \cdot Q_1 \cdots Q_k) \tag{3.4}$$

It is easy to see that all T_k are tridiagonal. More importantly, it can be shown (cf. [49], [91] for details of the related proof and for a mathematically precise notion of convergence in this context) that the matrices T_k generated by Algorithm 3.7 converge to an eigendecomposition $\Lambda = Q^* T Q$, symbolically

$$T_k \rightarrow \Lambda = \mathrm{diag}(\lambda_1, \ldots, \lambda_n), \quad \prod_{i=0}^{k} Q_i \rightarrow Q \quad (k = 0, 1, \ldots) \tag{3.5}$$

We conclude with some remarks:

Remark 3.4

- *The QR decomposition in Line 3 of Alg. 3.7 can be computed by applying a sequence of $n - 1$ Givens rotations (cf. Section 2.3.1.2).*

- *The QR algorithm is closely related to the inverse iteration (Alg. 3.2), to the power method (Alg. 3.1) and to the Rayleigh quotient iteration (Alg. 3.3). For a detailed discussion of this matter we refer to [49],[91] and [122].*

- *In practical implementations one uses an implicitly shifted version of Algorithm 3.7, in which the matrix $T_k - \sigma_k I$ is not explicitly formed. It can be shown that this essentially leads to the same sequence of matrices Q_k, and thus, does not affect the convergence behavior. For full technical details on implicitly shifted QR iterations and the resulting bulge chase mechanism see [49], [91] and [122]*

- *A popular choice for the shift σ_k in Line 2 of Alg. 3.7 is the so-called Wilkinson shift, which is the eigenvalue of*

$$T_k(n - 1 : n, n - 1 : n) = \begin{bmatrix} d_{n-1} & e_{n-1} \\ e_{n-1} & d_n \end{bmatrix} \tag{3.6}$$

closer to d_n and given by

$$\sigma_k = d_n + t - \mathrm{sign}(t)\sqrt{t^2 - e_{n-1}^2}, \quad t = \frac{d_{n-1} - d_n}{2} \tag{3.7}$$

- To obtain eigenvectors of T, it is necessary to accumulate the Q_k in Alg. 3.7.

- The implicitly shifted QR method is part of the LAPACK [2] software in the form of the routine xSTEQR, and it is used by the standard driver DSYEV for computing eigensystems of arbitrary symmetric matrices. However, for future releases it is expected that the QR algorithm will be superseded by the RRR method (which is in general faster and requires less work space) as algorithm for the standard driver.

$\square$

3.2.2.2. Bisection Method and Inverse Iteration

In practice, one is often interested in eigenpairs that lie in an interval $[\alpha, \beta]$ or in specific subsets, say eigenpairs related to the range from the i-th to the j-th eigenvalue. In these cases, bisection combined with subsequent *inverse iteration* may be an interesting choice. The idea behind *bisection methods* for finding eigenvalues of a tridiagonal matrix T is based on two observations:

1. The characteristic polynomial of a tridiagonal matrix T (3.1) may be expressed recursively using the following relation

$$p_r(x) = (d_r - x)p_{r-1}(x) - e_{r-1}^2 p_{r-2}(x), \qquad p_0(x) = 1, \quad p_1(x) = d_1 - x \quad (3.8)$$

   where $p_r(x) = \det(T_r - xI)$ is the characteristic polynomial of T's leading principal $(r \times r)$-submatrix T_r (verification by induction).

2. The number $N(\mu)$ of T's eigenvalues being less than μ is characterized by the following theorem:

Theorem 3.5 (Sturm sequence property)
If the tridiagonal matrix T in (3.1) has no zero subdiagonal entries, then the eigenvalues of T_{r-1} strictly separate the eigenvalues of T_r:

$$\lambda_1(T_r) < \lambda_1(T_{r-1}) < \lambda_2(T_r) < \lambda_2(T_{r-1}) < \ldots < \lambda_{r-1}(T_r) < \lambda_{r-1}(T_{r-1}) < \lambda_r(T_r) \quad (3.9)$$

Furthermore, the number $N(\mu)$ of T's eigenvalues that are less than μ is given by the number of sign changes in the sequence

$$\{p_0(\mu), p_1(\mu), \ldots, p_n(\mu)\} \qquad (3.10)$$

where p_r are the polynomials defined by (3.8). If $p_r(\mu)$ happens to be zero, then this is counted as a sign change.

Proof: The fact that the eigenvalues of T_{r-1} weakly separate those of T_r follows immediately by applying the inclusion principle (Lemma 2.18). To see that the separation is also strict, suppose that $p_r(\mu) = p_{r-1}(\mu)$ for some r and μ. As T is unreduced, it follows from (3.8) that $p_0(\mu) = p_1(\mu) = \ldots = p_r(\mu)$, which is clearly a contradiction. A proof for the assertion on $N(\mu)$ may be found in [132] $\square$

By Geršgorin's disc theorem, Theorem 2.8, we know that all eigenvalues $\lambda_k(T)$ are guaranteed to lie in the interval $[y, z]$ where

$$y = \min_{i=1,\ldots,n} d_i - |e_i| - |e_{i-1}| \qquad z = \max_{i=1,\ldots,n} d_i + |b_i| + |b_{i-1}| \qquad (3.11)$$

This directly gives rise to the following surprisingly simple bisection procedure for determining an eigenvalue λ_k ($k \in \{1, \ldots, n\}$) of T:

Algorithm 3.8: Bisection for λ_k

---

1 **function** $\mu = \mathbf{bisec}(T, k, \varepsilon)$
2 **while** $|z - y| \geq \varepsilon$ **do**
3 $\quad$ $\mu = (y + 2)/2$
4 $\quad$ **if** $N(\mu) < k$ **then**
5 $\quad\quad$ $z = \mu$
6 $\quad$ **else**
7 $\quad\quad$ $y = \mu$
8 $\quad$ **end if**
9 **end while**
10 **return** μ

However, Algorithm 3.8 as presented above is prone to underflow. Careful implementations, e.g. the LAPACK [2] routines xSTEBZ (cf. [7]) use recurrences different from (3.8) to avoid possible instabilities. The so-called LAPACK [2] *expert driver* xSYEVX, which allows for the computation of selected eigenpairs, uses the bisection routine xSTEBZ to determine the sought-after eigenvalues, and it subsequently calls xSTEIN to compute the associated eigenvectors. A feature which makes bisection attractive for the computation of partial eigensystems is the fact that only $\mathcal{O}(nk)$ flops are needed for the computation of k eigenvalues. In case that the eigenvalues are well separated, the application of *inverse iteration* will also cost $\mathcal{O}(nk)$ flops, which is the *best case*. However, when dealing with clustered eigenvalues, additional work in the form of explicit Gram-Schmidt orthogonalization has to be invested in order to make sure not to repeatedly obtain the same eigenvectors. This involves additional $\mathcal{O}(nk^2)$ flops and is the *worst case*. In the forthcoming release of LAPACK [2] the expert driver will make use of the routine xSTEGR (an implementation of the *RRR method* discussed below) instead of *inverse iteration*, which will lead to a gain in performance as the RRR algorithm (see below) uses a strategy to avoid explicit orthogonalization of eigenvectors.

3.2.2.3. Divide-and-conquer Method

The *Divide-and-conquer method* for computing eigensystems of a tridiagonal matrix T was proposed by CUPPEN [28] in 1981. The basic idea can be described as follows:

1. *Divide*:
   "Tear" $T \in \mathbb{R}^{n \times n}$ in two halves $T_1 \in \mathbb{R}^{n_1 \times n_1}$, $T_2 \in \mathbb{R}^{n_2 \times n_2}$ ($n = n_1 + n_2$) and compute the corresponding eigensystems $D_1 = Q_1^* T_1 Q_1$ and $D_2 = Q_2^* T_2 Q_2$

2. *Conquer*:
   "Glue" together the results for the smaller matrices T_1 and T_2 computed in the divide step to obtain the eigensystem of the original matrix T.

Using the notation of (3.1) we first motivate how the *divide* step is accomplished. Define v (where $m = n_1$, $\mathbf{e}_m^{(n_1)}$ denotes the mth unit vector of $\mathbb{R}^{n_1}$ and $\mathbf{e}_1^{(n_2)}$ the first unit vector of $\mathbb{R}^{n_2}$) as follows

$$v = \begin{bmatrix} \mathbf{e}_m^{(n_1)} \\ \theta \mathbf{e}_1^{(n_2)} \end{bmatrix} \tag{3.12}$$

and observe that $\tilde{T} = T - \rho v v^*$ is identical to T, except for the 4 entries given by the submatrix

$$\tilde{T}(m : m+1, m : m+1) = \begin{bmatrix} d_m - \rho & e_m - \rho\theta \\ e_m - \rho\theta & d_{m+1} - \rho\theta^2 \end{bmatrix} \tag{3.13}$$

Setting $\rho\theta = e_m$ we obtain

$$T = \begin{bmatrix} T_1 & 0 \\ 0 & T_2 \end{bmatrix} + \rho v v^* \tag{3.14}$$

Notice, that T_1 and T_2 are almost, but not quite submatrices of T, as $T_1(m, m) = \tilde{d}_m = d_m - \rho$ and $T_2(1, 1) = \tilde{d}_{m+1} = d_{m+1} - \rho\theta^2$. Now we are in the desired situation that eigendecompositions $Q_1^* T_1 Q_1 = D_1$ and $Q_2^* T_2 Q_2 = D_2$ can be computed independently where Q_i ($i = 1, 2$) are orthogonal and D_i ($i = 1, 2$) are diagonal matrices. If we now set $U = \mathrm{diag}(Q_1, Q_2)$ and denote $D = \mathrm{diag}(D_1, D_2) = \mathrm{diag}(d_1, d_2, \ldots, d_n)$, we obtain

$$U^* T U = U^* \left(\begin{bmatrix} T_1 & 0 \\ 0 & T_2 \end{bmatrix} + \rho v v^* \right) U = D + \rho z z^* \tag{3.15}$$

where

$$z = U^* v = \begin{bmatrix} Q_1^* \mathbf{e}_m^{(n_1)} \\ \theta Q_2^* \mathbf{e}_1^{(n_2)} \end{bmatrix} \tag{3.16}$$

Unfortunately, this is still not what we are looking for.

To *conquer* the desired eigendecomposition of T from (3.15), we need to construct an orthogonal matrix V such that

$$V^*(D + \rho zz^*)V = \Lambda = \text{diag}(\lambda_1, \ldots, \lambda_n) \qquad (3.17)$$

The following theorem tells us how to proceed:

Theorem 3.6 (Eigensystems of rank-1-modified diagonal matrices)
If $d_1 < d_2 \cdots < d_n$ and the components of z are nonzero, then the eigenvalues λ_i of $D + \rho zz^*$ satisfy the secular equation

$$f(\lambda) = 1 - \rho \sum_{i=1}^{n} \frac{z_i^2}{\lambda - d_i} = 0 \qquad (3.18)$$

The eigenvector associated with λ_i lies along the direction of $(\lambda_i I - D)^{-1}z$, such that the normalized eigenvector is given by

$$v_i = \frac{(\lambda_i I - D)^{-1}z}{\|(\lambda_i I - D)^{-1}z\|_2} \qquad (3.19)$$

Proof: see [49] or [122]   □

Hence, starting from (3.15) the *conquer* part of the algorithm comprises the following three steps:

1. Solve the secular equation (3.18) to compute eigenvalues λ_i

2. Compute eigenvectors v_i according to (3.19) (rescale the components z_i of z)

3. Compute $Q = U \cdot V$ to recover the eigenvectors of the original eigenproblem

Remark 3.7
- *The secular equation (3.18) can be solved efficiently, e.g. by means of a Newton-like method.*

- *The hypotheses in Theorem 3.6 seem rather restrictive. However, it can be shown that the suggested principle also works if there are repeated d_i and/or zero z_i (cf. [49]).*

- *The implementation in a computer code is quite subtle and technical, as several aspects like accuracy, deflation, just to mention a few, have to be taken into account. See [122] for more details.*

- The divide-and-conquer method (DAC) is attractive for problem sizes larger than a certain point of break even `mindac`. For smaller problems, the QR method is in general superior. Hence, DAC it is applied recursively, as long as the size of the matrices $T_i^{(k)}$ is greater than `mindac`. If it falls below this size, the eigensystem is solved by means of an alternative eigensolver (e.g. the QR algorithm).

- The conquer part involves a memory overhead in Step 3 of the conquer part, as additional $\mathcal{O}(n^2)$ workspace is needed to store the matrix V.

- A state-of-the-art implementation of the Divide-and-Conquer method is available in LAPACK [2, 7] in the form of the routine xSTEDC and the corresponding driver routine xSYEVD.

□

3.2.2.4. RRR Algorithm

The RRR method (*Relatively Robust Representation*) was introduced by DHILLON in his PhD thesis [32] in 1997 and has since then become the champion method in many cases, as it is almost always faster than the other methods, and as it requires the least workspace. A detailed explanation of the ideas is far beyond the scope of this thesis, we can only give a very rough sketch. Essentially, the RRR algorithm can be regarded as an improvement over *inverse iteration*. In a first step of the algorithm the spectrum of T is analyzed, and existing clusters of eigenvalues are identified. Then the algorithm uses an LDL^* factorization (called *representation*) for a number of translates $T - \sigma I$ of T, where σ is a shift near each cluster of eigenvalues. This procedure may be applied recursively to each cluster, which leads to a *representation tree*. The decisive advantage over *inverse iteration* is now that the RRR algorithm is *guaranteed* to consume only $\mathcal{O}(kn)$ for determining k eigenvectors, as the outlined approach avoids Gram-Schmidt orthogonalization when computing eigenvectors related to clustered eigenvalues.

The algorithm is implemented in the LAPACK [2, 7] routine xSTEGR and the corresponding driver routine xSYEVR. For future LAPACK [2] releases it is expected that the *RRR method* will replace the QR algorithm as eigensolver for the tridiagonal problem in the standard driver DSYEV. Although particularly well suited for the computation of partial eigensystems, LAPACK's [2] current implementation of xSYEVR only uses xSTEGR for complete eigensystems and switches to the combination of bisection (xSTEBZ) and *inverse iteration* (xSTEIN) when partial eigensytems are desired by the user. This weakness will be fixed in the release to come such that the expert driver is also expected to make use of the *RRR method*. For a thorough description and the theory behind the RRR algorithm we refer to DHILLON's PhD-thesis [32] and the related survey papers [33, 34].

3.2.3. Jacobi's Method

Jacobi's method belongs to the oldest algorithms for symmetric eigenvalue problems and is named after its inventor who proposed the algorithm in his famous article from 1846 [58]. As opposed to the previously presented methods, one typically does not transform the matrix A into tridiagonal form in a preprocessing step, but directly applies the algorithm to the unreduced matrix. The basic strategy of the algorithm is to construct a sequence of similarity transformations which successively reduces the "norm" of the off-diagonal elements represented by

$$\text{off}(A) = \sqrt{\sum_{i=1}^{n} \sum_{\substack{j=1 \\ j \neq i}} a_{ij}^2} \tag{3.20}$$

The smaller this quantity becomes, the better the eigenvalue approximations represented by the diagonal entries of A will be, which is an obvious consequence from Geršgorin's disc theorem (Theorem 2.8). JACOBI's basic idea was to construct a similarity transformation that annihilates the matrix entries a_{pq} and a_{qp} associated with a given index pair (p, q). The tools for this purpose are the rotation matrices which have already been introduced in Section 2.3.1.2 and which are called *Jacobi rotations* in our context (see also Remark 2.34). The essential element in a Jacobi eigenvalue procedure thus comprises the following steps:

- choose an index pair (p, q), $1 \leq p < q \leq n$

- compute a cosine-sine pair (c, s) such that

$$\begin{bmatrix} b_{pp} & b_{pq} \\ b_{qp} & b_{qq} \end{bmatrix} = \begin{bmatrix} c & s \\ -s & c \end{bmatrix}^T \begin{bmatrix} a_{pp} & a_{pq} \\ a_{qp} & a_{qq} \end{bmatrix} \begin{bmatrix} c & s \\ -s & c \end{bmatrix} \tag{3.21}$$

  is diagonal, i.e. $b_{pq} = b_{qp} = 0$

- overwrite A with $B = J^T A J$, where $J = J(p, q, \theta)$ is the *Jacobi rotation* related to the cosine-sine pair (c, s)

B agrees with A, except in rows and columns p and q, and since the Frobenius norm is left invariant under orthogonal similarity transformations, we have

$$a_{pp}^2 + a_{qq}^2 + 2a_{pq}^2 = b_{pp}^2 + b_{qq}^2 \tag{3.22}$$

and, hence,

$$\begin{aligned} \text{off}(B)^2 &= \|B\|_F - \sum_{i=1}^{n} b_{ii}^2 \\ &= \|A\|_F - \sum_{i=1}^{n} a_{ii}^2 + (a_{pp}^2 + a_{qq}^2 - b_{pp}^2 - b_{qq}^2) \\ &= \text{off}(A)^2 - 2a_{pq}^2 \end{aligned} \tag{3.23}$$

Let us now turn to the computation of the rotational parameter pair (c, s): For B in (3.21) to be diagonal we obviously require

$$0 = b_{pq} = a_{pq}(c^2 - s^2) + (a_{pp} - a_{qq})cs \tag{3.24}$$

If $a_{pq} = 0$, then we simply set $(c, s) = (1, 0)$. Otherwise define

$$\tau = \frac{a_{qq} - a_{pp}}{2a_{pq}} \quad \text{and} \quad t = s/c \tag{3.25}$$

Some straightforward algebra turns (3.24) into the quadratic equation

$$t^2 - 2\tau t - 1 = 0 \tag{3.26}$$

It is advantageous to select the smaller of the two roots (see [49]). We now obtain (c, s) using the well-known formulae

$$c = 1/\sqrt{1 + t^2} \quad \text{and} \quad s = tc \tag{3.27}$$

The computational steps in order to determine the parameters for a *Jacobi rotation* are summarized in Algorithm 3.9.

Algorithm 3.9: Generation of a Jacobi rotation

---

1 **function** $[c, s] = \mathbf{jacobi}(A, p, q)$
2 **if** $a_{pq} \neq 0$ **then**
3 $\quad$ $\tau = (a_{qq} - a_{pp})/(2a_{pq})$
4 $\quad$ **if** $\tau \geq 0$ **then**
5 $\quad$ $\quad$ $t = 1/(\tau + \sqrt{1 + \tau^2})$
6 $\quad$ **else**
7 $\quad$ $\quad$ $t = -1/(-\tau + \sqrt{1 + \tau^2})$
8 $\quad$ **end if**
9 $\quad$ $c = 1/\sqrt{1 + t^2}$
10 $\quad$ $s = tc$
11 **else**
12 $\quad$ $c = 1, s = 0$
13 **end if**
14 **return** $[c, s]$

We finally have to specify how to actually choose the indices p and q of the matrix element to be zeroed. Equation (3.23) motivates to choose the off-diagonal element with the largest modulus $|a_{pq}|$ because then the reduction of the quantity off(A) is maximal. This directly leads to the classical Jacobi algorithm:

Algorithm 3.10: Classical Jacobi

---

1 $V = I_n$; $eps = tol\|A\|_F$
2 **while** $off(A) > eps$ **do**
3 $\quad$ Choose (p,q) such that $|a_{pq}| = \max_{i \neq j} |a_{ij}|$
4 $\quad$ (c,s)=jacobi(A,p,q)
5 $\quad$ $A = J(p,q,\theta)^T A J(p,q,\theta)$
6 $\quad$ $V = V J(p,q,\theta)$
7 **end while**

Remark 3.8

- *Matrix elements that have been annihilated once may re-obtain non-zero values in later cycles of the **while**-loop. However, this does not affect the convergence because the quantity off(A) is reduced in every cycle.*

- *Alg. 3.10 accumulates the transformations carried out in the matrix V, and thus, also yields eigenvector approximations when the desired accuracy has been reached.*

- *With $N = n(n-1)/2$ it easily follows by induction that*

$$off(A^{(k)})^2 \leq \left(1 - \frac{1}{N}\right)^k off(A^{(0)})^2 \tag{3.28}$$

which implies that the classical Jacobi procedure converges linearly. However, it can be shown that the asymptotic convergence rate is even quadratic. See [49] and the references therein for more details.

- *Searching for the largest off-diagonal element in modulus (Line 3 of Alg. 3.10) is time-consuming (costs $\mathcal{O}(n^2)$), especially if n is large. A cheaper alternative, which also leads to quadratic convergence, is to annihilate the matrix elements in a row-by-row fashion instead. This referred to as a cyclic Jacobi procedure (see [49] for further details).*

- *In general, Jacobi's method is not competitive with the previously discussed algorithms in terms of floating point operations. Jacobi needs $2sn^3$ multiplications for s sweeps (N Jacobi updateds are customarily referred to as a sweep and s is usually a number between 3 and ten), and this is clearly more than the $4/3n^3$ flops required for tridiagonal reduction. Nonetheless, in the recent time the interest in the method has re-aroused because it can deliver eigenvalue approximations with a small error in the relative sense. This is an advantage over the methods based on tridiagonalization, which only guarantee that the error is bounded relative to the norm of the matrix (see [31]). Another interesting feature is the inherent parallelism of the method which may be exploited in the implementation.*

$\square$

3.2.4. Assessment and Summary

Let us now briefly summarize the pros and cons of the reviewed methods for the symmetric tridiagonal eigenproblems and the tridiagonalization approaches:

3.2.4.1. Tridiagonalization Approaches

Notice, that the discussion on the pros and cons comprises also the back-transformation of (partial) eigensystems of tridiagonals:

   1. standard approach (LAPACK [2, 7], name of the routine: xSYTRD)

      Pros: − in general superior, if more than one third of the eigenpairs is requested

      Cons: − in spite of blocking techniques about one half of the operations are still matrix-vector multiplications, relatively small level 3 fraction

              − exploitation of memory hierarchies not satisfactory

              − inferior to SBR, if a small fraction of selected eigenpairs is sought after

   2. two-stage approach (SBR Toolbox [13], name of the routine: xSYBTRD)

      Pros: − often leads to a considerable gain in performance for small partial eigensystems

              − takes advantage of memory hierarchy and performs well, especially on computers with fast processors and relatively slow memory

      Cons: − back-transformation twice as expensive as for the standard approach

              − not competitive for the computation of a medium or large fraction (more than one third) of eigenpairs

3.2.4.2. Tridiagonal Eigensolvers

   1. *QR method*

      Pros: − well-tried and reliable

      Cons: − in general slower than RRR and DAC

   2. *Divide-and-conquer method* (DAC)

      Pros: − faster than QR for larger problems (problem size greater than certain point of break even)

              − inherent parallel structure

      Cons: memory overhead

   3. *Bisection / inverse iteration* (BISECT / INVIT)

Pros: – designed for computation of partial eigensystems

– best case time complexity $\mathcal{O}(nk)$ for computation of k selected eigenpairs

Cons: – problems with clustered eigenvalues: requires explicit Gram-Schmidt orthogonalization

– may lead to worst case time complexity $\mathcal{O}(nk^2)$

4. RRR method

Pros: – in general superior to all other methods (it is faster and requires the least workspace)

– particularly well-suited for the computation of partial eigensystems

– improvement over INVIT, because time-complexity of $\mathcal{O}(nk)$ guaranteed even for clustered eigenvalues

Cons: – in some situations other methods (especially Jacobi's method, cf. [31]) may deliver results with a higher relative accuracy

Table 3.1.: Direct methods for the symmetric tridiagonal eigenproblem

Method	Costs		Part. eigsys	LAPACK [2]	
	Performance	Storage		Routine	Driver
QR	+	+	No	xSTEQR	xSYEV
DAC	++	−	No	xSTDEC	xSYEVD
BISECT / INVIT		+	Yes	xSTEBZ / xSTEIN	xSYEVX
RRR	+++	++	Yes	xSTEGR	xSYEVR

3.2.4.3. Summary

There is available a couple of methods for both, the solution of the tridiagonal eigenproblem and the tridiagonalization of symmetric matrices. We are primarily interested in computing partial eigensystems with a rather small fraction of eigenpairs (typically about 7-10 percent of all eigenpairs are sought-after). Thus, for our purposes the two-stage approach (SBR Toolbox [13]) as a tridiagonalization method and the RRR Algorithm [32],[7] as a tridiagonal eigensolver turns out to be the most favorable combination in the generic eigensolver (Alg. 3.4). This is confirmed by the numerical results in [87], where the eigensolver is applied to Hamiltonian matrices arising in the computation of rovibronic energy levels for triatomic molecules which exhibit the *Double Renner effect*.

3.3. Iterative Projection Methods

Direct methods are *black-box* methods, as one is neither required to possess any additional knowledge about the properties of the matrix (apart from its dimension n and the fact that it is Hermitian), nor does one need theoretical knowledge on how the chosen method works. The only thing the non-expert user has to care about is to provide enough work space for the storage of A and to call the routine properly in a computer code (choice of parameters etc.). Hence, for small and medium-sized problems direct algorithms are the methods of choice, as they are available in technically mature software libraries (most notably in LAPACK [2, 7]), work reliably and have a predictable convergence behavior. However, for larger problem sizes n (say $n > 10000$) the time complexity $\mathcal{O}(n^3)$ increasingly becomes perceivable and as the workspace of a computer is limited, one inevitably reaches the point, where eigenvalue computations are no more feasible by means of direct methods. For this reason, iterative projection methods come into play and may be a viable alternative. The fundamental difference with the direct methods lies in the fact that iterative projection methods only implictly access information of a given matrix A by matrix-vector multiplications Av. This feature very often – especially for large-scaled problems – turns out to be a key advantage over direct methods, as explicit storage of all matrix entries in an $(n \times n)$ array is *not* necessary, and as the sparsity and structure of A can be exploited in efficient matrix-vector multiplication subroutines. As a consequence – and this is what one hopes for in practice – iterative projection methods may scale much more favorably, i.e. like $\mathcal{O}(n)$ or $\mathcal{O}(n^2)$, at least for the computation of a few eigenpairs. The following example, which we will discuss in detail later on, impressively illustrates the advantages of exploiting sparsity:

Example 3.9 (see also Fig. 6.4 and the exposition in Section 6.4)
For the rotational quantum number $J = 13/2$ and a "big basis" (see Table 6.5) the variational computation of rovibronic energy levels for the **MgNC**-*molecule leads to a symmetric matrix of the size $n = 83328$, which implies that (storage requirement for a* `double` *variable in the C programming language is 8 bytes) $(83328 \times 83328 \times 8)/1024^3 = 51.73$ GB workspace is needed to store the complete matrix in memory. This is clearly beyond the amount of 32 GB available to us on a SUN^{TM} Fire machine. If, by contrast, we exploit sparsity and only store the non-zero entries, we require 3.78 GB workspace.*

The single vector iterations presented in Section 3.1 try to generate a sequence of iterates $q^{(i)} \in \mathbb{R}^n$, $q^{(i)} \to q$ (q is an eigenvector of A) and an iterate $q^{(k)}$ is computed from its predecessor $q^{(k-1)}$, whereas the basic idea behind iterative projection methods is to take into account *all* iterates generated so far and to look for an eigenvector approximation u in the subspace

$$\mathcal{K} = \mathrm{span}\{q^{(0)}, q^{(1)}, \ldots, q^{(k-1)}\} \subset \mathbb{R}^n, \qquad \dim(\mathcal{K}) = k < n \qquad (3.29)$$

This idea is also referred to as *subspace acceleration* in the literature, since it often leads to faster convergence as compared to the corresponding single vector iterations. Typically, after computing the residual $r = Au - \theta u$ a new search direction $q^{(k)} = f(r)$

is derived from r and the subspace is expanded from $\mathcal{K}$ to $\mathcal{K} + \text{span}\{q^{(k)}\}$. One hopes to obtain increasingly better approximations to the sought-after eigenvector q by iterating this procedure. Hence, any successful iterative projection method will rely on the efficient interplay of the two algorithmic components

1. *information extraction*

2. *subspace expansion*

These basic ideas are formalized in the following generic algorithmic template, where the basis V of $\mathcal{K}$ is chosen to be orthogonal:

Algorithm 3.11: Generic iterative projection method for Hermitian matrices

---

1 **function** (λ, q)=eigpair(A, v_0)
2 $V = [\,]$, choose starting vector $t = v_0$
3 **for** $m = 1, 2, \ldots, \nu$ **do**
4      $t = \text{orth}(V, t)$;   /* orthonormalization by means of Alg. 2.5 */
5      $V = [V, t]$
6   compute approximate eigenvector
         $u \in \text{span}\{V\}$, $\|u\|_2 = 1$ (Information Extraction)
7   compute approximate eigenvalue
         via Rayleigh quotient $\theta = u^* A u$
8   compute residual $r = Au - \theta u$
9   **if** $\|r\|_2 \leq \epsilon$ **then**
10   (θ, u)
11   **end if**
12   compute update $t = f(r)$ (Subspace Expansion)
13 **end for**

Essentially, any simple iterative projection method is based on this scheme, and so are the basic instances of the algorithms discussed in this thesis. The more sophisticated variants (see the related discussion in Sections 3.3.2.2 and 4.3) deviate in that they make use of restart and deflation techniques in order to compute more than one approximate eigenpair, and they keep the size of the search space $\mathcal{K}$ bounded. In fact, it will turn out that the success of such restarted schemes depends to a great deal upon how well information of interest can be "compressed" into a subspace $\mathcal{K}$ whose dimension m is considerably smaller than n. In the following, we will explain in more detail how the extraction of approximate eigenvectors from a subspace $\mathcal{K}$ in Line 6 and the construction of suitable subspace expansions $f(r)$ in Line 12 of the above template may be accomplished. The matrices we will be concerned with in our considerations are Hermitian, and for this reason, we will develop and discuss algorithms taking advantage of this specific property. All algorithms discussed in the following also have counterparts for the non-Hermitian case, a thorough survey and detailed description for both, the Hermitian and the non-Hermitian case, may be found in [8], [100] and [125]. For a general overview of existing software for iterative projection methods see [52].

3.3.1. Information Extraction

Basically, we can distinguish between three general approaches to find an approximation (θ, u) to an eigenpair (λ, x) of A in a prescribed subspace $\mathcal{K} \subset \mathbb{C}^n$

1. *orthogonal projection methods*
   Here one imposes the *Galërkin condition* on the resdiual

$$Au - \theta u \perp \mathcal{K} \qquad (3.30)$$

   i.e. the residual is required to be orthogonal to the subspace $\mathcal{K}$.

2. *oblique projection methods*
   Olique projection methods employ an additional subspace $\mathcal{L} \subset \mathbb{C}^n$, which leads to the *Petrov-Galërkin condition*

$$v^* \left(A - \theta I \right) u = 0 \qquad \forall v \in \mathcal{L} \qquad (3.31)$$

   For the choice $\mathcal{L} = \mathcal{K}$ (3.31) coincides with the *Galërkin condition* (3.30), which shows that this definition is a generalization. We will be mainly concerned with the case $\mathcal{L} = A\mathcal{K}$, so-called *harmonic extraction methods*.

3. *refined projection methods*
   These methods attempt to find an improved ("refined") approximation $\hat{u}$ of an approximate eigenpair (θ, u) coming from an orthogonal projection method by solving the least squares problem

$$\|(A - \theta I)\hat{u}\|_2 = \min_{\substack{v \in \mathcal{K} \\ \|v\|_2 = 1}} \|(A - \theta I)v\|_2 \qquad (3.32)$$

   The solution $\hat{u}$ minimizes the residual over all possible $v \in \mathbb{C}^n$ of norm unity.

Normally, orthogonal projection is the approach of choice to extract eigeninformation from a subspace. However, sometimes the approximations obtained are poor or even completely misleading. This may be the case, when the eigenpairs of interest are clustered or lie in the interior of the spectrum. The straightforward way out of this difficulty is to apply orthogonal projection methods to an operator obtained by a shift-and-invert spectral transformation of the original matrix, which is advocated in [110]. However, storage requirements and the costs for solving the arsing linear systems can be prohibitive. Then oblique or refined projection methods are valuable alternatives and, although slightly more expensive, they often yield better results than the orthogonal projection methods. We will give a brief survey and discussion of the problems arising with orthogonal projection methods at the end of the following paragraph. The issue of information extraction in the context of iterative projection methods is subject of intense current research, which is reflected in the large number of both theoretical and practical investigations on this topic (cf. [122] and refs. therein). Note that recently presented approaches even go beyond the above ideas, e.g. *refined harmonic extraction methods* (cf. [64]), which attempts to combine the advantages of oblique and refined projection methods.

3.3.1.1. Standard Extraction

Let $A \in \mathbb{C}^{n \times n}$ and $\mathcal{K}$ be a subspace of $\mathbb{C}^{n \times n}$ where $\dim \mathcal{K} = m < n$. Let us furthermore assume that the columns of $V \in \mathbb{C}^{n \times m}$ are an orthonormal basis of $\mathcal{K}$, i.e. $V^*V = I_m$ and span$\{V\} = \mathcal{K}$. We consider the Hermitian eigenvalue problem

$$Ax = \lambda x \tag{3.33}$$

and we are looking for approximate eigenpairs (θ, u) where $u \in \mathcal{K}$. The key idea is now to impose the *Galërkin condition* on the resdiual, i.e. we demand

$$Au - \theta u \perp \mathcal{K} \tag{3.34}$$

or, equivalently,

$$v^*(Au - \theta u) = 0 \qquad \forall v \in \mathcal{K} \tag{3.35}$$

Since u is supposed to be in $\mathcal{K}$, we can express it as a linear combination of the basis vectors

$$u = Vy \tag{3.36}$$

As (3.35) is especially valid for the basis vectors v_j of $\mathcal{K}$, plugging (3.36) into (3.35) yields

$$v_j^*(AVy - \theta Vy) = 0, \qquad j = 1, \ldots, m \tag{3.37}$$

and by collecting these equations $(j = 1, \ldots, m)$ we obtain the following equivalent matrix notation

$$M_m y = \theta y \tag{3.38}$$

where

$$M_m = V^*AV \tag{3.39}$$

is an $(m \times m)$-matrix. The orthogonality requirement (3.35) thus leads to a lower dimensional Hermitian eigenvalue problem. Before we formalize the approach derived above, we give some essential definitions:

Definition 3.10
Let $A \in \mathbb{C}^{n \times n}$ and let $V \in \mathbb{C}^{n \times m}$ be orthonormal, i.e. $V^*V = I_m$ and let (θ, y) an eigenpair of $M_m = V^*AV$ Then we call

- M_m the interaction matrix

- θ a Ritz value

- $u = Vy \in \mathbb{R}^n$ a Ritz vector

- (θ, u) a Ritz pair

of A with respect to $\mathcal{K} = $ span$\{V\}$.
One often also finds the terms Rayleigh-Ritz approximation, Galërkin approximation or Ritz-Galërkin approximation in the literature.

The following algorithm is known under the name *Rayleigh-Ritz procedure*, and it summarizes how to apply the above ideas in order to extract approximate eigenpairs from $\mathcal{K}$:

Algorithm 3.12: Standard extraction (Rayleigh-Ritz procedure)

---

1 compute an orthonormal basis $\{v_i\}_{i=1,\ldots,m}$ of the subspace $\mathcal{K}$ and let
$$V = [v_1, v_2, \ldots, v_m]$$

2 compute
$$M_m = V^*AV \qquad \qquad \qquad \textit{(interaction matrix)}$$

3 compute the eigenvalues
$$\theta_i \text{ of } M_m \ (i = 1, \ldots, m) \qquad \qquad \textit{(Ritz values)}$$

4 compute the eigenvectors
$$y_i \text{ of } M_m \ (i = 1, \ldots, m)$$

5 compute as approximate eigenvectors of A
$$u_i = Vy_i \qquad \qquad \qquad \qquad \textit{(Ritz vectors)}$$

Let us first note the following trivial but important corollary, which states that eigenvectors present in the subspace $\mathcal{K}$ are retrieved by the *Rayleigh-Ritz procedure*:

Corollary 3.11
Let (λ, x) be an eigenpair of A with $x = Vy$. Then (λ, Vy) is a Ritz pair.

Proof: Straightforward verification.   □

We now collect some essential properties of the *Rayleigh-Ritz procedure*:

Remark 3.12 (Basic properties of Rayleigh-Ritz procedure)
- *Line 2 of Algorithm 3.12 shows that information on A is only needed in the form of matrix-vector products $A \cdot v_j$.*

- *By the Poincaré separation theorem (Corollary 2.19) it is clear, that a Ritz value θ_i is an upper bound to the corresponding eigenvalue λ_i, if we order both eigenvalues $\lambda_j \ (j = 1, \ldots, n)$ and Ritz values $\theta_j \ (j = 1, \ldots, m)$ by ascending magnitude, i.e.*
$$\lambda_i \leq \theta_i \qquad (i = 1, \ldots, m)$$

- *Any Ritz value θ (related to the Ritz vector u) is as a convex combination of the eigenvalues of A. To see this, express θ by the corresponding Rayleigh quotient*
$$\theta = \frac{u^*M_m u}{u^*u} = \frac{u^*V^*AVu}{u^*u} = \frac{(Vu)^*AVu}{(Vu)^*Vu} \qquad (3.40)$$

Letting $x = Vu$ we are in the situation of Remark 2.16 and can re-use equation (2.23) such that
$$\theta = \sum_{k=1}^{n} \beta_k \lambda_k, \quad 0 \leq \beta_k = 1, \quad \sum_{i=1}^{n} \beta_i = 1 \qquad (3.41)$$

- The interaction matrix $M_m = V^*AV$ can be viewed as a generalized *Rayleigh quotient* and in analogy to Proposition 2.14 one can show that

$$\|R(M_m)\| \leq \|R(B)\| \qquad \forall B \in \mathbb{C}^{n \times n}$$

  where $R(B) = BV - VB$ is the residual matrix and $\| \cdot \|$ the spectral norm. For the related proof and the discussion in what sense the Rayleigh-Ritz procedure is optimal we refer to the profound exposition in [91].

- The quality of a Ritz approximation (θ, u) may be assessed by the error bounds (2.31), (2.34) and (2.35) presented in the introductory part of this thesis. However, in general there is no information available in advance about the distribution of the sought-after eigenvalues (separation and clusters), such that only the estimate (2.31) is of practical use. Without any additional knowledge no reasonable error bounds can be placed on the eigenvector approximations.

- The computational costs in Alg. 3.12 amount to

  1. $\mathcal{O}(m^2 \cdot n)$ for the orthogonalization of the basis vectors (e.g. by means of modified Gram-Schmidt)

  2. $\mathcal{O}(\ell \cdot m \cdot n)$ for computing $W = AV$ (ℓ is the avarage number of nonzero elements per row of A

  3. $\mathcal{O}(m^2 \cdot n)$ for computing $M_m = V^*W$

  4. $\mathcal{O}(m^3)$ for computing the eigensystem of M_m

  5. $\mathcal{O}(m^2 \cdot n)$ for computing m Ritz vectors $u_i = Vy_i$

  In practical situations A is often sparse and $\ell \ll n$, such that the Rayleigh-Ritz procedure in total is essentially an $\mathcal{O}(m^2 \cdot n)$-process.

- The naming of the procedure is an acknowledgement of the fact that the approach dates back to Rayleigh [77] and Ritz [99] who proposed it independently of each other at the beginning of the last century.

$\square$

As already indicated in the introductory part, the *Rayleigh-Ritz procedure* may produce bad approximations or even miserably fail. A phenomenon that one often encounters in this context, is that a certain *Ritz value* may be a good approximation to a corresponding eigenvalue of the matrix A, whereas the associated *Ritz vector* has little in common with the sought-after eigenvector. Such *Ritz values* are referred to as *spurious eigenvalues*, *ghost values*, *imposters* or *phantom values* in the literature. Unfortunately, it is difficult to figure out when this situation arises, because it is known from (2.35) that a small residual alone does not guarantee that an eigenvector approximation is appropriate. We cite the following example from [47] to illustrate the problem:

Example 3.13
Let

$$A = \begin{bmatrix} -1 & 0 & 0 \\ 0 & 0.1 & 0 \\ 0 & 0 & 1 \end{bmatrix} \tag{3.42}$$

and the subspace $\mathcal{V} = \text{span}\{v_1, v_2\}$ be given by

$$v_1 = \begin{bmatrix} 0.01 \\ -0.90 \\ -0.01 \end{bmatrix} \quad and \quad v_2 = \begin{bmatrix} -1 \\ 0 \\ 1 \end{bmatrix} \tag{3.43}$$

Then the subspace $\mathcal{V}$ obviously contains a suitable approximation (v_1) to the eigenvector

$$x = \begin{bmatrix} 0 \\ 1 \\ 0 \end{bmatrix}$$

related to the eigenvalue $\lambda = 0.1$ of the matrix A. The Rayleigh-Ritz procedure (Algorithm 3.12) then yields (after orthonormalization of the basis vectors) the Ritz values $\theta_1 = 0.1$ and $\theta_2 = 0.0$, as well as the Ritz vectors

$$u_1 = \begin{bmatrix} 0.0007697 \\ 1.0000000 \\ -0.0007697 \end{bmatrix} \quad and \quad u_2 = \begin{bmatrix} -1.0000000 \\ 0.0096224 \\ 1.0000000 \end{bmatrix} \tag{3.44}$$

Both, θ_1 and θ_2, are good approximations to the eigenvalue $\lambda = 0.1$, but only the Ritz vector u_1 is an appropriate approximation to the eigenvector x. Thus, θ_2 is a typical imposter. □

This example may seem somewhat artificial, but it reflects the general problem when it comes to looking for eigenvalue approximations in the interior of the spectrum of a given matrix A. There are different attempts to explain the possible failure of the *Rayleigh-Ritz procedure* (an in-depth discussion of this issue may be found in [47]):

1. SCOTT [110] explains the possible problems by the fact that *Ritz values* are convex combinations of the eigenvalues (cf. Remark 3.12). For *Ritz values* approximating interior eigenvalues this has the consequence that there may be also major contributions from exterior eigenvalues and that the resulting convex combination only accidentially approximate the sought-after eigenvalue. As this also carries over to the corresponding *Ritz vectors*, we are in the typical situation described above: The *Ritz value* is a good approximation, whereas the corresponding *Ritz vector* is completely erroneous. Clearly, for eigenvalues at the exterior this cannot happen as, owing to the convexity, there are only relevant contributions from exterior eigenvalues.

2. SLEIJPEN and VAN DER VORST [114] point out that *Ritz values* converge monotonically to exterior eigenvalues of A. Hence, a *Ritz value* may temporarily approximate an eigenvalue in the interior, but is well on its way to converge to an exterior eigenvalue. This has the consequence that the corresponding *Ritz vector* already has strong contributions from exterior eigenvectors.

3. STEWART [122] stresses that apart from the convergence of the desired eigenvalues one also needs their separation in order to have convergence of the *Ritz space*. To illustrate this, let us come back to the example in Remark 2.7 where we discussed the possible discontinuity of eigenvectors (e.g. in a neighborhood of multiple eigenvalues): We have seen that

$$A = \begin{bmatrix} \nu + \delta & \gamma \\ \gamma & \nu - \delta \end{bmatrix}, \qquad \delta, \gamma, \nu \in \mathbb{R} \tag{3.45}$$

has the eigenvalues $\lambda_1 = \nu - \delta$ and $\lambda_2 = \nu + \delta$ where $\sigma^2 = \delta^2 + \gamma^2$. Applying the *Rayleigh-Ritz procedure* with respect to the subspace $\mathcal{K} = \text{span}\{V\}$ where $V = [e_1]$ we obtain

$$M = V^* A V = \nu + \delta \tag{3.46}$$

i.e. the *Ritz value* is $\theta = \nu + \delta$. Furthermore, the "true" eigenvector x and the related *Ritz vector* u are

$$x = \begin{bmatrix} 1 \\ \pm\gamma/(\sigma + |\delta|) \end{bmatrix} \qquad \text{and} \qquad u = \begin{bmatrix} 1 \\ 0 \end{bmatrix} \tag{3.47}$$

As per Prop. 2.28, we can now compute the cosine of the acute angle ϕ between x and u to measure the error:

$$\cos \phi = (1 + \gamma^2/(\sigma + |\delta|^2)^{-1/2} \longrightarrow \begin{Bmatrix} 1 & \text{if } \delta \neq 0 \\ 1/\sqrt{2} & \text{if } \delta = 0 \end{Bmatrix} \text{ as } \gamma \to 0 \tag{3.48}$$

Obviously, one gets into trouble for $\delta = 0$, because then the "true" eigenvector is $x = (1, \pm 1)^*$ for *all* nonzero γ and the error angle is $\pi/4$. As explained in Remark 2.7, the problem is that for $(\delta, \gamma) = (0, 0)$ the eigenvalues λ_1 and λ_2 coincide. Consequently, it is not sensible to ask for the error angle ϕ, because the *Ritz vector* u cannot approximate both, $x_1 = (1, 1)$ and $x_2 = (1, -1)$, simultaneously. Therefore, to ensure convergence, it is necessary that the eigenvalues be separated sufficiently well from each other.

For a detailed convergence analysis of the Rayleigh-Ritz method we refer to [65], [122] and [124]. The oblique and refined projection methods presented in the following may be used as alternatives when facing difficulties with the *Rayleigh-Ritz procedure*.

3.3.1.2. Harmonic Extraction

We consider two subspaces $\mathcal{K} \subset \mathbb{C}^n$ and $\mathcal{L} \subset \mathbb{C}^n$ of equal dimension m, where the columns of $V \in \mathbb{C}^{n \times m}$ form a basis of $\mathcal{K}$ and the columns of $W \in \mathbb{C}^{n \times m}$ are a basis of $\mathcal{L}$.

Imposing the *Petrov-Galërkin condition*

$$v^* (A - \theta I) u = 0 \qquad \forall v \in \mathcal{L} \tag{3.49}$$

on the residual and expressing u as linear combination $u = Vy$ of the basis vectors in V, some algebra leads us to the generalized eigenvalue problem

$$W^* AV \, y = \theta \, W^* V y \tag{3.50}$$

Note that one may derive a standard eigenproblem from (3.50) by demanding $W^* V = I_m$ (then the bases for $\mathcal{K}$ and $\mathcal{L}$ are biorthogonal).

In the following, we will focus on the particular case $\mathcal{L} = A\mathcal{K}$, i.e. $W = AV$. Then (3.50) turns into

$$V^* A^* AVy = \theta V^* A^* Vy \tag{3.51}$$

As we are especially interested in finding approximations to interior eigenvalues of A near a target value σ, we we generalize (3.51) by replacing A with $(A - \sigma I)$ with some shift $\sigma \geq 0$ which gives rise to the following definition:

Definition 3.14 (Harmonic Ritz pairs)
Let $\mathcal{K}$ be a subspace and V be an orthonormal matrix whose columns build a basis for $\mathcal{K}$. Then $(\sigma + \delta, Vy)$ is a harmonic Ritz pair with shift σ, if

$$V^*(A - \sigma I)^*(A - \sigma I)V \, y = \delta \, V^*(A - \sigma I)^* V \, y \tag{3.52}$$

$\theta = \sigma + \delta$ *is called harmonic Ritz value and* $u = Vy$ *harmonic Ritz vector.*

First of all, it is important to note that eigenvectors that are exactly present in the subspace $\mathcal{K}$ are retrieved by a harmonic extraction method:

Corollary 3.15
Let (λ, x) be an eigenpair of A with $x = Vy$. Then (λ, Vy) is a harmonic Ritz pair.

Proof: Straightforward verification by comparing left-hand and right-hand side in (3.52). □

Furthermore, we can immediately derive the following interesting

Corollary 3.16
Let $(\sigma + \delta, Vy)$ be a harmonic Ritz pair of A with respect to $\mathcal{K}$ where $\|y\|_2 = 1$. Then the following inequality holds

$$\| (A - \sigma I) \, Vy \|_2 \leq \delta \tag{3.53}$$

Proof: On pre-multiplying (3.52) by y^*, we obtain

$$\| (A - \sigma I)Vy \|_2^2 \le |\delta| \cdot \| (A - \sigma I)Vy \|_2 \tag{3.54}$$

which proves the assertion.

$\square$

Basically, this inequality says that for any *harmonic Ritz pair* around the shift σ (within the radius δ) the pair must have a residual norm (with respect to σ) which is bounded by $|\delta|$. If, for instance, we consider a shift σ near a sougth-after eigenvalue λ and a tiny δ, then we see that imposters, in principle, cannot occur. This property is one of the reasons to favor harmonic projection methods when looking for eigenvalues in the interior of the spectrum.

However, it is still not clear, how to compute *harmonic Ritz pairs* of a given matrix A with respect to a prescribed subspace $\mathcal{K}$. To this end, we will now derive a characterization, which is more suitable for this purpose, and we will turn (3.52) into a standard eigenproblem by requiring $W = (A - \sigma I)V$ to be orthonormal, which implies

$$
\begin{aligned}
y &= \delta V^*(A - \sigma I)^* Vy \\
&= \delta V^*(A - \sigma I)^*(A - \sigma I)^{-1}(A - \sigma I)Vy \\
&= \delta W^*(A - \sigma)^{-1}Wy
\end{aligned}
$$

or, equivalently,

$$W^*(A - \sigma I)^{-1}Wy = \frac{1}{\delta}y \tag{3.55}$$

For this relation to hold, we have to ensure that the matrix V be transformed according to the orthonormalization of W. Thus, computing a QR-decomposition $W = QR$ and letting $\widetilde{W} = Q$ we have to use $\widetilde{V} = VR^{-1}$ to maintain the relation $\widetilde{W} = (A - \sigma I)\widetilde{V}$. (3.55) reveals that the specific choice of $\mathcal{L} = (A - \sigma I)\mathcal{K}$ implicitly leads to an orthogonal projection of $(A - \sigma I)^{-1}$ onto W. It is important to note that the inverse $(A - \sigma I)^{-1}$ is *not* explicitly computed. However, as W depends on A, and thus, cannot be chosen arbitrarily, we cannot expect (3.55) to work equally well as an explicit inversion of $(A - \sigma I)$. Numerical experiments show that harmonic extraction improves the convergence behavior considerably as compared to the Rayleigh-Ritz method. This is also confirmed by theoretical investigations in [89], where it is shown that for Hermitian A the *harmonic Ritz values* converge monotonically to the the smallest non-zero eigenvalue in absolute value. The above ideas are summarized in the following

Definition 3.17 (Harmonic Ritz pairs, alternative definition)
Let $\mathcal{K}$ be a subspace and let the columns of V form a basis of $\mathcal{K}$ such that the columns of $W = (A - \sigma I)V$ are an orthonormal basis of $\mathcal{L} = (A - \sigma I)\mathcal{K}$. Then $(\sigma + \delta, Vy)$ is a harmonic Ritz pair with shift σ, if

$$W^*V = W^*(A - \sigma I)^{-1}Wy = \frac{1}{\delta}y \tag{3.56}$$

$\theta = \sigma + \delta$ *is called* harmonic Ritz value *and* $u = Vy$ *harmonic Ritz vector.*

This definition gives rise to Algorithm 3.13 in which the computation of *harmonic Ritz pairs* is described.

Algorithm 3.13: Harmonic extraction (harmonic Rayleigh-Ritz procedure)

---

1 compute an orthonormal basis W of the subspace
$$\mathcal{L} = (A - \sigma I)\mathcal{K}$$
and a basis V of the subspace $\mathcal{K}$, such that
$$W = (A - \sigma I)V.$$
2 compute
$$M_m = W^*V \qquad\qquad\qquad\qquad\qquad (interaction\ matrix)$$
3 compute the eigenvalues
$$\tfrac{1}{\delta_i} \text{ of } M_m \ (i = 1, \ldots, m)$$
4 compute eigenvalue approximations
$$\theta_i = \sigma + \delta_i \qquad\qquad\qquad\qquad\quad (harmonic\ Ritz\ values)$$
5 compute the eigenvectors
$$y_i \text{ of } M_m \ (i = 1, \ldots, m)$$
6 compute approximate eigenvectors of A as
$$u_i = V y_i \qquad\qquad\qquad\qquad\qquad (harmonic\ Ritz\ vectors)$$

We conclude with some remarks:

Remark 3.18

- *The appellation* harmonic *can be explained by the following observation: According to Remark 3.12,* Ritz *values are convex combination of the eigenvalues, hence in our case we have (for shift $\sigma = 0$):*

$$\frac{1}{\delta_i} = \sum_{k=1}^{n} \beta_k \frac{1}{\lambda_k}, \qquad \sum_{i=1}^{n} \beta_i = 1$$

  or, equivalently,

$$\theta_i = \delta_i = \frac{1}{\sum_{k=1}^{n} \beta_k \frac{1}{\lambda_k}}, \qquad \sum_{i=1}^{n} \beta_i = 1$$

  This shows that the harmonic Ritz values *are weighted harmonic means of the eigenvalues of A.*

- *In practice, one replaces the harmonic Ritz values computed in Alg. 3.13 by the Rayleigh quotients $\rho(u_i) = u_i^* A u_i$ which are known to be optimal with respect to given eigenvector approximations (see Prop. 2.14).*

- *The computation of harmonic Ritz pairs is only slightly more expensive than the computation of their unharmonic counterparts in Algorithm 3.12, the essential difference being, that apart from the orthonormalization of W additional work has to be invested in V in order to maintain the relation $W = (A - \sigma I)V$. This can be realized e.g. by a Gram-Schmidt procedure that simultaneously operates on W and V.*

- *A formal introduction of harmonic Ritz values was given in [89]. The references [81],[113] especially point out their use as a device to find appropriate approximations to interior eigenvalues of symmetric matrices. Interesting theoretical investigations on the properties of harmonic Ritz values along with their relation to refined Ritz values may be found in [113]. The characterization of harmonic Ritz values by the Petrov-Galërkin condition*

$$x \in \mathcal{K}, \quad (A - \sigma I)x - \delta x \perp (A - \sigma I)\mathcal{K}$$

  and the alternative Definition 3.17 is due to SLEIJPEN *and* VAN DER VORST *[114], whereas* STEWART *[122] prefers the equivalent Definition 3.14, which also leads to a different computation technique. A recent convergence analysis of the harmonic extraction may be found in [64].*

- *An example from our experiments comparing harmonic and standard extraction will be given later on in Section 4.4.3, where we discuss the use of the different extraction method in the context of the Jacobi-Davidson variants presented in Chapter 4.*

□

3.3.1.3. Refined Extraction

We have already seen that it is difficult to assess the quality of a *Ritz vector* computed in Algorithm 3.12 and especially the associated norm of the residual is not guaranteed to be minimal. Example 3.13 impressively illustrates the potential dangers arising in the use of standard extraction. A straightforward remedy to this problem is to enforce that the residual norm related to a *Ritz value* be minimal, which leads to an improved ("refined") eigenvector approximation. The following definition gives a concise formulation of the arising optimization problem:

Definition 3.19 (Refined Ritz vector)
Let $A \in \mathbb{C}^{n \times n}$ be Hermitian, $\mathcal{K}$ be a subspace and V be an orthonormal basis for $\mathcal{K}$. Furthermore, let (θ, u) be a Ritz approximation obtained by Algorithm 3.12. Then we call the solution $\hat{u}$ of the minimization problem

$$\|(A - \theta I)\hat{u}\|_2 = \min_{\substack{v \in \mathcal{K} \\ \|v\|_2 = 1}} \|(A - \theta I)v\|_2 \tag{3.57}$$

a refined Ritz vector and $\rho(\hat{u}) = \hat{u}^* A \hat{u}$ the associated refined Ritz value.

To solve the least squares problem (3.57) we exploit that v can be written as a linear combination of vectors in V, i.e. $v = Vy$ with an appropriate $y \in \mathbb{C}^m$. Then

$$\|(A - \theta I)\hat{u}\|_2 = \min_{\substack{v \in \mathcal{K} \\ \|v\|_2 = 1}} \|(A - \theta I)v\|_2 = \min_{\substack{y \in \mathbb{C}^m \\ \|y\|_2 = 1}} \|(A - \theta I)Vy\|_2 = \min_{\substack{y \in \mathbb{C}^m \\ \|y\|_2 = 1}} \|\widehat{W}y\|_2 \tag{3.58}$$

By computing a singular value decomposition

$$\widehat{W} = (A - \theta I)V = \widehat{R}\Sigma\widehat{S}^* \tag{3.59}$$

and making use of the variational characterization (2.66) for the smallest singular value σ_{min} we can now derive a minimizer $\hat{u} = V\hat{y}$ from the formulation (3.58):

$$\|(A - \theta I)\hat{u}\|_2 = \min_{\substack{y \in \mathbb{C}^m \\ \|y\|_2 = 1}} \|\widehat{W}y\|_2 = \sigma_{min}(\widehat{W}) \tag{3.60}$$

The minimum $\hat{y}$ is taken for the right singular vector $\hat{s}_{min}$, and we obtain a solution $\hat{u}$ of the least squares problem (3.57) by:

$$\hat{u} = V\hat{s}_{min} \tag{3.61}$$

Note that this solution need not necessarily be unique, as σ_{min} may have a multiplicity greater than one, such that any of the related right singular vectors can be employed to construct a minimizer of (3.57). The above argumentation shows that we only need the right singular vector $\hat{s}_{min}$ of the matrix $\widehat{W} = (A - \theta I)V$ in order to determine the *refined Ritz vector*. Extending the *Rayleigh-Ritz procedure* (Alg. 3.12) with the above recipe for computing a *refined Ritz vector* we obtain the following

---

Algorithm 3.14: Refined extraction (refined Rayleigh-Ritz procedure)

---

1 compute Ritz pairs (θ_i, u_i) of A with respect to the basis vectors $V \in \mathbb{C}^{n \times m}$ of the subspace $\mathcal{K}$ using the Rayleigh-Ritz procedure (Alg. 3.12)

2 choose index
$$r \in \{1, \ldots, m\} \text{ of the Ritz pair to be refined}$$

3 compute
$$\widehat{W}_r = (A - \theta_r I)V$$

4 compute SVD
$$\widehat{W}_r = \widehat{R}\Sigma\widehat{S}^*$$

5 compute
$$\hat{u}_r = V\hat{s}_{min} \qquad\qquad\qquad\qquad\qquad (\textit{refined Ritz vector})$$

6 compute
$$\hat{\theta}_r = \hat{u}_r^* A\hat{u}_r \qquad\qquad\qquad\qquad\qquad (\textit{refined Ritz value})$$

Remark 3.20

- *The above algorithmic template comprises the standard Rayleigh-Ritz procedure and, hence, it will be more expensive in total. Fortunately, as the costs for computing the SVD of an $n \times m$-matrix amount to $\mathcal{O}(n \cdot m^2)$, the order of magnitude will remain the same, since the Rayleigh-Ritz procedure is also an $\mathcal{O}(n \cdot m^2)$ process. Of course, it is possible to compute refined approximations to all, rather than only to one Ritz pair. However, then m SVDs (one for each shifted matrix $\widehat{W}_r$) have to be computed and, consequently, $m \cdot \mathcal{O}(n \cdot m^2) = \mathcal{O}(n \cdot m^3)$ additional work has to be invested, which is often too expensive. We shall see later on, that is sufficient for our purposes to compute only one refined Ritz pair.*

- To obtain the right singular vector associated with σ_{min}, one can compute an SVD of $\widehat{W}$, for instance by means of the LAPACK [2, 7] routine xGESVD. An alternative is to compute the eigensystem of the cross-product matrix $\widehat{W}^*\widehat{W}$ and to use its relation to the SVD of $\widehat{W}$ according to Corollary 2.45. This is advocated in [122] as a means to save computational effort when several refined approximations are of interest. The analysis in [122] shows that in general there are no stability problems using the cross-product approach in our context, because suitable criteria allow to anticipate possible inaccuracies and to avoid difficulties.

- Refined extraction methods in the sense of the above definition were introduced by JIA and examined in the context of different iterative projection methods ([63], [36]). Related approaches are treated in [47], e.g.

$\square$

3.3.2. Subspace Expansion

The subspace expansion in the generic template (Alg. 3.11) is the characteristic algorithmic component which actually distinguishes the different iterative projection methods from each other. Rather surprisingly, there are only a few possibilites discussed in the literature and for a long time, until about 1975, when DAVIDSON proposed his method [29], the only approach to be considered was the subspace expansion leading to the Lanczos method.

3.3.2.1. Lanczos Method

The simplest conceivable way to expand a given subspace $\mathcal{K}$ is to choose the residual

$$t = f(r) = r = Au - \theta u \tag{3.62}$$

with respect to the current eigenvector approximation u as a new search direction in Algorithm 3.11. Starting from scratch with an arbitrary initial guess $r \neq 0$, it follows by induction, that this choice leads to a Krylov space

$$\mathcal{K}_m = \{r, Ar, A^2r, \dots, A^{m-1}r\} \tag{3.63}$$

In the following, however, we do not employ the generic scheme (Alg. 3.11) as a framework, but we pursue a slightly different approach, since it is more appropriate to use the Lanczos procedure (Alg. 2.7) to compute an orthonormal basis of the Krylov space $\mathcal{K}_m$. Comparing (2.59) with (3.39) reveals that the interaction matrix M_m with respect to the Lanczos vectors V_m in the *Rayleigh-Ritz procedure* (Alg. 3.12) coincides with the tridiagonal matrix T_m. A clever combination of both algorithms results in the following simple algorithm for computing several eigenpairs:

Algorithm 3.15: Lanczos method for Hermitian eigenproblems

1 choose a vector v_1 such that $\|v_1\|_2 = 1$. Set $\beta_1 \equiv 0$, $v_0 \equiv 0$
2 **for** $j = 1, 2, \ldots, m$ **do**
3 $\quad$ compute $r_j = Av_j - \beta_j v_{j-1}$
4 $\quad$ compute $\alpha_j = (r_j, v_j)$
5 $\quad$ $r_j = r_j - \alpha_j v_j$
6 $\quad$ reorthogonalize if necessary
7 $\quad$ $\beta_{j+1} = \|r_j\|_2$.
8 $\quad$ compute approximate eigenvalues of A from eigendecomposition of
   $\quad$ $T_j = S\Theta^{(j)}S^*$ (T_j is defined as per Notation 2.39)
9 $\quad$ test bounds (residual norms) for convergence by means of (3.65) (see below)
10 $\quad$ $v_{j+1} = w_j / \beta_{j+1}$
11 **end for**
12 compute approximate eigenvectors $X = V_j S$

Using the Lanczos relation (2.57) the residual $r_i^{(j)}$ (cf. Line 9) of a Ritz pair $(\theta_i^{(j)}, x_i^{(j)})$ in the jth pass of the **for**-loop in Algorithm 3.15 can be expressed as

$$r_i^{(j)} = Ax_i^{(j)} - \theta_i^{(j)}x_i^{(j)} = AV_j s_i^{(j)} - V_j s_i^{(j)}\theta_i^{(j)} = (AV_j - V_j T_j)s_i^{(j)} = v_{j+1}\beta_{j+1}s_{j,i}^{(j)} \quad (3.64)$$

and, hence, the residual norm is determined from

$$\|r_i^{(j)}\|_2 = |\beta_{j+1}s_{i,j}^{(j)}| \quad (3.65)$$

This shows that it is not necessary to compute the *Ritz vectors* $x_i = V_j s_i$ during the **for**-loop (Lines 2-11) in order to obtain the residual norms of the *Ritz pairs*. These time-consuming matrix-vector multiplications can be postponed until the the end of the algorithm when the *Ritz values* of interest have converged (Line 12).

Remark 3.21

- *As pointed out in Section 2.3.3.2 one has to take into account that the Lanczos vectors lose their global orthogonality. There are different approaches to deal with this difficulty:* PARLETT *and his co-workers have proposed reorthogonalization strategies (partial and selective orthogonalization), which are described e.g. in [91].* CULLUM *and* WILLOUGHBY *[27], [26] suggest to do no reorthogonalization at all, which results in multiple copies of already detected eigenvalues (so called spurious eigenvalues). However they are able to state clever criteria, which allow for filtering out these multiple copies after one run of the Lanczos algorithm. A general survey of the Lanczos method and the pros and cons of the different strategies may be found in [8].*

- *The Lanczos algorithm 3.15 can be interpreted as a subspace accelerated power method and indeed it can be shown that it is superior to the single vector power method discussed in Section 3.1.1 (for the related proof see [91] or [49]).*

- *The Lanczos method, which was originally introduced in 1950 [70] as a means to reduce the input matrix A to tridiagonal form, has been known for more than half a century and a lot of research has been done to analyze its convergence properties. In contrast to the algorithms to be discussed in the following there is a beautiful and mature convergence theory available along with compelling quantitative results, the so-called Kaniel-Paige-Saad bounds (for details see e.g. [91], [100] or [49]). Essentially, one can say that convergence to eigenvalues at the ends of the spectrum will be faster, i.e. these eigenvalues are in general the first ones to be detected. The better these eigenvalues are separated from the rest of the spectrum, the faster the convergence will be. Unfortunately, rather often this is not true in practice, and in these cases the Lanczos method is slow to converge and the number of required iterations can be high. As a consequence, the costs for storing all Lanczos vectors in memory and keeping them orthogonal may become prohibitive. For this reason, we derive a strategy which tries to remedy these disadvantages in the following.*

□

3.3.2.2. Implicitly Restarted Lanczos Method (IRLM)

A general approach to limit storage requirements and computational effort is the incorporation of so-called *restarts* and a particularly simple way of doing so are *explicit restarts*, in which one retains the most recent Lanczos vector v_k, discards all previously computed basis vectors $v_1, \ldots, v_{k-1}$ and launches a new Lanczos process with $\tilde{v}_1 = v_k$ as a starting vector. Clearly, this meets our requirements as the memory consumption is bounded. On the other hand, one often loses valuable information, for which much work had to be invested. A more sophisticated alternative is due to LEHOUCQ [73] and SORENSEN [117] who suggest an *implicit restart* scheme. Their basic idea can be described as follows: The result of applying $m = k + p$ steps in the basic Lanczos method (Algorithm 3.15) is characterized by the following algebraic top-level formulation (cf. (2.57)):

$$AV_m = V_m T_m + r_m e_m^* \tag{3.66}$$

The objective is now to compress this $(k + p)$-step factorization to a k-step Lanczos factorization containing the most interesting information. This can be realized by applying p implicitly shifted QR steps on T_m and leads to

$$AV_m^+ = V_m^+ T_m^+ + r_m e_m^* Q \tag{3.67}$$

where

$$V_m^+ = V_m Q, \qquad T_m^+ = Q^* T_m Q \quad \text{and} \quad Q = Q_1 Q_2 \cdots Q_p \tag{3.68}$$

Each of the orthogonal matrices Q_j is associated with the shift μ_j during the shifted QR algorithm. Since all Q_j exhibit Hessenberg structure, the first $k - 1$ entries of the vector $e_m^* Q$ are zero. This in turn shows that the k leading columns in (3.67) are still in Lanczos relation and represent what we are looking for, an updated k-step Lanczos factorization

$$AV_k^+ = V_k^+ r_k^+ e_k^*$$ (3.69)

where $V_k^+ = V_m^+(1:k)$ is formed from the first k columns of V_m^+. The updated residual is of the form

$$r_k^+ = V_m^+ e_{k+1}\beta_k + r_m Q(m,k)$$ (3.70)

where $Q(m,k)$ is the element which is located at the mth row and kth column of the matrix Q defined in (3.68). The above ideas provide a general recipe for shrinking a $(k+p)$-step Lanczos relation to a k-step factorization. Iterating this process eventually leads to an algorithm called the implicitly restarted Lanczos method (IRLM) and is described in Alg. 3.16.

Algorithm 3.16: IRLM for Hermitian eigenproblems

---

1 start with $v_1 = v/\|v\|_2$ where $v \neq 0$ is an arbitrary initial guess
2 compute an m-step Lanczos factorization
$$AV_m = V_m T_m + r_m e_m^*$$
3 **repeat**
4   compute $\sigma(T_m)$ and select p shifts $\mu_1, \mu_2, \ldots, \mu_p$
5   initialize $Q = I_m$
6   **for** $j = 1, 2, \ldots, p$ **do**
7   QR-factorize $Q_j R_j = T_m - \mu_j I$
8   update $T_m = Q_j^* T_m Q_j$, $Q = Q Q_j$
9   **end for**
10     $r_k = v_{k+1}\beta_k + r_m\sigma_k$, with $\beta_k = T_m(k+1,k)$ and $\sigma_k = Q(m,k)$
11     $V_k = V_m Q(:, 1:k)$, $T_k = T_m(1:k, 1:k)$
12   beginning with the k-step Lanczos factorization
$$AV_k = V_k T_k + r_k e_k^*$$
    apply p additional steps of the Lanczos process to obtain
$$AV_m = V_m T_m + r_m e_m^*$$
13 **until** *convergence, i.e. $T_k = D_k$ diagonal*

Remark 3.22

- *To compute the initial $(k+p)$-step Lanczos factorization in Line 1 one can make use of Algorithm 3.15.*

- *The shifts μ_j in Line 4 are chosen according to the eigenvalues of interest. To this end, the user typically specifies a "wanted set" of eigenvalues. Possible and sensible choices are e.g. the k smallest or the k largest eigenvalues of the matrix A.*

- *A popular choice for the shifts μ_j that has proven successful in practice are so-called exact shifts. To this end the Ritz values, i.e. the eigenvalues θ_i of T_m are partitioned in two disjoint sets of k wanted and p unwanted eigenvalues. The p*

unwanted eigenvalues are then used as shifts $\mu_1, \ldots, \mu_p$. Repeating this strategy successively filters out the unwanted eigenvalues.

- An interesting theoretical property lies in the fact that one shift-cycle in Lines 6-9 of the IRLM (Alg. 3.16) can be viewed as the implicit application of a polynomial ψ in A to the starting vector v_1, i.e.

$$v_1 \leftarrow \psi(A)v_1 \qquad (3.71)$$

where the zeros of ψ are the the p shifts $\mu_1, \ldots, \mu_p$ used in the QR process

$$\psi(\lambda) = \prod_{j=1}^{p}(\lambda - \mu_j) \qquad (3.72)$$

For full details on this relation see [117]. This interpretation as a polynomial filter also motivates other interesting choices of the shifts μ_j, the roots of Čebyšëv[2] polynomials, the roots of Leja polynomials, the roots of least squares polynomials or harmonic Ritz values. For the corresponding references see the related discussion in [8].

- In analogy to the QR-algorithm discussed in Section 3.2.2.1 the application of p shifted QR iteration steps should be done using implicitly shifted QR factorizations.

- All of the above ideas related to Lanczos factorizations (2.57) directly carry over to Arnoldi factorizations (2.50) and, hence, they result in a straightforward generalization of the IRLM to a method for non-Hermitian matrices, the implicitly restarted Arnoldi method (IRAM).

- There are careful implementations of both, the IRLM and the IRAM, freely available in the form of the state-of-the-art software package ARPACK [75], a detailed description of which is given in the related user's guide [76]. We will use the ARPACK software later on, apply it to the eigenproblems we are interested in and make comparsions with the results obtained by the Jacobi-Davidson methods to be disussed in the following.

- We shall see that a key problem in the application of the ARPACK software is how large to choose the maximal subspace dimension $m = k + p$ relative to the number of sought-after eigenvalues k_{max}. It is clear that m must be at least equal to k_{max}, in [76] it is recommended to choose $m \geq 2 \cdot k_{max}$. Unfortunately there is no general recipe for a succesful adjustment of the parameter m and, even worse, the computational effort to be invested in order to obtain k_{max} converged eigenpairs depends rather sensitively on this choice.

□

---

[2]see also remarks on transliteration of Russian names in Appendix A.2

3.3.2.3. Davidson's Method

The *Davidson method* is named after its inventor, a quantum chemist, who proposed the algorithm in 1975 [29]. He designed the method for the application to eigenvalue computations in electronic structure calculations where the arising Hermitian matrices are typically very large and almost diagonal, i.e. they are strongly diagonal dominant. To take advantage of this specific property, he suggested to expand the search space $\mathcal{K}$ by

$$t = f(r) = (D_A - \theta)^{-1} r \tag{3.73}$$

where D_A is the diagonal matrix whose entries are the diagonal entries of A and

$$r = Au - \theta u \tag{3.74}$$

is the residual r with respect to a given approximation (θ, u). Using (3.73) as *subspace expansion* and the *Rayleigh-Ritz procedure* (Alg. 3.12) as a method for *information extraction* the, generic Algorithm 3.11 turns into a basic version of Davidson's method:

Algorithm 3.17: Davidson's method for $\lambda_{min}(A)$

---

1   **function** $(\lambda, q) = $ **davidson**(A, v_0)
2   choose a starting vector $t = v_0$
3   $M = [], V = [], W = []$
4   **for** $m = 1, 2, \ldots, \nu_1$ **do**
5      $v_m = \text{orth}(V, t)$;   /* orthonormalization by means of Alg. 2.5 */
6      $w_m = Av_m$
7      $M = \left[\begin{array}{c|c} M & V^* w_m \\ \hline w^* V & v_m^* w_m \end{array} \right]$
8      $V = [V, v_m], W = [W, w_m]$
9   compute smallest eigenpair (θ, s) of M $(\|s\|_2 = 1)$
10     $u = Vs$
11     $w = Ws$
12     $r = w - \theta u$
13   **if** $\|r\|_2 \leq \varepsilon$ **then**
14 | **return** (θ, u)
15   **end if**
16   solve t from:
17     $(D_A - \theta I)t = r$
18 **end for**

Remark 3.23
- *Note, that in contrast to the Rayleigh-Ritz procedure (Alg. 3.12) only the Ritz vector u (Line 10) related to the smallest eigenvalue approximation θ (Line 9) is of interest and is actually computed in one pass of the **for**-loop.*

- As we consider Hermitian matrices A, we can adopt the Rayleigh-Ritz procedure to this situation: It is sufficient to build up and store the upper triangular part of the interaction matrix M. Thus, the update of M in every step of the Davidson iteration can be realized by adding one column and an additional bottom element.

- In practice, one restricts the size of the search space by incorporating so called restart techniques. Furthermore, computing more than one eigenpair may be achieved by means of deflation. Further details on both, restart and deflation, will be discussed in the corresponding Section 4.3 on the Jacobi-Davidson method and can be applied without any significant modification to extend the Davidson method.

- Further theoretical and practical investigations on the original version of Davidson's method along with an explanation for its success when dealing with strongly diagonal dominant matrices may be found in [30].

$\square$

The algorithm is reported to work well in the context described above which explains its popularity among quantum chemists and it is easy to implement. However, for a long time it was not well understood, why *Davidson's method* is successful and, until recently, there was no satisfactory convergence theory available. The obvious and intuitive idea to regard D_A as a preconditioner $K \approx A$ is problematic. This becomes evident when trying to generalize *Davidson's method* to arbitrary approximations $K \approx A$. In the asymptotic case $K = A$ we have

$$t = (K - \theta I)^{-1}r = (A - \theta I)^{-1}r = (A - \theta I)^{-1} \cdot (A - \theta I)u = u \qquad (3.75)$$

and, clearly, this leads to stagnation, as no new information enters into the subspace. Hence, the approximation K must not be too good, which is counterintuitive and shows that the notion of K as a preconditioner for A is misleading in this context. Nonetheless, there are also approaches with K different from D_A, that yield satisfactory results (cf. [103], [25]). Notice that for general choices of K a factorization of $B := K - \theta I$ (line 17) is required in every pass of the **for**-loop in Alg. 3.17, which is often rather time-consuming. Another interesting particluar case is the choice $K = I$, which leads to a process that generates the same sequence of vectors as the Lanczos method. Hence, *Davidson's method* could be also viewed as a generalization of the Lanczos algorithm.

The above discussion reveals that *Davidson's method* heavily relies on the diagonal dominance of A and is only applicable, when this requirement is satisfied. Unlike e.g. the Lanczos method, it is thus no multi-purpose eigensolver. Theoretical results on the convergence of *Davidson's method*, which also make more precise statements on how to choose the approximation K, may be found in [103] and [25]. For a list of available software for *Davidson's method* see [52] and the references therein.

4. The Jacobi-Davidson Method and its Variants

The *Jacobi-Davidson method* was proposed by SLEIJPEN and VAN DER VORST [114] in 1996 as an improvement over *Davidson's method*, which explains the second part of the method's name. The first part is an acknowledgement of its relation to a method described by JACOBI in his famous article from 1846 [58], which we will call *JOCC method* (acronym for *Jacobi's Orthogonal Complement Correction*) in the following in order to distinguish it from the well-known diagonalization method (cf. Section 3.2.3) introduced in the same article. Like *Davidson's method* or the Lanczos algorithm it belongs to the family of iterative projection methods, and hence, from a systematic point of view, the following description should have been placed into a corresponding subsection of the previous chapter. However, as the *Jacobi-Davidson method* is in the center of interest of this thesis, it is appropriate to dedicate it a chapter of its own right. We will first give a motivation and we will present a basic version of the *Jacobi-Davidson method* for computing one eigenpair. Besides, we will briefly comment on its convergence properties as well as on its relation to other methods. Finally, we will develop more sophisticated variants of the algorithm which are suited for the practical use and which we will actually apply to the eigenvalue computations we are concerned with in this thesis.

4.1. Motivation of the Algorithm

4.1.1. JOCC Method

Originally, JACOBI did not use his plane rotations the same way as we are used to these days. He only went half the way and employed them as a means to make the matrix diagonally dominant. For the final diagonalization he used the JOCC method, which we will now briefly describe:

Let $A \in \mathbb{C}^{n \times n}$ be diagonally dominant and let us without loss of generality assume that $a_{11} = \alpha$ is the largest diagonal element of A. Then α is an approximation to A's largest eigenvalue λ and e_1 an approximation for the corresponding eigenvector. This leads to the following notation of the eigenvalue problem

$$A \begin{bmatrix} 1 \\ z \end{bmatrix} = \begin{bmatrix} \alpha & c^T \\ b & F \end{bmatrix} \begin{bmatrix} 1 \\ z \end{bmatrix} = \lambda \begin{bmatrix} 1 \\ z \end{bmatrix} \tag{4.1}$$

where α is a scalar, $F \in \mathbb{C}^{(n-1) \times (n-1)}$ is a square matrix and b, c and z vectors of appropriate size. Denoting the eigenvector related to the sought-after eigenvalue λ by

$u = (1, z)^T$ we can transform (4.1) into the equivalent system

$$\lambda = \alpha + c^* z \tag{4.2}$$
$$(F - \lambda I)z = -b \tag{4.3}$$

Jacobi's idea was now two solve (4.3) by turning the above equations into a related two-step iteration scheme where D_F is formed by the diagonal entries of F:

Algorithm 4.1: Jacobi's orthogonal complement correction (JOCC)

$$\left.\begin{array}{rcl} \theta_k &=& \alpha + c^* z_k \\ (D_F - \theta_k I)z_{k+1} &=& (D_F - F)z_k - b \end{array}\right\} \quad k = 0, 1, 2, \ldots \tag{4.4}$$

Alogrithm 4.1 belongs to the family of single vector iterations discussed in Section 3.1. The main feature of the JOCC iteration scheme is that eigenvector corrections in the orthogonal complement of the first unit vector (i.e. span$\{e_1\}^T$) are constructed in *every* iteration step, which also explains the name of the method. SLEIJPEN and VAN DER VORST [114] point out, that this is not quite the algorithm that Jacobi actually made use of, but the above version is better suited for our purposes as we shall see in the following.

4.1.2. Davidson's Method Revisited

Let us now briefly review *Davidson's method* (Alg. 3.17) in the context of the above notation (4.1), where an iterate u_k is again assumed to be scaled such that its first coordinate is 1, i.e. $u_k = (1, z_k^*)^*$. Furthermore, let θ_k be the associated eigenvalue approximation. Then the residual is given by

$$r_k = (A - \theta_k I)u_k = \begin{bmatrix} \alpha - \theta_k + c^* z_k \\ (F - \theta_k I)z_k + b \end{bmatrix} \tag{4.5}$$

In *Davidson's method*, a new search direction t_k is now computed from

$$(D_A - \theta_k I)t_k = -r_k \tag{4.6}$$

With $t_k = (\gamma, y_k^*)^*$, where γ is a scalar and $\hat{y}_k = (0, y_k^*)^*$, the component of t orthogonal to e_1, we conclude from (4.6) and (4.5)

$$(D_F - \theta_k I)y_k = -(F - \theta_k I)z_k - b = (D_F - F)z_k - (D_F - \theta_k I)z_k - b \tag{4.7}$$

or, equivalently,

$$(D_F - \theta_k I)(z_k + y_k) = (D_F - F)z_k - b \tag{4.8}$$

A comparison of (4.8) and (4.4) now reveals that $z_k + y_k$ is the z_{k+1} one would have obtained by one step of JOCC applied to z_k. However, after this point the iterates of *Davidson's method* differ from the JOCC scheme, as it does not take $t_k = (1, (z_k + y_k)^*)^*$ as the next eigenvector approximation, but expands the subspace $\mathcal{K}$ built up so far by t_k and then extracts the new approximation from span$\{\mathcal{K}, t_k\}$ by means of the *Rayleigh-Ritz procedure*. Clearly, this is an improvement over the JOCC method, as now more information enters into the computation of the eigenvalue. However, *Davidson's method* also only constructs eigenvector corrections in the orthogonal complement of e_1.

4.2. The Basic Jacobi-Davidson Method for Computing one Eigenpair

The close relation between JOCC and *Davidson's method* was the starting point for the considerations of Sleijpen and van der Vorst [114], whose idea was to go beyond the approach of Davidson by computing orthogonal corrections to the *current iterate*, rather than to the unit vector e_1 in every step of the iteration. This consequently leads to a different kind of subspace expansion, which we will derive now:
Let us again consider the Hermitian eigenvalue problem

$$Ax = \lambda x \tag{4.9}$$

and suppose that an approximation (θ, u) to an eigenpair (λ, x) where $\theta = \rho(u) = \tilde{u}^* Au$ and $\|u\|_2 = 1$ is available. We are now looking for an *exact* orthogonal correction t of u, i.e.

$$A(u + t) = \lambda(u + t), \qquad t \perp u$$

or, equivalently,

$$(A - \lambda I)t = -(A - \lambda I)u$$

By pre-multiplying with the projector $P = I - uu^*$, we obtain

$$(I - uu^*)(A - \lambda I)t = -(I - uu^*)(A - \lambda I)u$$

Because of $t \perp u$ the projector $P = (I - uu^*)$ acts as identity on t and we can write

$$\begin{aligned}
(I - uu^*)(A - \lambda I)(I - uu^*)t &= -(I - uu^*)(A - \lambda I)u \\
&= -(A - \lambda I)u + uu^*(A - \lambda I)u \\
&= -(A - \lambda I)u + u\,(u^* Au) - \lambda u \\
&= -(A - \theta I)u \\
&= -r \tag{4.10}
\end{aligned}$$

Solving t from (4.10) yields the desired correction of u. However, in practical situations one does not know λ beforehand, and hence, it is straightforward to replace λ by the current eigenvalue approximation $\theta = u^*Au$. This finally leads to the following relation, which is referred to as the *Jacobi-Davidson correction equation*:

$$(I - uu^*)(A - \theta I)(I - uu^*)t = -(A - \theta I)u = -r \tag{4.11}$$

Using a possibly approximate solution t of (4.11) as subspace expansion and the Rayleigh-Ritz method (Alg. 3.12) as a device for information extraction, the algorithmic framework (Alg. 3.11) yields a basic version of the *Jacobi-Davidson method*:

Algorithm 4.2: Jacobi-Davidson method for $\lambda_{min}(A)$ (JD)

---

1 **function** $(\lambda, q) = $ **jacobi-davidson**(A, v_0)
2 choose a starting vector $t = v_0$
3 **for** $m = 1, 2, \ldots, \nu_1$ **do**
4     $\quad v_m = \text{orth}(V, t)$;   /* orthonormalization by means of Alg. 2.5 */
5     $\quad w_m = Av_m$
6     $\quad M = \left[\begin{array}{c|c} M & V^*w_m \\ \hline w_m^*V & v_m^*w_m \end{array} \right]$
7     $\quad V = [V, v_m], W = [W, w_m]$
8     $\quad$compute smallest eigenpair (θ, s) of M ($\|s\|_2 = 1$)
9     $\quad u = Vs$
10    $\quad w = Ws$
11    $\quad r = w - \theta u$
12    $\quad$**if** $\|r\|_2 \leq \varepsilon$ **then**
13    $\quad\quad$**return** (θ, u)
14    $\quad$**end if**
15    $\quad$(approximately) solve the correction equation (4.11) for $t \perp u$:
16    $\quad\boxed{(I - uu^*)(A - \theta I)(I - uu^*)t = -r}$
17 **end for**

Of course, it is also possible to employ the above algorithm to compute eigenpairs other than the smallest one. We will come back to this issue later on. The correction equation in Line 16 of the above template is the core of the *Jacobi-Davidson method* and plays a key role for the success of the algorithm. In practice, (4.11) is not solved exactly, and even rather crude approximations may lead to success. Before we discuss in detail how to solve the correction equation in practical situations, we make some comments on the existence of solutions and on related methods in order to have a solid basis to build upon.

4.2.1. Consistency of the Correction Equation

First of all, to put things precisely, it should be mentioned that – although it virtually never leads to problems in practice – it cannot be guaranteed that (4.11) always has unique a solution, as is highlighted in the following theorem from [36]:

Theorem 4.1
Assume that (ρ, u) is an approximate eigenpair of the matrix A with $u \in \mathcal{K}$ and $\rho = u^* A u$, and select a matrix $U_\perp \in \mathbb{C}^{n \times (n-1)}$ such that $[u, U_\perp]$ is unitary. Then the columns of $U_\perp$ form an orthonormal basis of $\mathrm{span}\{u\}^\perp$. Set $r = (A - \rho I)u$. Then $r \perp u$, and there exists a unique b such that $r = U_\perp b$. For the linear system

$$(I - uu^*)(A - \rho I)(I - uu^*)t = -r \qquad t \perp u \qquad (4.12)$$

the following results hold:

1. *equation (4.12) has no solution if $b \notin \mathrm{Im}(U_\perp^* A U_\perp - \rho I)$*

2. *equation (4.12) has at least one solution if $b \in \mathrm{Im}(U_\perp^* A U_\perp - \rho I)$*

3. *equation (4.12) has a unique solution, if and only if ρ is not an eigenvalue of $U_\perp^* A U_\perp$*

Proof: $[u, U_\perp][u, U_\perp]^* = I$, hence

$$I - uu^* = U_\perp U_\perp^*$$

and (4.12) can be written as

$$(U_\perp U_\perp^* A U_\perp U_\perp^* - \rho I)t = -r$$

Setting $t = U_\perp x$ we obtain

$$U_\perp (U_\perp^* A U_\perp - \rho I)x = -U_\perp b$$

Pre-multiplying $U_\perp^*$ from the left then gives

$$(U_\perp^* A U_\perp - \rho I)x = -b \qquad (4.13)$$

The assertions of the theorem now directly follow from (4.13) and the well-known criteria for the existence of solutions of linear systems. $\qquad \square$

It is also interesting to see what happens when an eigenvalue approximation θ is getting close to an eigenvalue λ of A, as it is the case when the iteration converges. A common fear is that the shifted matrix $A - \theta I$ becomes nearly singular and this might lead to problems in the solution of the related linear systems. PARLETT [91] could refute these objections for the *inverse iteration* (Alg. 3.2) and the *Rayleigh quotient iteration* (Alg.

3.3), where one is in a similar situation (see Remark 3.2). In case of the Jacobi-Davidson iteration, a new aspect comes into play: The role of the projectors $P = I - uu^*$. In [44] a heuristic analysis of the asymptotic case that the eigenvector x coincides with the approximation u is made. To see what happens upon convergence $(u \to x)$ we make a heuristic analysis (cf. [44]), i.e. we assume $u = x$ in the left-hand side and $u \neq x$ in the right-hand side of (4.11). Then we have

$$(I - xx^*)(A - \lambda I)(I - xx^*)t = -r \tag{4.14}$$

and $(A - \lambda I)$ is singular. However, the eigenvector x, which is the singular direction, is projected out by $(I - xx^*)$, so that the solution t is in the orthogonal complement of x. Thus, equation (4.14) is well-defined. Of course, the above argumentation is also valid for the asymptotic case $u = x$ in the right-hand side (cf. [114] for a related discussion). The considerations in this paragraph and the experiences made in the practical use of Alg. 3.2 show that there are no stability problems solving the correction equation (4.11).

4.2.2. Relation to Other Methods

The *Jacobi-Davidson method* exhibits interesting relations to other methods, as we shall see in the following. To this end, we consider the more general situation that an approximation $M \approx (A - \theta I)$ is available. Then using $t \perp u$ we obtain the following instance of (4.11)

$$
\begin{aligned}
(I - uu^*)Mt &= -r \\
\Leftrightarrow Mt - uu^* Mt &= -r \\
\Leftrightarrow Mt - \alpha u &= -r
\end{aligned}
\tag{4.15}
$$

where $\alpha = u^* Mt$.

4.2.2.1. JD and RQI

Let us first consider the case that we do not use any approximation at all, but work with the exact correction equation, i.e. $M = A - \theta I$. Solving t from (4.15) then leads to

$$
\begin{aligned}
t &= \alpha(A - \theta I)^{-1}u - (A - \theta I)^{-1}r \\
&= \alpha(A - \theta I)^{-1}u - u
\end{aligned}
\tag{4.16}
$$

In the subsequent step of Algorithm 4.2 t is made orthogonal to the subspace $\mathcal{K}$ built so far and since $u \in \mathcal{K}$, the new search direction is actually

$$\tilde{t} = (A - \theta I)^{-1}u \tag{4.17}$$

This shows that the *Jacobi-Davidson method* is related to *inverse iteration* (Algorithm 3.2) and especially to the *Rayleigh quotient iteration* for Hermitian matrices (Algorithm 3.3), provided that (4.11) is solved *exactly*. Notice, however, that the *Jacobi-Davidson method* is *not* identical, but actually an improvement over RQI, as it is a subspace accelerated process, i.e. *all* iterates generated so far are assembled in $\mathcal{K}$ and the new eigenvector approximation is computed from $\mathcal{K}$. Finally, the discussion in this paragraph also makes more precise in what sense the JD algorithm can be viewed as an improvement over *Davidson's method*: As opposed to *Davidson's method* (cf. discussion in Section 3.3.2.3) there is no danger of stagnation when using an approximation M which is very close or possibly even identical to $A - \theta I$. On the contrary, the better the approximation is, the more satisfactory the convergence will be, and this shows that the JD-approach removes the conceptual weakness of *Davidson's method* in the context of general approximations $K \approx A$. Notay [84] directly compares the performance of the JD and the *Davidson method* in a couple of numerical experiments, and in fact JD is faster in most cases, except for matrices that are almost diagonal.

4.2.2.2. JD and Davidson

Using $M = K - \theta I$ as an approximation and choosing $\alpha = 0$ in (4.15) yields the subspace expansion

$$t = (K - \theta I)^{-1} r \qquad (4.18)$$

which leads to a (generalized) instance of *Davidson's method* (Algorithm 3.17).

4.2.2.3. JD and Olsen

In many cases the exact solution of (4.11) is much too expensive. Therefore, one may attempt to use a fixed approximation $M \approx (A - \theta I)$ for different values of θ instead. Then solving for t in (4.15) leads to

$$t = \alpha M^{-1} u - M^{-1} r, \qquad t \perp u \qquad (4.19)$$

We can now determine the parameter α by exploiting the orthogonality constraint $t \perp u$:

$$
\begin{aligned}
u^* t &= \alpha u^* M^{-1} u - u^* M^{-1} r = 0 \\
\Longleftrightarrow \qquad \alpha &= \frac{u^* M^{-1} r}{u^* M^{-1} u}
\end{aligned} \qquad (4.20)
$$

This approach was proposed by OLSEN and his co-authors in 1990 [88] in the context of FCI calculations (*Full Configuration Interaction*) arising in quantum chemistry. Thus, we will refer to the subspace expansion defined by (4.19) and (4.20) as *Olsen's method* in the following. Note that the derivation of the parameter α is typical of our following considerations and will come up repeatedly in different disguise.

4.2.2.4. JD and Lanczos

The simplest choice $M = I$ and $\alpha = 0$ results in a subspace expansion by the residual $r = Au - \theta u$. Obviously, this results in a subspace with Krylov structure, i.e.

$$\mathcal{K} = \{v_0, Av_0, A^2 v_0, \ldots, A^{m-1} v_0\} \tag{4.21}$$

where v_0 is the starting vector. In exact artihmetic the process obtained by using $-r$ as expansion in our framework, Alg. 3.11, generates the same sequence of approximations as the Lanczos method.

4.2.3. Solving the Correction Equation

We have already pointed out that the efficiency of the *Jacobi-Davidson method* crucially depends on whether one succeeds in solving the correction equation (4.11) appropriately. According to the above discussion, the exact solution leads to a subspace accelerated *Rayleigh quotient iteration* which has attractive convergence properties, but is in general not feasible because the involved costs are prohibitive. The natural way out of this difficulty is to resort to cheaper approximate solutions instead, e.g. obtained by means of one-step approximations, like Olsen's method previously discussed in Section 4.2.2.3, or multi-step approximations using iterative solvers for linear systems which we will present in the following. Before we go into detail, let us collect some general guidelines which are borne out of experience and have proven successful in practice (for a detailed motivation and discussion on this issue see e.g. [114], [8], [55]):

- The correction equation should not be solved too accurately. Especially in the beginning of the iteration process relatively crude approximations are often sufficient, and it would be even counterproductive to be particularly exact there. More precisely speaking, the demanded accuracy should be gradually increased, e.g. by imposing some rule on the norm of the residual like

$$\|r_m\|_2 = \|t_m - r\|_2 \leq 0.7^m, \tag{4.22}$$

  where m denotes the number of the JD iteration step (`for`-loop in Algorithm 4.2).

- When working with an iterative Krylov solver, the number of solver steps should be limited above by a small constant `itsolvermax` (e.g. `itsolvermax = 5`), regardless of whether the demanded accuracy is reached or not.

- Any other approximation may be sensible and is admissible, provided that the computational effort is limited. We actually make use of a mixture between Olsen's method and the two above principles in our computations. This means that we expand the subspace $\mathcal{K}$ by means of (4.19) and (4.20) in the first two or three steps of the JD-loop and then switch over to the solution of the correction equation using Krylov methods.

4.2.3.1. Iterative Krylov Methods for Linear Systems

We consider a linear system

$$Ax = b, \qquad A \in \mathbb{C}^{n \times n}, b \in \mathbb{C}^n \qquad (4.23)$$

where the coefficient matrix A is sparse, the size of the problem n is large and the solution $x \in \mathbb{C}^n$ is sought-after. Analogously to the situation of eigenvalue problems discussed in the sections 3.2 and 3.3, one can use either direct methods (LU factorisation, Gaussian elimination) or iterative methods to solve (4.23). We will make use of specific iterative methods, so called *Krylov subspace methods*, as they access information of the matrix A only implicitly by means of matrix-vector products which makes them particularly suited for our purposes. We can only describe the basic ideas and give a rough sketch of the underlying principles. For a detailed and thourough discussion we refer to the books by SAAD [101] and GREENBAUM [50], a brief survey for the practical use may be found in [9].

Given a suitable initial guess x^0 along with the related residual $r^0 = b - Ax^0$, the Krylov subspace methods we are interested in successively build up an orthonormal basis of the Krylov space

$$\mathcal{K}_m(A, r^0) = \text{span}\{r^0, Ar^0, A^2 r^0, \ldots, A^{k-1} r^0\} \qquad (4.24)$$

which results in a Lanczos or Arnoldi process (depending on whether A is Hermitian or not) and attempt to construct an approximate solution

$$x^m \in x^0 + \mathcal{K}_m(A, r^0) \qquad (4.25)$$

In the following, we will restrict ourselves to Krylov methods which minimize the corresponding residual (GMRES/MINRES), i.e.

$$\|b - Ax^m\|_2 = \min_{x \in x^0 + \mathcal{K}_m(A, r^0)} \|b - Ax\|_2 \qquad (4.26)$$

or try to minimize it in a certain sense (QMR/QMRS). Notice that there is a bewildering variety of methods ([101], [50], [9]) all of which having their pros and cons. In principle, any method which is designed for the use of general non-Hermitian matrices is admissible (other possible choices thus may be CGS, BiCGstab etc.).

4.2.3.2. GMRES

Let us consider the general case that $A \in \mathbb{C}^{n \times n}$ is not Hermitian. Then the orthonormal basis $V_m \in \mathbb{C}^{n \times m}$ of $\mathcal{K}_m$ is obtained by the Arnoldi procedure (Alg. 2.6), and any vector

$x \in x^0 + \mathcal{K}_m$ can be written as an affine linear combination

$$x = x^0 + V_m y, \qquad y \in \mathbb{C}^m \tag{4.27}$$

We are looking for a minimizer y^m of the functional J

$$J(y) = \|b - Ax\|_2 = \|b - A(x^0 + V_m y)\|_2 \tag{4.28}$$

Using (4.28) and the Arnoldi relation (2.51) gives

$$
\begin{aligned}
b - Ax &= b - A(x^0 + V_m y) \\
&= r^0 - AV_m y \\
&= \beta v_1 - V_{m+1}\bar{H}_m y \\
&= V_{m+1}(\beta e_1 - \bar{H}_m y) \tag{4.29}
\end{aligned}
$$

Since the columns of V_{m+1} are orthonormal, then

$$J(y) = \|b - A(x^0 + V_m y)\|_2 = \|\beta e_1 - \bar{H}_m y\|_2 \tag{4.30}$$

To obtain the solution $x^m = x^0 + V_m y^m$ we have to solve the least squares problem

$$y_m = \operatorname*{argmin}_{y \in \mathbb{C}^n} \|\beta e_1 - \bar{H}_m y\|_2 \tag{4.31}$$

This leads to the GMRES method (*Generalized Minimal Residual*) which was proposed by SAAD and SCHULTZ [102] in 1986:

Algorithm 4.3: GMRES

1 compute $r^0 = b - Ax^0$, $\beta := \|r^0\|_2$ and $v_1 = r^0/\beta$
2 **for** $j = 1, 2, \ldots, m$ **do**
3 $\quad$ compute $w_j := Av_j$
4 $\quad$ **for** $i = 1, \ldots, j$ **do**
5 $\quad\quad$ $h_{ij} := (w_j, v_i)$
6 $\quad\quad$ $w_j := w_j - h_{ij} v_i$
7 $\quad$ **end for**
8 $\quad$ $h_{j+1,j} = \|w_j\|_2$
9 $\quad$ **if** $h_{j+1,j} = 0$ **then**
10 $\quad\quad$ set $m := j$, goto 14
11 $\quad$ **end if**
12 $\quad$ $v_{j+1} = w_j/h_{j+1,j}$
13 **end for**
14 define the $(m+1) \times m$ Hessenberg matrix $\bar{H}_m = \{h_{ij}\}_{1 \le i \le m+1, 1 \le j \le m}$
15 compute y^m, the minimizer of $\|\beta e_1 - \bar{H}_m y\|_2$ and $x^m = x^0 + V_m y^m$

Remark 4.2

- In practical implementations the least squares problem in Line 15 is solved by means of a QR decomposition of the upper Hessenberg matrix $\bar{H}_m \in \mathbb{C}^{(m+1)\times m}$

$$\bar{H}_m = Q_m \bar{R}_m \tag{4.32}$$

where $Q_m \in \mathbb{C}^{(m+1)\times(m+1)}$ is obtained by the product

$$Q_m = J_1 \cdot J_2 \ldots \cdot J_m \tag{4.33}$$

of the m Givens rotations $J_i \in \mathbb{C}^{(m+1)\times(m+1)}$ (cf. Section 2.3.1.2) that successively annihilate the subdiagonal elements $\bar{h}_{i+1,i}$, and $\bar{R}_m \in \mathbb{C}^{(m+1)\times m}$. Since Q is unitary, one can transform the least squares problem into

$$\min \|\beta e_1 - \bar{H}_m y\|_2 = \min \|\bar{g}_m - \bar{R}_m y\|_2 \tag{4.34}$$

and the sought-after minimizer y^m can now be readily determined from the triangular system

$$Ry^m = g_m \tag{4.35}$$

where $R = \bar{R}(1:m, 1:m)$ and $g_m = \bar{g}_m(1:m)$.

- The storage requirement for the m Arnoldi vectors in V_m and the Hessenberg matrix $\bar{H}_m$ amounts to $\mathcal{O}(mn+m^2)$ which is a disadvantage when one is interested in accurate solutions, and thus, a large number m of solver steps may be required. In our situation, however, this is no problem, as modest accuracy of the solutions is in general sufficient, and as we restrict the number of solver steps m a priori.

$\square$

4.2.3.3. MINRES

The MINRES algorithm (*Minimal Residual*) was introduced by PAIGE and SAUNDERS [90] in 1975 and it can be regarded as a simplification of GMRES for the particular case that $A \in \mathbb{C}^{n\times n}$ is Hermitian. Then the Arnoldi process (Alg. 2.6) for the computation of the orthonogonal basis V reduces to the Lanczos procedure (Alg. 2.7) and the orthonormalization of a new basis vector v_{j+1} can be done by means of three-term recurrences which also implies that it is no more necessary to keep all basis vectors in memory. Furthermore, the Hessenberg Matrix $\bar{H}_m$ reduces to a tridiagonal matrix $\bar{T}_m$, the solution of the minimization problem (4.31) simplifies, and the update $x^m = x^0 + V_m y^m$ in Line 15 of Alg. 4.3 can be accomplished using short recurrences. For a more in-depth description on how to derive MINRES from GMRES and the algorithmic details we refer to [42], [50] and [9].

4.2.3.4. QMR and QMRS

The main idea behind the QMR method (*Quasi Minimal Residual*) introduced by FREUND and NACHTIGAL [40] in 1991 is to use short recurrences also for non-Hermitian matrices. To this end, they propose to pursue the same approach as for MINRES, but to use an unsymmetric Lanczos process which produces bi-orthogonal bases V_m and W_m for the Krylov spaces $\mathcal{K}_m(A, r_0)$ and $\mathcal{K}_m(A^*, r_0)$, i.e. $V_m^* W_m = I$ resp. $v_i^* w_j = \delta_{ij}$. Analogous to the standard Lanczos procedure (Alg. 2.7) the result is a relation of the form

$$AV_m = V_{m+1} \bar{T}_m \tag{4.36}$$

in which $\bar{T}_m$ is the $(m+1) \times m$ tridiagonal matrix

$$\bar{T}_m = \begin{bmatrix} T_m \\ \delta_{m+1} e_m^* \end{bmatrix} \tag{4.37}$$

T_m is unsymmetric and V_m is in general no more orthogonal, but by requiring $\|v_i\|_2 = 1$ $(i = 1, \ldots, m)$ we have

$$
\begin{aligned}
\|b - Ax^m\|_2 &= \|b - A(x^0 + V_m y^m)\|_2 \\
&= \|r^0 - AV_m y^m\|_2 \\
&= \|V_{m+1}(\beta e_1 - \bar{T}_m y^m)\|_2 \\
&\leq \|V_{m+1}\|_2 \cdot \|(\beta e_1 - \bar{T}_m y)\|_2 \\
&\leq \sqrt{m+1} \cdot \|(\beta e_1 - \bar{T}_m y)\|_2
\end{aligned}
\tag{4.38}
$$

Following the same ideas as for GMRES/MINRES, one can now try to minimize the second factor in (4.38), i.e.

$$J(y) = \|(\beta e_1 - \bar{H}_m y)\|_2 \tag{4.39}$$

which leads to a quasi minimal residual and explains the name of the method.

The resulting QMR algorithm is twice as expensive as the corresponding MINRES method, as now in every step two matrix-vector multiplications (involving A and A^*) are required.

Fortunately, the unsymmetric Lanczos process simplifies considerably, if it is applied to a J-Hermitian matrix B.

Definition 4.3
A matrix $B \in \mathbb{C}^n$ is called J-Hermitian, iff there is a matrix $J \in \mathbb{C}^{n \times n}$ such that $B^* J = BJ$

Then only one matrix-vector operation involving the operator B is needed in the unsymmetric Lanczos process (including one additional operation involving J) (cf. [41]) and this leads to a simplified version of the QMR method for J-symmetric matrices called QMRS (*Quasi Minimal Residual Simplified*). This variant is also due to FREUND and NACHTIGAL [41] and was proposed in 1994. As we shall see in the following discussion on preconditioners, the QMRS method is particularly well suited in our context.

4.2.3.5. Preconditioners

Unfortunately, Krylov subspace methods are often rather slow to converge and they may exhibit an irregular behavior. On the other hand, it is known from the theory (see [101], [50]) that the convergence rate of these methods strongly depends on the spectral properties of the coefficient matrix $A \in \mathbb{C}^{n \times n}$. Hence, a possible way to improve the convergence speed is to transform A into a better-suited matrix. To this end, it is common to construct a matrix $M \approx A$ which is referred to as a *preconditioner* and to consider the equivalent linear system

$$M^{-1}Ax = M^{-1}b \qquad (4.40)$$

Effectively, the Krylov method is now applied to the preconditioned operator $B :=$ $M^{-1}A$, and consequently, a Krylov subspace

$$\mathcal{K}(M^{-1}A, r^0) = \text{span}\{r^0, M^{-1}Ar^0, (M^{-1}A)^2 r^0, \dots, (M^{-1}A)^{m-1} r^0\} \qquad (4.41)$$

related to the preconditioned residual $r^0 = M^{-1}(b - Ax^0)$ is constructed. The approach in (4.40) is referred to as *left preconditioning*. Other possibilities are *right preconditioning*, i.e.

$$AM^{-1}u = b, \qquad u = Mx \qquad (4.42)$$

or *split preconditioning* where a preconditioner is given in factored form $M = M_1 M_2$.

$$M_1^{-1}AM_2^{-1}u = M_1^{-1}b, \qquad x = M_2^{-1}u \qquad (4.43)$$

It is natural to ask whether possible savings in the number of iteration steps (which are not guaranteed) compensate for the costs of setting up M once and multiplying M^{-1} in every step of the iteration. Hence, the general difficulty in devising a preconditioner is to find a matrix M that approximates A reasonably well on the one hand, and which is both, cheap to construct and cheap to invert, on the other hand. In most cases, one does not invert M explicitly and realizes the application of the operator M^{-1} by means of a user-supplied sub-routine. Preconditioners have emerged as a key ingredient for the success of Krylov solvers, and it is no surprise that they represent a wide area for research. For a general state-of-the-art survey of preconditioning techniques we refer to [11] and [101]. Unfortunately, there is no general recipe on how to find a good preconditioner, and often one is most successful in taking advantage of the problem specific situation. As we employ Krylov solvers for the solution of the *Jacobi-Davidson correction equation*, the construction and application of suitable preconditioners is of central importance in this thesis and we will come back to this aspect later on when we analyze the properties of the matrices we have to deal with.

When applying a Krylov method to a preconditioned operator $B = M^{-1}A$ one has to take into account the requirements of the method. In case of MINRES the trouble is that the preconditioned operator B will be in general no more Hermitian, even if A and

M are. Hence, the method is only applicable, when a positive definite preconditioner $M = LL^*$ is available and a split-preconditioning technique according to (4.43) with $B = L^{-1}A(L^{-1})^*$ is employed. Clearly, this is a prohibitive restriction for the case that A is Hermitian and highly indefinite.

In our situation QMRS is to be preferred over MINRES as the matrices we are dealing with are in general indefinite (due to the involved shift θ) and so are the preconditioners. Then working with a Hermitian left preconditioner K and defining $B = K^{-1}A$, $J = K^*$ implies

$$B^*J = A^*(K^{-1})^*K^* = A^* = K^*(K^*)^{-1}A^* = K^*K^{-1}A = JB \qquad (4.44)$$

Hence, according to Definition 4.3 $B = K^{-1}A$ is J-Hermitian with $J = K^*$ and the QMRS method is applicable to the preconditioned operator B.

4.2.3.6. Preconditioning the Correction Equation

We will now discuss in more detail how to use Krylov solvers along with preconditioners to solve the *Jacobi-Davidson correction equation* (4.11). To this end, we define the operator

$$\widetilde{A} = (I - uu^*)(A - \theta I)(I - uu^*) \qquad (4.45)$$

We also have to take into account the projectors for the preconditioner

$$\widetilde{K} = (I - uu^*)K(I - uu^*) \qquad (4.46)$$

where K is an approximation to A such that we arrive at the preconditioned correction equation

$$\widetilde{K}^{-1}\widetilde{A}v = \widetilde{K}^{-1}r \qquad (4.47)$$

Before we proceed, it is appropriate to make the following remarks:

- $\widetilde{A}$ and $\widetilde{K}$ are singular operators due to the projections involved. Hence, at first glance, one has to be careful with the notion preconditioner in this context. Fortunately, we are looking for solutions in span$\{u\}^T$, i.e. we are operating in the orthogonal complement of the singular direction u, such that our considerations in the following are sensible and all operations are well-defined.

- From (4.46) one recognizes that A is first approximated by K and then the projections are carried out. As pointed out in [37], this need not necessarily lead to the best approximation and one could also try to approximate the deflated operator $\widetilde{A}$ directly. However, in general there is no straightforward way to do so.

Solving $z \perp u$ from

$$\widetilde{K}z = \widetilde{A}v \qquad v \perp u \qquad (4.48)$$

seems rather complicated due to the involved projections at first glance. However, it turns out, that everything amounts to a surprisingly simple computational scheme. First of all, given an appropriate approximaton K for A let us assume that $v \perp u$ (see remark below). Then the right projector in the first step of the computation of $\widetilde{A}v$ can be omitted:

$$p = (A - \theta I)v \qquad (4.49)$$

To obtain the final result of the multiplication, the remaining projector is pre-multiplied

$$\widetilde{A}v = (I - uu^*)p \qquad (4.50)$$

such that

$$\widetilde{K}z = (I - uu^*)p \qquad (4.51)$$

Computing the scalar products and collecting the resulting terms now leads to

$$
\begin{aligned}
Kz &= (I - uu^*)p = p - \alpha u \\
z &= K^{-1}p - \alpha K^{-1}u
\end{aligned}
\qquad (4.52)
$$

and we can use the already well-known trick (cf. (4.19), (4.20)) and exploit the orthogonality requirement $z \perp u$ to determine the parameter α:

$$
\begin{aligned}
u^*z &= u^*K^{-1}p - \alpha u^*K^{-1}u = 0 \qquad &(4.53) \\
\alpha &= \frac{u^*K^{-1}p}{u^*K^{-1}u} \qquad &(4.54)
\end{aligned}
$$

Applying the preconditioner $\widetilde{K}$ to the right-hand side of (4.11), i.e. solving $\widetilde{r}$ from

$$\widetilde{K}\widetilde{r} = r \qquad (4.55)$$

can be done along similar lines, i.e.

$$
\begin{aligned}
(I - uu^*)K\widetilde{r} &= r \\
\widetilde{r} &= K^{-1}r - \alpha'K^{-1}u
\end{aligned}
\qquad (4.56)
$$

Following the familiar principle, we can obtain the paramter α' by exploiting the orthogonality condition $\widetilde{r} \perp u$:

$$\alpha' = \frac{u^*K^{-1}r}{u^*K^{-1}u} \qquad (4.57)$$

We summarize the steps (4.49) – (4.57) in the following template for the solution of the preconditioned correction equation (4.47):

Algorithm 4.4: Solution of the JD correction equation (left-preconditioning)

1 define operators

$$\widetilde{K} \equiv (I - uu^*)K(I - uu^*)$$
$$\widetilde{A} \equiv (I - uu^*)(A - \theta I)(I - uu^*)$$

2 solve y from $\boxed{Ky = u}$, $\mu = u^*y$

3 compute $\widetilde{r} \equiv \widetilde{K}^{-1}r$ as

4   (a) solve $\widehat{r}$ from $\boxed{K\widehat{r} = r}$

5   (b) $\widetilde{r} = \widehat{r} - \frac{u^*\widehat{r}}{\mu}y$

6 apply a Krylov solver with $t_0 = 0$, operator $\widetilde{K}^{-1}\widetilde{A}$,

7 and right-hand side $-\widetilde{r}$

8 compute $z = \widetilde{K}^{-1}\widetilde{A}v$ as:

9   (a) $p = (A - \theta I)v$

10   (b) solve $\widehat{p}$ from $\boxed{K\widehat{p} = p}$

11   (c) $z = \widehat{p} - \frac{u^*\widehat{p}}{\mu}y$

Remark 4.4

- *To make sure that we operate in the orthogonal complement of u, it is easiest to choose $t_0 = 0$ as starting vector for a Krylov solver, as is suggested in the above template.*

- *The arising preconditioner operations are highlighted by boxes. Suppose we want to carry out n_{LS} steps of a Krylov solver (e.g. GMRES) to (approximately) solve the preconditioned correction equation. Then the number of involved preconditioner solves amounts to*

$$
\begin{array}{lll}
1 & \text{solve for} & y = K^{-1}u \\
1 & \text{solve for} & \widehat{r} = K^{-1}r \\
n_{LS} & \text{solves for} & \widehat{p} = K^{-1}(A - \theta I)v \\
\hline
n_{LS} + 2 & \text{solves} & \text{altogether}
\end{array}
\tag{4.58}
$$

- *Strictly speaking, one actually has to construct a new preconditioner K in every pass of the JD-loop as θ varies and so does $(A - \theta I)$. In practice, this is often too time-consuming and luckily it turns out that in many cases it is possible to work with a fixed approximation K for several shifts θ.*

- *One may also derive a similar template for the case that one wants to apply a right preconditioner i.e. $\widetilde{A}K^{-1}$ (see [8] for details). We will restrict ourselves to the template above as it is sufficient for our purposes.*

$\square$

4.2.4. Convergence of the Jacobi-Davidson Method

We have already seen that the *Jacobi-Davidson method* can be interpreted as a subspace accelerated instance of the *Rayleigh quotient iteration* (cf. (4.16) and (4.17)), which explains the rapid convergence of the method if the correction equation (4.11) is solved exactly. However, as we have pointed out in the related discussion on the RQI in Section 3.1.3, this is only of little practical value, as the exact solution of the related system is in general much too expensive. Hence, to understand and analyze the convergence behavior of the JD process as it is actually used in practice, one has to take into account the following aspects:

- influence of the inexact solution of the correction equation (4.11) involving variable accuracy requirements

- influence of the preconditioners K

- influence of the subspace acceleration on the convergence behavior

It is not difficult to imagine that this turns out to be a rather complicated matter, and the convergence theory available up to now (about 10 years after the introduction of the JD method) is far from mature. For this reason, one can often only give heuristic motivations for the success of the method. This is especially true of the JDQR method (an extension of the JD method for computing several eigenpairs) to be discussed in the sequel. Recent investigations on this topic try to reduce the problem to particular cases that are easier to deal with. In [123] the *Jacobi-Davidson method* without subspace acceleration by means of Rayleigh-Ritz (Alg. 3.12) is analyzed and related to inexact *Rayleigh quotient iterations*, for which convergence results are available.

Another interesting observation lies in the fact that the *Jacobi-Davidson method* can be interpreted as an inexact (subspace accelerated) Newton scheme, as pointed out in [115]: The eigenvalue problem is nonlinear. For almost any fixed scaling vectors $\widetilde{u}$ and w, the eigenvector x (scaled such that $\widetilde{u}^*x = 1$) is the solution of the equation

$$F(x) = 0 \quad \text{where} \quad F(u) = Au - \theta u \quad \text{and} \quad \theta = \theta(u) = \frac{w^*Au}{w^*u} \qquad (4.59)$$

Clearly, the function F is nonlinear and maps the hyper-plane $\{u \mid \widetilde{u}^*u = 1\}$ to the hyperspace $w^\perp$. Particularly, this means that all residuals $r = F(u)$ are orthogonal to w. For a given approximation u_k (step k) the next Newton approximate u_{k+1} is now obtained by

$$u_{k+1} = u_k + t, \quad \text{where} \quad t \perp \widetilde{u} \quad \text{satisfies} \quad \left(\frac{\partial F}{\partial u}\bigg|_u\right) t = -r = -F(u_k) \qquad (4.60)$$

where the Jacobi matrix of F operates on $\widetilde{u}^\perp$ and is given by

$$\left(\frac{\partial F}{\partial u}\bigg|_{u_k}\right) t = \left(I - \frac{u_k w^*}{w^*u_k}\right)(A - \theta_k I)t \qquad t \perp \widetilde{u} \qquad (4.61)$$

This leads to the following correction equation for a Newton step

$$t \perp \tilde{u} \quad \text{and} \quad \left(I - \frac{u_k w^*}{w^* u_k}\right)(A - \theta_k I)t = -r \qquad (4.62)$$

which is already somewhat reminiscent of the *Jacobi-Davidson correction equation* (4.11), the difference being that in our derivation $\tilde{u}$ and w are kept fixed throughout the Newton process. From the theory it is known that the iterates u_k obtained from (4.62) converge asymptotically quadratically towards x if the initial guess u_1 ($\tilde{u}^* u_1 = 1$) is sufficiently close to x. Now that the error contracts quadratically if u_k is close enough to x, the non-stationary choice $\tilde{u} = u_k$ and $w = u_k$ also results in asymptotic quadratic convergence, and the Newton correction equation (4.62) coincides with the *Jacobi-Davidson correction equation* (4.11). Hence, the *Jacobi-Davidson method* is an inexact Newton scheme with subspace acceleration. This interpretation also gives additional motivation for the guidelines on how to solve the correction equation in Section 4.2.3, as solving the correction equation with increasing accuracy has proven useful for inexact Newton methods. A more profound and systematic discussion on the relation between Newton schemes and the *Jacobi-Davidson method* is given in the book by STEWART [122].

4.3. The JDQR Variants for Computing Several Eigenpairs

The *Jacobi-Davidson method* (Alg. 4.2) has attractive convergence properties, but it is not well-suited for the practical use for the following reasons:

- It is only capable of computing *one* eigenpair. In practice, and especially in our situation, we are interested in computing *several* eigenpairs.

- The dimension of the search space $\mathcal{K}$ is not bounded, and hence, one cannot predict how much memory will be needed for the computation of one eigenpair with reasonable accuracy.

In the following, we will derive a generalized version of the JD method which was introduced in [37] to remedy these drawbacks.

4.3.1. Deflation

In order to compute more than one eigenpair, one may apply the *Jacobi-Davidson method* to a *deflated operator*. More precisely speaking, we consider

$$B = (I - QQ^*)A(I - QQ^*) \qquad (4.63)$$

where

$$Q = [q_1, \ldots, q_k] \in \mathbb{C}^{n \times k} \tag{4.64}$$

holds the eigenvector approximations q_i of A computed so far. When we now apply JD to B, these eigenvectors q_i are projected out and will no more appear in the computational process, provided that the approximations q_i are accurate enough. This technique is called *explicit deflation* and is recommended in [37], as it works reliably and is numerically stable. Consequently, plugging (4.63) into (4.11) results in a modified correction equation:

$$(I - uu^*)(B - \theta I)(I - uu^*)t = -r$$
$$(I - uu^*)(I - QQ^*)(A - \theta I)(I - QQ^*)(I - uu^*) = -r$$

or, equivalently,

$$(I - \widetilde{Q}\widetilde{Q}^*)(A - \theta I)(I - \widetilde{Q}\widetilde{Q}^*)t = -r \tag{4.65}$$

where we define

$$\widetilde{Q} := [Q, u] \tag{4.66}$$

Using (4.65) as correction equation and appending the detected approximations in Q now enables us to use the Jacobi-Davidson process in order to compute several eigenpairs. The obvious disadvantage of this strategy, however, is that the computation of subsquent approximations q_j $(j > k)$ is getting increasingly expensive the more approximate eigenvectors q_i $(i = 1, \ldots, k)$ have already been detected.

4.3.2. Restarts

The discussion in this section is somewhat reminiscent of the motivation for the implicitly restarted Lanczos method (IRLM) in Section 3.3.2.2, the difference being that we need not preserve any structure like the Lanczos relation (2.57), which is the backbone of the IRLM. This gives us more freedom on how to accomplish a restart in case of the *Jacobi-Davidson method*. As we shall see later on, when we compare the IRLM and restarted JD variants, the approach discussed in the following turns out to be superior in the context of our eigenvalue problems. In analogy to the related discussion on the Lanczos method, one could make use of explicit restart techniques, which employ the most recent approximation u in Alg. 4.2 as a new starting vector t_0 and discard the search space $\mathcal{K}$ built up so far, when the dimension of $\mathcal{K}$ reaches a pre-defined upper bound m_{max}. However, we may lose hard-won information, when we remove the complete subspace. In many cases the subspace $\mathcal{K}$ will already contain valuable information on nearby eigenvectors to be computed in subsequent steps. Hence, it might be wiser not to recompute $\mathcal{K}$ from scratch, but to impose a lower bound $m_{min} > 1$ on the dimension of $\mathcal{K}$ and to retain m_{min} vectors of $\mathcal{K}$ potentially containing the most interesting information. To this end, we will pursue a more sophisticated strategy, and to see how the restart is accomplished, let us assume that we have computed an orthogonal matrix V whose columns are the basis of the search space $\mathcal{K}$ where $\dim \mathcal{K} = m_{max}$. Let $W = AV$, M be the interaction matrix $M = W^*V = V^*AV$ and

$$S^*MS = \Theta = \text{diag}(\theta_1, \theta_2, \ldots, \theta_{m_{max}}), \qquad \theta_1 \leq \theta_2 \ldots \leq \theta_{m_{max}} \tag{4.67}$$

the eigendecomposition of M. If we now use the m_{min} leading columns of S, i.e. the eigenvectors related to the m_{min} smallest eigenvalues of M, and define

$$\widetilde{S} := S(:, 1 : m_{min}) \tag{4.68}$$

then we obtain a basis $\widetilde{V}$ of the shrinked subspace by

$$\widetilde{V} := V \cdot \widetilde{S} \tag{4.69}$$

and, correspondingly,

$$\widetilde{W} := W \cdot \widetilde{S} \tag{4.70}$$

In other words, we use the *Ritz vectors* obtained from the eigenvectors assembled in $\widetilde{S}$ as a new basis for the shrinked search space $\widetilde{\mathcal{K}}$. As the eigenvectors in S (and consequently the eigenvectors in $\widetilde{S}$) are stored according to the related eigenvalues which are presumed to be ordered by ascending magnitude, the resulting *Ritz vectors* in $\widetilde{V}$ actually contain the information of interest, i.e. the search directions related to the smallest eigenvalue approximations. Furthermore, the scheme given by (4.69), (4.70) has the advantage that we can easily compute the new interaction matrix $\widetilde{M}$:

$$\widetilde{M} = \widetilde{W}^*\widetilde{V} = \widetilde{S}^*W^*V\widetilde{S} = \widetilde{S}^*M\widetilde{S} = \text{diag}(\theta_1, \theta_2, \ldots, \theta_{m_{min}}) \tag{4.71}$$

This restart approach allows to control the dimension of the search space by imposing

$$m_{min} \leq \dim \mathcal{K} \leq m_{max} \tag{4.72}$$

As a by-product, the storage requirements for the basis vectors stored in V, the related matrix W and the interaction matrix M are kept limited as well. Unlike the IRAM or the IRLM methods implemented in the ARPACK software package [76] the JDQR variants of the *Jacobi-Davidson method* presented in the following are almost insensitive with regard to changes in these parameters. Another important difference is that the dimension of the search space $\mathcal{K}$ need not be chosen to be larger than the number of sought-after eigenvalues k_{max} as is the case for the IRLM/IRAM methods. In our numerical experiments it was sufficient to work with $m_{min} = 15$ and $m_{max} = 20$, and rather surprisingly, this choice worked constantly well, independent of the problem size n and the number of desired eigenvalues k_{max}.

4.3.3. Standard JDQR

We are now in a position to formulate the JDQR method (Alg. 4.5) which incorporates the deflation and restart techniques developped above and which is a straightforward extension (see [37]) of the basic *Jacobi-Davidson method* (Alg. 4.2) introduced in Section 4.2:

Algorithm 4.5: Jacobi-Davidson method for the k_{max} eigenpairs (JDQR)

---

1   choose suitable parameters $1 < m_{min} < m_{max} < n$, for example $m_{min} = 15$, $m_{max} = 20$
2   choose starting vector $t = v_0$ and target value τ
3   $k = 0$, $m = 0$, $Q = []$, $M = []$, $V = []$, $W = []$
4   **while** $k < k_{max}$ **do**
5      $m = m + 1$
6      $v_m = \operatorname{orth}(V, t)$;   `/* orthonormalization by means of Alg. 2.5 */`
7      $w_m = Av_m$
8      $M = \begin{bmatrix} M & V^*w_m \\ w_m^*V & v_m^*w_m \end{bmatrix}$
9      $V = [V, v_m]$, $W = [W, w_m]$
10   compute sorted eigendecomposition $M = S\Theta S^*$:
               $|\theta_i - \tau| \geq |\theta_{i-1} - \tau|$, $i = 1, \ldots, m$
11     $u = Vs_1$, $w = Ws_1$, $r = w - \theta_1 u$
12   **while** $\|r\|_2 \leq \varepsilon$ **do**
13        $\tilde{\lambda}_{k+1} = \theta_1$, $Q = [Q, u]$, $k = k + 1$
14   **if** $k = k_{max}$ **then**
15   stop
16   **end if**
17        $V = V \cdot S[\,:,2:m]$, $W = W \cdot S[\,:,2:m]$
18        $M = \operatorname{diag}(\theta_2, \ldots, \theta_m)$
19        $m = m - 1$
20   **for** $i = 1, \ldots, m$ **do**
21          $\theta_i = \theta_{i+1}$
22   **end for**
23        $u = v_1$, $r = w_1 - \theta_1 u$
24   **end while**
25   **if** $m \geq m_{max}$ **then**
26        $V = V \cdot S[\,:,1:m_{min}]$, $W = W \cdot S[\,:,1:m_{min}]$
27        $M = \operatorname{diag}(\theta_1, \ldots, \theta_{m_{min}})$
28        $m = m_{min}$
29   **end if**
30     $\theta = \theta_1$, $\tilde{Q} = [Q, u]$
31   solve $t \perp \tilde{Q}$ (approximately) from:
           $\boxed{(I - \tilde{Q}\tilde{Q}^*)(A - \theta I)(I - \tilde{Q}\tilde{Q}^*)t = -r}$
32 **end while**

To conclude, let us now briefly comment on some technicalities and details:

Remark 4.5

- *To realize the restart scheme, we require a sorted eigendecomposition, i.e. the eigenvalues have to be arranged by ascending magnitude and the eigenvectors must be stored correspondingly (Line 10 of Alg. 4.5).*

- *Upon detection of an approximate eigenpair, the following steps have to be taken: The detected approximate eigenvector u is appended to Q. Besides, the first col-*

umn of S, in which the related eigenvector of the interaction matrix M is stored, has to be removed, i.e. $\widetilde{S} = S[:, 2:m]$ and V and W have to be shrinked using (4.69) and (4.70). Finally, the interaction matrix M has to be updated correspondingly (Lines 26-4.5 of 4.5).

- Furthermore, we introduce a target value τ, which aims at the computation of k_{max} eigenpairs nearest τ. For $\tau = 0$ the k_{max} smallest eigenpairs in modulus are computed. Choosing some $\tau > 0$ amounts to computing interior eigenvalues. However, it will turn out that it is more advisable to choose the preconditioned variants along with refined or harmonic extraction for this purpose.

- The appellation JDQR (cf. [37]) is an abbreviation for Jacobi-Davidson QR and indicates that the method can be regarded as an iterative approach for the QR method (Algorithm 3.7).

- Using the same restart and deflation techniques one can derive a related scheme for the Davidson method (Alg. 3.17).

$\square$

4.3.4. Convergence of the JDQR Method

Similar to the basic *Jacobi-Davidson method* we can again only give a heuristic motivation for the sucess of the method when it comes to computing several subsequent eigenpairs, and to this end, we review the argumentation given in [37]. We consider the case that A is symmetric and for the sake of simplicity let us assume that the eigenvalues are all simple, ordered by ascending magnitude

$$\lambda_1 < \lambda_2 < \ldots < \lambda_n \tag{4.73}$$

and that the *Jacobi-Davidson correction equation* (4.11) is solved exactly. As discussed in Section 4.2.2.1, this leads to the following subspace expansion

$$t = \alpha(A - \theta I)^{-1} u - u \tag{4.74}$$

where the orthogonality constraint $t \perp u$ implies that

$$\alpha = \frac{1}{u^*(A - \theta I)^{-1} u} \tag{4.75}$$

Writing u as linear combination of the eigenvectors x_i

$$u = \sum_{i=1}^{n} \gamma_i x_i \tag{4.76}$$

we obtain

$$(A - \theta I)^{-1} u = \sum_{i=1}^{n} \frac{\gamma_i}{\lambda_i - \theta} x_i \tag{4.77}$$

Without loss of generality we may assume that $\gamma_i \neq 0$, because u is a *Ritz vector*, which means that from $\gamma_i = 0$ it follows that either $\angle(x_i, V) = 0$ or $\angle(x_i, V) = \frac{\pi}{2}$. The latter case is rather unlikely to happen due to rounding errors and the first case only occurs upon full convergence. It is plain to see from (4.77) that eigenvector components corresponding to eigenvalues closer to θ are more amplified. The component of the resulting vector which is orthogonal to u is used as a new search direction and basis vector for the *Ritz basis* V. When u is on its way to converge to x_1, it has a large component in direction of x_1, and thus, makes a small angle with it. This in turn has the consequence that other components than x_1 in t necessarily become dominant, in other words:

$$t \sim \sum_{i \neq 1}^{n} \frac{\gamma_i}{\lambda_i - \theta} x_i \tag{4.78}$$

This phenomenon is ilustrated in Figure 4.1 where the dash-dotted lines represent the amplification factors $1/|\lambda_i - \theta|$ for the components in the direction of x_i ($i = 2, 3, 4$). The same argumentation also carries over to the subsequent eigenpairs. If the angle $\angle(x_2, V)$ becomes very small, then the corresponding γ_2 will be tiny as well, and other components will become dominant due to orthogonalization. We see that convergence to one eigenpair enriches the search space with valuable information on nearby eigenpairs to be detected in subsequent steps of the iteration, which also justifies the need for a restarting scheme that tries to retain information of interest.

Figure 4.1.: Amplification factors of eigenvectors

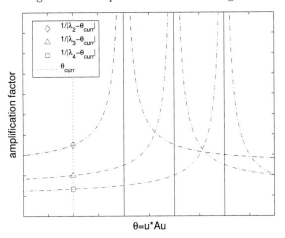

$$\theta = u^* A u$$

4.3.5. Preconditioning the Deflated Correction Equation

Again we need to discuss how a preconditioned Krylov solver like GMRES, QMR, MIN-RES etc. may be applied to the deflated correction equation (4.65). Basically, we can follow the same ideas as for the single vector *Jacobi-Davidson correction equation* presented in Section 4.2.3.6. However, things are now more complicated due to the more involved projections. We consider

$$\widetilde{A} = (I - \widetilde{Q}\widetilde{Q}^*)(A - \theta I)(I - \widetilde{Q}\widetilde{Q}^*) \tag{4.79}$$

along with a suitable approximation

$$\widetilde{K} = (I - \widetilde{Q}\widetilde{Q}^*)K(I - \widetilde{Q}\widetilde{Q}^*) \tag{4.80}$$

and the preconditioned correction equation

$$\widetilde{K}^{-1}\widetilde{A}v = \widetilde{K}^{-1}r \tag{4.81}$$

We first consider the left-hand side, i.e. we solve $z \perp \widetilde{Q}$ from

$$\widetilde{K}z = \widetilde{A}v, \qquad v \perp \widetilde{Q} \tag{4.82}$$

To obtain $\widetilde{A}v$ we first compute

$$p = (A - \theta I)v \tag{4.83}$$

where we can leave out the right projector (because $v \perp \widetilde{Q}$), and finally,

$$\widetilde{A}v = (I - \widetilde{Q}\widetilde{Q}^*)p \tag{4.84}$$

such that we arrive at:

$$\widetilde{K}z = (I - \widetilde{Q}\widetilde{Q}^*)p \tag{4.85}$$

The difference with the single vector case is now that we make use of an auxiliary vector $\vec{\alpha}$ instead of the corresponding scalar α in (4.52) to simplify the above equation:

$$\begin{aligned} Kz &= p - Q\vec{\alpha} \\ z &= K^{-1}p - K^{-1}Q\vec{\alpha} \end{aligned} \tag{4.86}$$

In analogy to (4.54) we can exploit the orthogonality constraint $z \perp \widetilde{Q}$ in order to determine $\vec{\alpha}$

$$\begin{aligned} \widetilde{Q}^*z &= \widetilde{Q}^*K^{-1}p - \widetilde{Q}^*K^{-1}Q\vec{\alpha} = 0 \tag{4.87} \\ \vec{\alpha} &= (\widetilde{Q}^*K^{-1}Q)^{-1}\widetilde{Q}^*K^{-1}p \tag{4.88} \end{aligned}$$

The same ideas also apply for the right-hand side, hence

$$\begin{aligned} \widetilde{K}\widetilde{r} &= r \\ \widetilde{r} &= K^{-1}r - K^{-1}Q\vec{\alpha'} \end{aligned} \tag{4.89}$$

and

$$\vec{\alpha'} = (\widetilde{Q}^*K^{-1}\widetilde{Q})^{-1}\widetilde{Q}^*K^{-1}r \tag{4.90}$$

We can now arrange the relations (4.83) – (4.90) in a scheme for the application of a preconditioned Krylov method in one pass of the JDQR loop to solve the preconditioned correction equation (4.81):

Algorithm 4.6: Solution of the deflated correction equation (left-preconditioning)

---

1 Define operators

$$\check{K} \equiv (I - \widetilde{Q}\widetilde{Q}^*)K(I - \widetilde{Q}\widetilde{Q}^*)$$
$$\check{A} \equiv (I - \widetilde{Q}\widetilde{Q}^*)(A - \theta I)(I - \widetilde{Q}\widetilde{Q}^*)$$

2 solve $\widetilde{Y}$ from $\boxed{K\widetilde{Y} = \widetilde{Q}}$

3 compute $\widetilde{H} = \widetilde{Q}^*\widetilde{Y}$

4 LU-factorize $\widetilde{H} = \mathcal{L}\mathcal{U}$

5 compute $\widetilde{r} \equiv \check{K}^{-1}r$ as

6 (a) solve $\widehat{r}$ from $\boxed{K\widehat{r} = r}$

7 (b) $\vec{\gamma} = \widetilde{Q}^*\widehat{r}$

8   solve $\vec{\beta}$ from $\mathcal{L}\vec{\beta} = \vec{\gamma}$

9   solve $\vec{\alpha}$ from $\mathcal{U}\vec{\alpha} = \vec{\beta}$

10 (c) $\widetilde{r} = \widehat{r} - \widetilde{Y}\vec{\alpha}$

11 apply a Krylov solver with $t_0 = 0$, operator $\check{K}^{-1}\check{A}$,

12 and right-hand side $-\widetilde{r}$

13 compute $z = \check{K}^{-1}\check{A}v$ as:

14 (a) $p = (A - \theta I)v$

15 (b) solve $\widehat{p}$ from $\boxed{K\widehat{p} = p}$

16 (c) $\vec{\gamma} = \widetilde{Q}^*\widehat{p}$

17   solve $\vec{\beta}$ from $\mathcal{L}\vec{\beta} = \vec{\gamma}$

18   solve $\vec{\alpha}$ from $\mathcal{U}\vec{\alpha} = \vec{\beta}$

19 (d) $z = \widehat{p} - \widetilde{Y}\vec{\alpha}$

Remark 4.6

- *Storing the approximate eigenvectors in the above computational scheme becomes twice as expensive, since for every detected approximate eigenvector a preconditioned counterpart is required.*

- *Applying n_{LS} steps of a Krylov solver for the approximate solution of the correction equation leads to the following number of preconditioning operations (highlighted by boxes in the above template):*

$$
\begin{array}{lll}
k + 1 & \text{solves for} & \widetilde{Y} = K^{-1}\widetilde{Q} \\
1 & \text{solve for} & \widehat{r} = K^{-1}r \\
n_{LS} & \text{solves for} & \widehat{p} = K^{-1}(A - \theta I)v \\
\hline
n_{LS} + k + 2 & \text{solves} & \text{altogether}
\end{array}
\tag{4.91}
$$

- *In comparison to the related single-vector scheme described in Algorithm 4.4 (cf. the related table in (4.58)) k additional operations for preconditioning the eigenvectors assembled in Q have to be invested. Furthermore, an LU factorization of*

$\widetilde{H}$ is required to carry out the multiplication by $\widetilde{H}^{-1}$. Fortunately, the arising costs $\mathcal{O}(k^3)$ for the latter are negligible, provided that $k \ll n$.

$\square$

4.3.6. Variants of JDQR Using a Fixed Preconditioner

Working with a *variable* preconditioner (for different shifts θ) implies that $\widetilde{Y}$ and $\widetilde{H}$ in Algorithm 4.6 have to be recomputed *each time* K changes, which is cumbersome and much too expensive. Hence, one will attempt to work with a *fixed* preconditioner as long as possible to avoid computational overhead. On the other hand, the preconditioner K cannot be expected to have constant quality, since it is constructed as an approximation to a matrix $A - \tau I$ with a fixed shift τ, and $A - \theta I$ varies in *every* pass of the JD loop such that the distance $|\theta - \tau|$ gradually increases. However, rather surprisingly, our experiments have shown that this approach enables us to compute several hundreds of eigenpairs, even if K is held constant for a long time. We shall see this later on in Section 7.5 (see Table 7.14 and Fig. 7.15), when we discuss the application of the algorithm in the context of our situation. Until now, no satisfactory theoretical explanation for this phenomenon is available, and again, we have to resort to a heuristic motivation (cf. [37]): Suppose that the preconditioner K is determined by the splitting

$$(A - \tau I) = K - R \tag{4.92}$$

Then applying the projections it follows

$$(I - \widetilde{Q}\widetilde{Q}^*)(A - \theta I)(I - \widetilde{Q}\widetilde{Q}^*) =$$
$$(I - \widetilde{Q}\widetilde{Q}^*)K(I - \widetilde{Q}\widetilde{Q}^*)$$
$$- (I - \widetilde{Q}\widetilde{Q}^*)R(I - \widetilde{Q}\widetilde{Q}^*) - (\tau - \theta)(I - \widetilde{Q}\widetilde{Q}^*) \tag{4.93}$$

On the one hand, the preconditioning error is enlarged by an additive part with small shift $(\tau - \theta)$, but on the other hand the projections reduce the error represented by R by filtering out detected eigenvectors. In other words, if R is large with respect to eigenvectors corresponding to eigenvalues near τ, then the projected error matrix $(I - \widetilde{Q}\widetilde{Q}^*)R(I - \widetilde{Q}\widetilde{Q}^*)$ will be significantly smaller. In these situations the only deterioration is a small shift due to $\tau - \theta$.

Thus, the observations made in experiments along with the theoretical motivation indicate that is rewarding to develop variants of the JDQR method working with a fixed preconditioner K. They turn out to be well-suited and flexible for the practical use and all our numerical experiments related to the Jacobi-Davidson approach will be based on one of these variants.

4.3.6.1. Preconditioned Jacobi-Davidson Correction Equation

Under the assumption that the approximation $K \approx A$ is held fixed throughout the JDQR algorithm it is possible to derive a preconditioned version of the deflated correction equation (4.65). For the sake of simplicity we introduce the short-hand notations (as in the definition of $\widehat{Q}$ in (4.66)), where matrices with a tilde denote temporarily expanded matrices whose last columns (and in case of $\widetilde{H}$ also the last row) change in every pass of the iteration scheme:

$$y = K^{-1}u \tag{4.94}$$
$$Y = K^{-1}Q \tag{4.95}$$
$$\widetilde{Y} = K^{-1}\widetilde{Q} = K^{-1}(Q, u) = (K^{-1}Q, K^{-1}u) = (Y, y) \tag{4.96}$$
$$H = Q^*Y \tag{4.97}$$
$$\widetilde{H} = \widetilde{Q}^*\widetilde{Y} \tag{4.98}$$

First of all, the straightforward application of the above definitions in (4.98) reveals how to determine $\widetilde{H}$ from H:

$$\begin{aligned}
\widetilde{H} &= \widetilde{Q}^*\widetilde{Y} = [Q, u]^* \, [K^{-1}Q, K^{-1}u] \\
&= \begin{bmatrix} Q^*K^{-1}Q & Q^*K^{-1}u \\ u^*K^{-1}Q & u^*K^{-1}u \end{bmatrix} \\
&= \left[\begin{array}{c|c} H & Q^*y \\ \hline u^*Y & u^*y \end{array} \right] \tag{4.99}
\end{aligned}$$

Thus, all one has to do is append one row, one column and the right bottom element which amounts to two matrix-vector operations and one inner product. Following the same principles, one can also update H when a new eigenvector approximation has been detected. Let us now again turn our attention to the computation of the *left-hand side* of (4.81): Plugging (4.83) and (4.88) into (4.86) and using the abbreviations (4.98), (4.96) yields:

$$\begin{aligned}
\widetilde{K}^{-1}\widetilde{A}v &= K^{-1}(A - \theta I)v - K^{-1}\widetilde{Q}(\widetilde{Q}^*K^{-1}\widetilde{Q})^{-1}\widetilde{Q}^*K^{-1}(A - \theta I)v \\
&= K^{-1}(A - \theta I)v - \widetilde{Y}\widetilde{H}^{-1}\widetilde{Q}^*K^{-1}(A - \theta I)v \\
&= (I - \widetilde{Y}\widetilde{H}^{-1}\widetilde{Q}^*)K^{-1}(A - \theta I)v \tag{4.100}
\end{aligned}$$

As $v \perp \widetilde{Q}$ we can again insert a skew projection, hence

$$\widetilde{K}^{-1}\widetilde{A}v = (I - \widetilde{Y}\widetilde{H}^{-1}\widetilde{Q}^*)K^{-1}(A - \theta I)(I - \widetilde{Y}\widetilde{H}^{-1}\widetilde{Q}^*)v \tag{4.101}$$

The right-hand side of (4.81) can be computed more explicitly as well, and combining (4.89), (4.90) with (4.98) and (4.96) gives:

$$\widetilde{K}^{-1}r = -(I - \widetilde{Y}\widetilde{H}^{-1}\widetilde{Q}^*) \, K^{-1} \, r \tag{4.102}$$

Equating the left-hand side (4.101) and the right-hand side (4.102) of (4.81), we finally arrive at the preconditioned version of the deflated correction equation (4.65):

$$(I - \widetilde{Y}\widetilde{H}^{-1}\widetilde{Q}^*) \, K^{-1} \, (A - \theta I) \, (I - \widetilde{Y}\widetilde{H}^{-1}\widetilde{Q}^*) \, t = -(I - \widetilde{Y}\widetilde{H}^{-1}\widetilde{Q}^*) \, K^{-1} \, r \tag{4.103}$$

4.3.6.2. Preconditioned Standard JDQR

We have to make the following extensions and changes in the template for the JDQR algorithm (Alg. 4.5) to obtain a variant working with a fixed preconditioner K which is described in Algorithm 4.7:

- in *every* pass of the loop one has to compute the temporarily expanded matrices $\widetilde{Y}$ by appending $\widetilde{y} = K^{-1}u$ and $\widetilde{H}$ by means of (4.99)

- whenever an eigenpair (λ, q) is detected, one has to update Y by appending $y = K^{-1}q$ and H has to be expanded following the principles in (4.99)

- the deflated correction equation (4.65) has to be replaced by its preconditioned counterpart (4.103)

4.3.6.3. Preconditioned Refined JDQR

Let us now come back to the computation of eigenvalues in the interior of the spectrum. In Section 3.3.1 we introduced alternatives to the *Rayleigh-Ritz procedure* (Alg. 3.12) which may lead to better results when looking for *interior eigenvalues* of A near a target value $\tau > 0$. In the following, we will briefly discuss how to adapt the previously derived preconditioned JDQR Algorithm 4.7 to a variant using refined extraction described in Alg. 4.8:

- According to Algorithm 3.14, a singular value decomposition $\widehat{W} = \widehat{R}^*\Sigma\widehat{S}$ of $\widehat{W} = W - \theta_1 V$ is required in order to obtain a *refined Ritz vector* $u = V\widehat{s}_1$ related to the smallest *Ritz value* θ_1.

- As per Definition 3.19 one uses the *Rayleigh quotient* $\theta = u^*Au/\|u\|_2^2$ as *refined Ritz value* and improved eigenvalue approximation.

- The main difference with Algorithm 4.7 now lies in the realization of restarts when the maximal dimension m_{max} of the search space $\mathcal{K}$ is reached, or when an approximate eigenpair is detected. To this end, one now uses the right singular vectors of $\widehat{W}$, which are assembled in $\widehat{S}$, instead of the eigenvectors of the interaction matrix M stored in S. Replacing all occurences of S by $\widehat{S}$ the restart technique developped in Section 4.3.2 and incorporated in Alg. 4.7 almost directly carries over to the refined variant. The only exception is the explicit re-computation of the updated interaction matrix by the matrix-matrix product $M = V^*W$ on restart (Line 31) resp. on detection of an approximate eigenpair (Line 25).

- The approach presented in Alg. 4.8 is novel and goes beyond the algorithm proposed in [36], in which a refined variant of the basic *Jacobi-Davidson method* (Alg. 4.2) without preconditioning is described.

Algorithm 4.7: Standard JDQR incorporating left preconditioning

1   choose suitable parameters $1 < m_{min} < m_{max} < n$
2   choose starting vector $t = v_0$ and target value τ
3   $k = 0$, $m = 0$, $Q = [\,]$, $Y = [\,]$, $H = [\,]$, $M = [\,]$, $V = [\,]$, $W = [\,]$
4   **while** $k < k_{max}$ **do**
5      $m = m + 1$
6      $v_m = \mathrm{orth}(V, t)$;   /* orthonormalization by means of Alg. 2.5 */
7      $w_m = A v_m$
8      $M = \left[\begin{array}{c|c} M & V^* w_m \\ \hline w_m^* V & v_m^* w_m \end{array} \right]$
9      $V = [V, v_m]$, $W = [W, w_m]$
10   compute sorted eigendecomposition $M = S\Theta S^*$:
         $|\theta_i - \tau| \geq |\theta_{i-1} - \tau|$
11     $u = V s_1$, $w = W s_1$, $r = w - \theta_1 u$
12   **while** $\|r\|_2 \leq \varepsilon$ **do**
13        $\widetilde{\lambda}_{k+1} = \theta_1$,
14        $y = K^{-1} u$
15        $H = \left[\begin{array}{c|c} H & Q^* y \\ \hline u^* Y & u^* y \end{array} \right]$
16        $Q = [Q, u]$, $Y = [Y, y]$, $k = k + 1$
17   **if** $k = k_{max}$ **then**
18   stop
19   **end if**
20        $V = V \cdot S[\,:, 2:m]$, $W = W \cdot S[\,:, 2:m]$
21        $M = \mathrm{diag}(\theta_2, \ldots, \theta_m)$
22        $m = m - 1$
23   **for** $i = 1, \ldots, m$ **do**
24          $\theta_i = \theta_{i+1}$
25   **end for**
26        $u = v_1$, $r = w_1 - \theta_1 u$
27   **end while**
28   **if** $m \geq m_{max}$ **then**
29        $V = V \cdot S[\,:, 1:m_{min}]$, $W = W \cdot S[\,:, 1:m_{min}]$
30        $M = \mathrm{diag}(\theta_1, \ldots, \theta_{m_{min}})$
31        $m = m_{min}$
32   **end if**
33     $\widetilde{y} = K^{-1} u$
34     $\widetilde{H} = \left[\begin{array}{c|c} H & Q^* \widetilde{y} \\ \hline u^* Y & u^* \widetilde{y} \end{array} \right]$
35     $\widetilde{Q} = [Q, u]$, $\widetilde{Y} = [Y, \widetilde{y}]$, $\theta = \theta_1$
36   solve $t \perp \widetilde{Q}$ (approximately) from:
37     $\boxed{ (I - \widetilde{Y} \widetilde{H}^{-1} \widetilde{Q}^*) \, K^{-1} \, (A - \theta I) \, (I - \widetilde{Y} \widetilde{H}^{-1} \widetilde{Q}^*) \, t = -(I - \widetilde{Y} \widetilde{H}^{-1} \widetilde{Q}^*) \, K^{-1} \, r }$
38   **end while**

Algorithm 4.8: Refined JDQR incorporating left preconditioning

---

1 choose suitable parameters $1 < m_{min} < m_{max} < n$
2 choose starting vector $t = v_0$ and target value τ
3 $k = 0$, $m = 0$, $Q = []$, $Y = []$, $H = []$, $M = []$, $V = []$, $W = []$
4 **while** $k < k_{max}$ **do**
5 $\quad$ $m = m + 1$
6 $\quad$ $v_m = \text{orth}(V, t)$;   /* orthonormalization by means of Alg. 2.5 */
7 $\quad$ $w_m = Av_m$
8 $\quad$ $M = \left[\begin{array}{c|c} M & V^* w_m \\ \hline w_m^* V & v_m^* w_m \end{array} \right]$
9 $\quad$ $V = [V, v_m]$, $W = [W, w_m]$
10 $\quad$ compute sorted eigendecomposition $M = S\Theta S^*$:
$\quad$ $|\theta_i - \tau| \geq |\theta_{i-1} - \tau|$
11 $\quad$ $\widehat{W} = W - \theta_1 V$
12 $\quad$ compute sorted singular value decomposition $\widehat{W} = \widehat{R}^* \Sigma \widehat{S}$
13 $\quad$ $\Sigma = \text{diag}(\sigma_1, \ldots, \sigma_m)$, $\sigma_1 \leq \sigma_2 \leq \ldots \sigma_m$
14 $\quad$ $u = V\hat{s}_1$, $w = W\hat{s}_1$
15 $\quad$ $\theta = u^* w / \|u\|_2^2$, $r = w - \theta u$
16 $\quad$ **while** $\|r\|_2 \leq \varepsilon$ **do**
17 $\quad\quad$ $\widetilde{\lambda}_{k+1} = \theta$,
18 $\quad\quad$ $y = K^{-1} u$
19 $\quad\quad$ $H = \left[\begin{array}{c|c} H & Q^* y \\ \hline u^* Y & u^* y \end{array} \right]$
20 $\quad\quad$ $Q = [Q, u]$, $Y = [Y, y]$, $k = k + 1$
21 $\quad\quad$ **if** $k = k_{max}$ **then**
22 $\quad\quad\quad$ stop
23 $\quad\quad$ **end if**
24 $\quad\quad$ $V = V \cdot \widehat{S}[:, 2 : m]$, $W = W \cdot \widehat{S}[:, 2 : m]$
25 $\quad\quad$ $M = V^* W$
26 $\quad\quad$ $u = v_1$, $\theta = u^* w_1 / \|u\|_2^2$, $r = w_1 - \theta u$
27 $\quad\quad$ $m = m - 1$
28 $\quad$ **end while**
29 $\quad$ **if** $m \geq m_{max}$ **then**
30 $\quad\quad$ $V = V \cdot \widehat{S}[:, 1 : m_{min}]$, $W = W \cdot \widehat{S}[:, 1 : m_{min}]$
31 $\quad\quad$ $M = V^* W$
32 $\quad\quad$ $m = m_{min}$
33 $\quad$ **end if**
34 $\quad$ $\widetilde{y} = K^{-1} u$
35 $\quad$ $\widetilde{H} = \left[\begin{array}{c|c} H & Q^* \widetilde{y} \\ \hline u^* Y & u^* \widetilde{y} \end{array} \right]$
36 $\quad$ $\widetilde{Q} = [Q, u]$, $\widetilde{Y} = [Y, \widetilde{y}]$, $\theta = \theta_1$
37 $\quad$ solve $t \perp \widetilde{Q}$ (approximately) from:
38 $\quad$ $\boxed{(I - \widetilde{Y}\widetilde{H}^{-1}\widetilde{Q}^*) K^{-1} (A - \theta I) (I - \widetilde{Y}\widetilde{H}^{-1}\widetilde{Q}^*) t = -(I - \widetilde{Y}\widetilde{H}^{-1}\widetilde{Q}^*) K^{-1} r}$
39 **end while**

Algorithm 4.9: Harmonic JDQR incorporating left preconditioning

---

1 choose suitable parameters $1 < m_{min} < m_{max} < n$
2 choose starting vector $t = v_0$ and target value τ
3 $k = 0$, $m = 0$, $Q = []$, $Y = []$, $H = []$, $M = []$, $V = []$, $W = []$,
4 **while** $k < k_{max}$ **do**
5 $w = (A - \tau I)t$
6   **for** $i = 1, \ldots, m$ **do**
7 $\quad \gamma = w_i^* w$, $w = w - \gamma w_i$, $t = t - \gamma v_i$
8   **end for**
9 $m = m + 1$, $w_m = w/\|w\|_2$, $v_m = t/\|w\|_2$
10 $M = \left[\begin{array}{c|c} M & W^* v_m \\ \hline v_m^* W & w_m^* v_m \end{array} \right]$
11 $V = [V, v_m]$, $W = [W, w_m]$
12   compute sorted eigendecomposition $M = S \widetilde{\Theta} S^*$:
          $\widetilde{\theta}_1 \leq \widetilde{\theta}_2 \leq \ldots$
13 $\widetilde{u} = V s_1$, $\mu = \|\widetilde{u}\|_2$, $u = \widetilde{u}/\mu$, $\vartheta = \widetilde{\theta}_1/\mu^2$
14 $w = W s_1$, $r = \widetilde{w}/\mu - \vartheta u$
15   **while** $\|r\|_2 \leq \varepsilon$ **do**
16 $\quad \widetilde{\lambda}_k = \vartheta + \tau$,
17 $\quad y = K^{-1} u$
18 $\quad H = \left[\begin{array}{c|c} H & Q^* y \\ \hline u^* Y & u^* y \end{array} \right]$
19 $\quad Q = [Q, u]$, $Y = [Y, y]$, $k = k + 1$
20 $\quad$ **if** $k = k_{max}$ **then**
21 $\quad\quad$ stop
22 $\quad$ **end if**
23 $\quad V = V \cdot S[\,:, 2 : m]$, $W = W \cdot S[\,:, 2 : m]$
24 $\quad M = \mathrm{diag}(\widetilde{\theta}_2, \ldots, \widetilde{\theta}_m)$
25 $\quad m = m - 1$
26 $\quad$ **for** $i = 1, \ldots, m$ **do**
27 $\quad\quad \widetilde{\theta}_i = \widetilde{\theta}_{i+1}$
28 $\quad$ **end for**
29 $\quad \mu = \|v_1\|_2$, $\vartheta = \widetilde{\theta}_1/\mu^2$, $u = v_1/\mu$, $r = w_1/\mu - \vartheta u$
30   **end while**
31   **if** $m \geq m_{max}$ **then**
32 $\quad V = V \cdot S[\,:, 1 : m_{min}]$, $W = W \cdot S[\,:, 1 : m_{min}]$
33 $\quad M = \mathrm{diag}(\widetilde{\theta}_1, \ldots, \widetilde{\theta}_{m_{min}})$
34 $\quad m = m_{min}$
35   **end if**
36 $\widetilde{y} = K^{-1} u$
37 $\widetilde{H} = \left[\begin{array}{c|c} H & Q^* \widetilde{y} \\ \hline u^* Y & u^* \widetilde{y} \end{array} \right]$
38 $\widetilde{Q} = [Q, u]$, $\widetilde{Y} = [Y, \widetilde{y}]$, $\theta = \vartheta + \tau$
39   solve $t \perp \widetilde{Q}$ (approximately) from:
40 $\boxed{(I - \widetilde{Y} \widetilde{H}^{-1} \widetilde{Q}^*) K^{-1} (A - \theta I) (I - \widetilde{Y} \widetilde{H}^{-1} \widetilde{Q}^*) t = -(I - \widetilde{Y} \widetilde{H}^{-1} \widetilde{Q}^*) K^{-1} r}$
41 **end while**

4.3.6.4. Preconditioned Harmonic JDQR

To obtain a harmonic version (Alg. 4.9) of the preconditioned JDQR method, the following changes have to be made in Algorithm 4.7:

- Using the harmonic Rayleigh-Ritz procdure (Alg. 3.13), one has now to compute shifted matrix-vector products $(A - \tau I)v$, and the columns of W are orthonormalized along with the corresponding transformation of V (lines 6-9). This makes the harmonic version slightly more expensive than the standard variant (Alg. 4.7). All other costs remain identical.

- One has to be careful with a *harmonic Ritz vector* $u = Vs$ computed in compliance with Definition 3.17. As the columns of V are not orthonormal, u has to be normalized in Line 13 of Algorithm 4.9 in order to make it compatible for the correction equation, where the current approximation u is required to have unit norm. Furthermore, one does not use the *harmonic Ritz value* $\tilde{\theta}_1$ computed in line 12, but the *Rayleigh quotient* ϑ related to the normalized *harmonic Ritz vector* u:

$$\vartheta = \frac{\tilde{u}^* A \tilde{u}}{\|\tilde{u}\|_2^2} = \frac{s_1^* V^* A V s_1^*}{\|\tilde{u}\|_2^2} = \frac{s_1^* W^* V s_1^*}{\|\tilde{u}\|_2^2} = \frac{s_1^* M s_1^*}{\|\tilde{u}\|_2^2} = \frac{\tilde{\theta}_1}{\|\tilde{u}\|_2^2} \qquad (4.104)$$

- When the residual r has reached the desired accuracy, then the actual eigenvalue approximation $\tilde{\lambda}$ is obtained by re-adding the target value τ, i.e. $\tilde{\lambda} = \vartheta + \tau$.

4.3.7. Storage Requirements and Computational Costs

In the following, a summary and a brief discussion of the arising costs for the previously derived preconditioned JDQR variants (Algorithms 4.7, 4.8 and 4.9) is given. Let us remind that we denote by

- n the problemsize (i.e. the dimension of the matrix A under consideration)

- k the number of eigenpairs detected so far, where $0 \leq k \leq k_{max}$

- m the current dimension of the search space, where $m_{min} \leq m \leq m_{max}$.

- ℓ the avarage number of nonzero elements per row of A

The following table gives a survey of the storage requirements, i.e. the amount of memory that has to be allocated in a computer code for the matrices and vectors listed below. Working with double-precision variables, for instance, implies that the proportionality constant in the $\mathcal{O}$-notation is 8, which is exactly the memory requirement in bytes for storing one double variable.

Table 4.1.: Storage requirements for the preconditioned JDQR methods

Matrix	Description / Formula	Costs	Std.	Ref.	Harm.
$\widetilde{Q}$	eigenvector approximations	$\mathcal{O}(k_{max} \cdot n)$	✓	✓	✓
$\widetilde{Y}$	preconditioned eigenvector approximations	$\mathcal{O}(k_{max} \cdot n)$	✓	✓	✓
$\widetilde{H}$	$\widetilde{H} = \widetilde{Q}^*\widetilde{Y}$	$\mathcal{O}(k_{max}^2)$	✓	✓	✓
$\widetilde{H}^{-1}$	LU decomposition of $\widetilde{H}$ is required	$\mathcal{O}(k_{max}^2)$	✓	✓	✓
V	basis vectors for $\mathcal{K}$	$\mathcal{O}(m_{max} \cdot n)$	✓	✓	✓
W	$W = AV$	$\mathcal{O}(m_{max} \cdot n)$	✓	✓	✓
M	$M = V^*W = V^*AV$ (interaction matrix)	$\mathcal{O}(m_{max}^2)$	✓	✓	✓
S	eigenvectors of M	$\mathcal{O}(m_{max}^2)$	✓	✓	✓
$\widehat{W}$	$\widehat{W} = W - \theta_1 V$	$\mathcal{O}(m_{max} \cdot n)$		✓	
$\widehat{S}$	right singular vectors of $\widehat{W}$	$\mathcal{O}(m_{max} \cdot n)$		✓	
u, v, w, y	several vectors	$\mathcal{O}(n)$	✓	✓	✓

Finally, it is appropriate to have a closer look at the computational process in the pre-conditioned standard JDQR method (Alg. 4.7). Most of what is stated in the following is also true of the refined and harmonic variants. However, as already emphasized in the preceding sections, the latter are slightly more expensive and we will point out the differences at the corresponding positions in the following itemization. First of all, the costs arising anyway in *every* pass of the JDQR loop amount to

- costs for updating the matrices V, W and M

    1. orthogonalization of new basis vector v_m against V $\qquad\qquad \mathcal{O}(nm)$
    2. update of matrix W by appending the vector $w_m = Av_m$ $\qquad \mathcal{O}(\ell n)$
    3. update of interaction matrix M by
        a) appending the column V^*w_m $\qquad\qquad\qquad\qquad\quad \mathcal{O}(nm)$
        b) appending the right bottom element $v_m^*w_m$ $\qquad\qquad\quad \mathcal{O}(n)$

    Note that in case of the harmonic JDQR variant w_m is made orthogonal to W and v_m is transformed accordingly, such that the corresponding costs in 1. are about $1\frac{1}{2}$ times higher, the order of magnitude of work to be invested remaining the same.

- costs for computing *Ritz vectors, Ritz values* and residual

    1. complete eigensystem of M $\qquad\qquad\qquad\qquad\qquad\qquad\qquad \mathcal{O}(m^3)$
    2. *Ritz vector* $u = Vs$ and related vector $w = Ws$ $\qquad\qquad\quad \mathcal{O}(nm)$
    3. residual $r = w - \theta u$ $\qquad\qquad\qquad\qquad\qquad\qquad\qquad\quad \mathcal{O}(n)$

In case of the refined JDQR variant, additional work for computing an SVD of $\widehat{W} = W - \theta_1 V$ (costs $\mathcal{O}(nm^2)$) and the computation of a *refined Ritz vector* (costs $\mathcal{O}(nm)$) has to be invested.

- costs for solving the preconditioned correction equation

  1. 2 preconditioner solves for $\widetilde{y} = K^{-1}u$ and $\widetilde{r} = K^{-1}r$
  2. update of auxiliary matrix $\widetilde{H}$ by
     - a) appending the column $Q^*\widetilde{y}$ $\mathcal{O}(nk)$
     - b) appending the row u^*Y $\mathcal{O}(nk)$
     - c) appending the right bottom element $u^*\widetilde{y}$ $\mathcal{O}(n)$
  3. LU factorization of $\widetilde{H}$ $\mathcal{O}(k^3)$
  4. applying n_{LS} steps of a Krylov solver
     - a) n_{LS} preconditioner solves
     - b) n_{LS} matrix-vector operations involving A $\mathcal{O}(n_{LS}\, \ell n)$
     - c) n_{LS} skew projections $I - \widetilde{Y}\widetilde{H}^{-1}\widetilde{Q}^*$ $\mathcal{O}((n_{LS}\,(k^2 + 2kn)\,))$
     - d) additional costs depending on the choice of solver

We see that keeping the preconditioner K fixed leads to the intended savings in the computational process, since preconditioned eigenvector approximations detected in previous steps can be saved in Y and re-used in the subsequent computations of the algorithm. In other words, only two new preconditioner operations in one pass of the loop have to be carried out, as it is the case for the single-vector scheme described in Alg. 4.4. However, it also becomes evident that the required operations for solving the correction equation depend on the number of detected approximations k. When k increases and is no more small in comparison to n, this can become rather time-consuming. Thus, for the sake of efficiency, the number of sought-after eigenpairs k_{max} should not be chosen to large.

In case of a restart, i.e. when $m \geq m_{max}$ we have to take into account

- costs for shrinking the matrices V and W

  1. $\widetilde{V} = V \cdot S[\,:\,,1:m_{min}]$ $\mathcal{O}(n\, m_{max}\, m_{min})$
  2. $\widetilde{W} = W \cdot S[\,:\,,1:m_{min}]$ $\mathcal{O}(n\, m_{max}\, m_{min})$

- update of M no costs
  Note that in the refined case M has to be explicitly
  re-computed as $M = V^*W$, involving $\mathcal{O}(nm^2)$ additional work.

Finally, on detection of an approximate eigenpair (θ, u) one has to consider

- costs for shrinking the matrices V and W

  1. $\widetilde{V} = V \cdot S[\,:\,,2:m]$ $\mathcal{O}(n\, m^2)$
  2. $\widetilde{W} = W \cdot S[\,:\,,2:m]$ $\mathcal{O}(n\, m^2)$
  3. M is updated by shifting the diagonal entries by one position $\mathcal{O}(m)$
     Using the refined variant M has to be re-computed
     as $M = V^*W$, the costs being $\mathcal{O}(nm^2)$

4.4. Summary and Guidelines for the Practical Use

The previously discussed JDQR variants working with a fixed preconditioner K (Algorithms 4.7, 4.8 and 4.9) are the algorithms of choice for the practical use as they are most flexible and contain all other variants discussed so far as particular cases. For $K = I$ the preconditioned JDQR algorithm simplifies to the basic JDQR method (Alg. 4.5) and looking for only one eigenpair amounts to the originally introduced basic *Jacobi-Davidson method* (Alg. 4.2). In principle, preconditioning need not necessarily be incorporated, but all numerical experiments have shown that it is mandatory in our situation to make the method converge at all. Thus, to apply the method the user has to supply

- a subroutine for matrix-vector multiplication

- a subroutine for a preconditioner solve

and we will see that the proper implementation and efficient design of these algorithmic ingredients is crucial for the success in practice.

4.4.1. Choice of the Parameters

Apart from the choice of a possible preconditioner and the implementation of the matrix-vector multiplication, the user can influence the behavior of the preconditioned JDQR methods by a couple of parameters and switches which are listed below along with the choices that we actually employ in our numerical experiments:

1. required accuracy for JD-residual $tol = 10^{-10}$

2. minimal dimension of search space $\mathcal{V}$: $m_{min} = 15$

3. maximal dimension of search space $\mathcal{V}$: $m_{max} = 20$

4. maximal number of solver steps : $l_{max} = 5$

5. use subspace expansion according to Olsen when $\|r\|_2 \geq 10^{-3}$, otherwise use Krylov solver and demand $tol_{LGS} = 0.7^k$ in the kth pass of the JD-loop

6. Krylov solvers: MINRES, GMRES, QMRS (see the following discussion)

7. extraction methods: standard, harmonic, refined (see the following discussion)

The parameter choices 1.-5. regarding subspace dimension, solver steps and accuracy have proven of value in practice. For a more detailed discussion on how changes may affect the convergence behavior see [114], [44]. In the following, we comment slightly in more detail on the choice of the Krylov solver and the extraction method.

We will give numerical evidence to the following recommendations by means of some example computations. They all refer to the following medium-sized model problem, which is an anticipation of our discussion in the Chapters 6 et seqq., the construction will be explained in Section 6.4 and systematic numerical experiments will be presented in Chapter 7.

Problem 4.7 (Model problem)
We consider an **MgNC** *molecule and want to compute its rovibronic engery levels related to the smallest rotational quantum number* $J = 1/2$ *and the parameters* $S = 1/2$ *and* $\Gamma_{rve} = A'$. *To obtain results with reasonable accuracy we employ a large vibrational basis set ("big basis", cf. Table 6.5) whose size is determined by the parameters*

$$N_r^{(lim)} = 16, N_R^{(lim)} = 6, (v_2^a)^{(lim)} = 31, (v_2^b)^{(lim)} = 31 \tag{4.105}$$

In Section 6.3 (cf. Alg. 6.3) we will see that this choice leads to a real-symmetric matrix $\mathbf{H}^{(\mathbf{J},\mathbf{S},\mathbf{\Gamma_{rve}})}$ *(called Hamiltonian matrix) with the dimension* $n = 11904$. *It consists of three dense square blocks on the diagonal (two* (2976×2976)*-blocks and one* (5952×5952) *block) whose Frobenius norms are large as compared to the off-diagonal blocks which are either sparse or even zero. Full details on the structure of the Hamiltonian matrices (sparsity pattern, distribution of information, etc.) arising in our computations will be given in Chapter 6 and Section 7.1.*

In the following examples we employ a simple block Jacobi preconditioner which is constructed from the three dense blocks on the diagonal (a precise specification will be given in Section 7.5.1.1, Def. 7.13).   □

4.4.2. Choice of the Krylov Solver

Formally, the MINRES method is suited for our purposes, because we are interested in solving linear systems with symmetric indefinite coefficient matrices. However, as we have already exposed in Section 4.2.3.5, a positive definite preconditioner is required in order to assure the symmetry of the preconditioned operator. This is rather restrictive, and only meets our situation when one wants to compute k_{max} smallest eigenvalues as the initial fixed preconditioner is then positive definite. Our experiments have shown that MINRES is applicable in this specific situation, even for the computation of more than one hundred eigenpairs, but in general its performance is inferior in comparison to GMRES or QMRS. Clearly, when interior eigenvalues of A, i.e. eigenvalues near a target value $\tau > 0$ are sought-after, then preconditioned MINRES cannot be used anymore as $A - \tau I$ is indefinite, and any reasonable preconditioner K will be as well. In general, QMRS is best suited for our demands, as the preconditioner is allowed to be indefinite and as it employs short recurrences which makes it slightly more efficient than GMRES (although the advantage is in general only marginal). PUTTIN [93] reports stability problems using QMR variants for the solution of the correction equation. We cannot confirm these difficulties from our observations, but to play it safe one has the option to resort to GMRES.

Result 4.8

*We applied the preconditioned standard JDQR method to the computation of the 100 smallest eigenvalues of the matrix **H** described in Problem 4.7 and obtained the following timings depending on which method was employed for the solution of the correction equation:* □

Table 4.2.: Influence of the Krylov solver

Krylov Solver	Time (secs)	MV-ops
MINRES	3800.680000	13852
GMRES	3485.450000	11602
QMRS	3475.840000	11516

4.4.3. Choice of the Extraction Method

The choice of the extraction method primarily depends on what part of the spectrum one is interested in. For exterior eigenvalues, i.e. the k_{max} smallest eigenvalues in our case, the use of standard extraction (Alg. 3.12) is recommended, since it yields satisfactory results and exhibits a smooth and regular convergence behavior, as the following example shows:

Result 4.9 (Exterior eigenvalues)

Computing the ten smallest eigenvalues of $\mathbf{H}^{(\mathbf{J},\mathbf{S},\mathbf{\Gamma}_{\mathbf{rve}})}$ (see Problem 4.7) by means of the preconditioned standard JDQR variant (Alg. 4.7) produces the convergence history as depicted in Figure 4.2: One can also recognize from the plot, that slightly more

Figure 4.2.: Exterior eigenvalues obtained by standard extraction

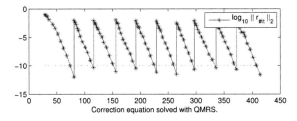

Correction equation solved with QMRS.

matrix-vector operations are needed for the detection of the first eigenpair than for the computation of the subsequent eigenpairs, which is also typically true in general. □

When computing eigenpairs in the interior of the spectrum things are more involved, and the discussion in Section 3.3.1 anticipates that the use of standard extraction (Alg.

3.12) may lead to problems. In order to illustrate the phenomena arising in the use of the
different extraction methods, we applied the corresponding variants of the preconditioned
JDQR algorithm to Problem 4.7:

Result 4.10 (Interior eigenvalues)

*We are now interested in 10 eigenvalues nearest the target $\tau = 0.0196$. Then 200
eigenvalues lie to the left of τ and the remaining part of the spectrum lies to its right.
Note that the block Jacobi preconditioner is now constructed from the shifted matrix
$\mathbf{H}^{(J,S,\Gamma_{rve})} - \tau I$. Application of the three different extraction methods then yields the
results as shown in the Figures 4.3, 4.4 and 4.5.*

Figure 4.3.: Interior eigenvalues obtained by standard extraction

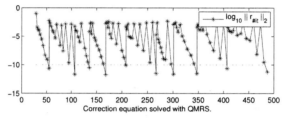

Figure 4.4.: Interior eigenvalues obtained by refined extraction

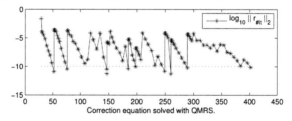

Figure 4.5.: Interior eigenvalues obtained by harmonic extraction

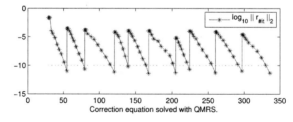

Standard extraction exhibits a highly oscillatory convergence behavior, and as a consequence, considerably more matrix-vector operations are required to obtain the 10 sought-after eigenvalues. By contrast, harmonic extraction leads to the best results as the convergence is smooth and regular and the least number of matrix-vector multiplications is needed. Refined extraction is also better than the standard variant in terms of matrix vector multiplications, however, the convergence is still somewhat oscillatory. Note, that harmonic extraction need not always be better than the refined variant. The potential danger in the use of harmonic extraction is its tendency to stagnate, which sometimes could be observed in our experiments. It is important to notice that the quality of the preconditioner also has a great influence on the convergence behavior as we shall see later on. □

4.4.4. Related Approaches and Software Availability

For the sake of completeness, we give a brief survey of available software for JD methods and comment on some approaches that have been suggested as generalizations of the *Jacobi-Davidson method* (see also [54] for the state-of-the-art situation):

1. The MATLAB® [5] code by SLEIJPEN [112] provides routines for the preconditioned standard and harmonic JDQR variants (Algorithms 4.7 and 4.9). It is user friendly and provides a couple of switches to experiment with. Unfortunately, we could not use it for our purposes, as it is too slow and as the currrent MATLAB® software [5] on the Sun Fire machine is only able to adress 4 GB memory.

2. STATHOPOULOS and McCOMBS [118, 120] have developped a recent software package called PRIMME (acronym for **pr**econditioned **i**terative **m**ulti**m**ethod **e**igensolver) which comprises robust implementations of Arnoldi's/Lanczos' method, the (generalized) Davidson method and the Jacobi-Davidson/JDQR methods. It is written in the C programming language (a FORTRAN 77 [85] interface is available, too) and may be obtained from [119]. The design of the code aims at an improved interplay between the outer iteration (JD loop) on the one hand, and the inner loop (Krylov method) on the other hand. To this end, more sophisticated and reliable stopping criteria, as well as variants of Krylov solvers specialized for the use in the context of *Jacobi-Davidson methods* are developped. In the current version of the software, there is no option to use refined/harmonic extraction as an alternative to the standard *Rayleigh-Ritz procedure*.

3. The software package JADAMILU (acronym for **Ja**cobi-**Da**vidson method with **m**ultilevel **ILU** preconditioning) is due to BOLLHÖFER and NOTAY and may be obtained from [16],[17] (including a related technical report and a user's guide). It is written in FORTRAN 77 [85] and particularly suited for the computation of interior eigenvalues of sparse matrices. The authors' intention is to provide a user-friendly black-box solver, i.e. it automatically constructs an ILU-type precondi-

tioner for the given matrix which based on the ILUPACK software by BOLLHÖFER [15].

4. NOTAY [82] attempts to use the *CG method (Conjugate Gradients)* as a Krylov solver for the correction equation (4.11). The CG method (see [50], [101], [43] for theory and algorithmic details) is actually designed for the application to linear systems with positive definite coefficient matrices, such that, at first glance (see also the related discussion in Section 4.2.3.5), it does not seem to be suited for solving the correction equation. However, a closer analysis reveals that due to the effect of the projectors the method becomes applicable and leads to gain in convergence speed. The MATLAB® [5] code of his approach is available from [83], and comparing it with the corresponding MATLAB® code for the standard preconditioned JDQR variant by SLEIJPEN [112] shows that NOTAY's JDCG approach is only superior for a very small number (about 5-10) of eigenpairs. For a larger number JDCG is clearly inferior. We thus do not pursue this approach in this thesis either.

5. GOEKE [47] and GEUS [45] experiment with block variants, in which they attempt to simultaneously compute n_b approximations (θ_i^k, u_i^k) (where n_b is an appropriately chosen block size and k the number of the iteration step) and search directions t_i^k. This can be done using the following procedure

for $i = 1, \ldots, n_b$ **do**
    solve approximately

$$\left(I - \sum_{j=1}^{n_b} u_j^k (u_j^k)^*\right) (A - \theta_i^k) \left(I - \sum_{j=1}^{n_b} u_j^k (u_j^k)^*\right) t_i^k = -\left(I - \sum_{j=1}^{n_b} u_j^k (u_j^k)^*\right) r_i^k$$

end for

The use of block variants may be advantageous when dealing with clustered eigenvalues. In our experiments we could not observe any gain in performance, sometimes their use even resulted in slowing down convergence. We thus do not pursue this approach in our experiments. The software package JDBSYM by GEUS [46] written in the programming language C [67] contains the preconditioned standard variant (Alg. 4.7) as a special case for the choice $n_b = 1$. However, the software is not well suited for the computation of interior eigenvalues , since the information extraction is only implemented in the form of the *Rayleigh-Ritz procedure* (Alg. 3.12).

6. GENSEBERGER [44] examines a modified correction equation in which the new search direction t is required to be orthogonal to the *complete* search space $\mathcal{K}$ built up so far spanned by the basis vectors V_m:

$$(I - V_m V_m^*)(A - \theta I)(I - V_m V_m^*) = -r \qquad (4.106)$$

Depending on the dimension of the search space $\mathcal{K}$, the involved projections are now much more costly, and in practical situations the approach is even slower to converge than the basic Algorithm 4.2. Furthermore, if one tries to incorporate preconditioners using the techniqes and ideas discussed in Section 4.3.5, it becomes evident that all vectors of V_m have to be preconditioned, which results in a further increase of storage and computational costs. In general this is too expensive, and hence, not attractive for the practical use.

Part II.

Quantum Chemistry

5. Eigenvalue Problems in Theoretical Spectroscopy

In this chapter a brief and general survey of some important basic aspects in theoretical spectroscopy will be given, including some elements from functional analysis, which are required for the proper mathematical treatment. In particular, we will explain how symmetric matrix eigenvalue problems arise from the theoretical framework, and what factors may have an impact on their structure and complexity. Finally, we will outline four general solution strategies, and in this context we will come back to the dichotomy between direct and iterative projection methods for eigenvalue problems, which we have already explained in Chapter 3.

5.1. Motivation and Introduction

First of all, we give a brief introduction to what theoretical spectroscopy is concerned with, and towards this end, we review the motivation given in [61]: Let us consider the simple experiment described in Figure 5.1

Figure 5.1.: Simple spectroscopic experiment

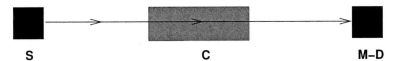

S   **C**   **M–D**

where **C** is a transparent cell containing some substance to be analyzed in the form of a gas. The source of radiation **S** is emitting light, which is sent through the cell **C**. The **M − D** apparatus of the experiment, a monochromator-detector, now receives the incoming radiation from the cell **C**, evaluates the intensities $I(\nu)$ related to the frequencies ν and plots the corresponding diagram Fig. 5.3. Repeating the same experiment with an empty cell **C** , i.e. without the molecular gas, results in the diagram depicted in Fig. 5.2, the so-called *reference spectrum*. By a *spectrum* in this context, we mean the intensity $I(\nu)$ depending on the frequency ν (one could also use the wavelength λ or the wavenumber $\bar{\nu}$ as physical units instead, see below). A comparison of the Figures 5.3 and 5.2 shows that at certain discrete frequencies the related intensities are noticably reduced, because the molecular gas in **C** obviously absorbs these parts of the reference spectrum. For this reason, the resulting diagram Fig. 5.3 is referred to as *absorption spectrum*. The physical explanation for the observation made in the experiment is that

Figure 5.2.: Reference spectrum

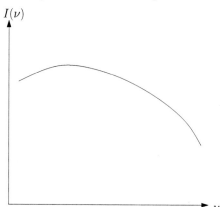

Figure 5.3.: Absorption spectrum

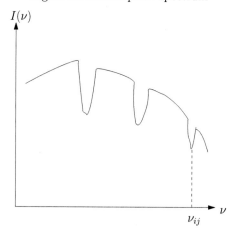

light (or radiation in general) exhibits both wave and particle properties, which is well-known as *wave-particle duality* in the literature (see [109] for instance). This means that monochromatic light with a certain frequency ν (wave property) can also be interpreted as a stream of *photons* (particle property) each of which having an energy $E = h\nu$ (where h is Planck's constant). Many of the photons related to characteristic frequencies ν_{ij} are obviously "swallowed" by the molecules M of the gas in the cell $\mathbf{C}$

$$M + h\nu_{ij} \rightarrow M^* \qquad (5.1)$$

where the asterisk in the above notation symbolizes that the molecule M is in an excited state, the energy being increased by $h\nu_{ij}$ after absorbing the light quantum. We also see that absorption does not occur for arbitrary frequencies ν and is only possible for two allowed energy states E_i and E_j of the molecule M for which the condition

$$h\nu_{ij} = E_j - E_i \qquad (5.2)$$

is satisfied. The main concern of *theoretical spectroscopy* is now to predict these discrete

Figure 5.4.: Absorption of a Light quantum

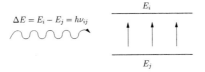

$$\Delta E = E_i - E_j = h\nu_{ij}$$

energy levels E_i of a given molecule M (as far as it is possible) by purely theoretical and computational means, the exception being the use of some physical constants (cf. Table 5.1) which are only attainable by means of experiments. For this reason, one often also speaks of *ab initio theory* in this context. The underlying physical concept in the following considerations is *quantum mechanics*, which provides a theoretical framework for the quantitative description of systems at atomic and subatomic levels. For an introduction and a general account on the topic we refer to standard textbooks, such as [109], for instance. For the case that atoms or molecules are the physical systems of interest, the scientific discipline is called *quantum chemistry*, a branch of which is the field of *theoretical spectroscopy* (see [21], [61]). In the following sections we will see that using the principles of quantum mechanics the characteristic energy levels E_i of a molecule can be characterized as the eigenvalues of the Hamiltonian (5.52) to be defined and introduced in Section 5.3. It will turn out, that only in very rare cases it is possible to explicitly state eigenvalues and eigenfunctions in terms of closed analytic expressions. Consequently, one has to resort to numerical techniqes, and to this end, general strategies will be derived.

Table 5.1.: Some constants of nature

Description	Name	Value	Unit
Planck's constant	h	$6.626176 \cdot 10^{-34}$	Js
reduced Planck's constant	$\hbar = \frac{h}{2\pi}$	$1.05457168 \cdot 10^{-34}$	Js
speed of light	c	$299\,792\,458$	m/s
elementary charge	e	$1.602\,176\,53 \cdot 10^{-19}$	C
mass of a proton	m_p	$1.672\,621\,71 \cdot 10^{-27}$	kg
mass of a neutron	m_n	$1.674\,927\,29 \cdot 10^{-27}$	kg
mass of an electron	m_e	$9.109\,3826 \cdot 10^{-31}$	kg
vacuum permittivity	ϵ_0	$8.854\,187\,8176 \cdot 10^{-12}$	C^2/Jm

5.2. Prerequisites from Functional Analysis

The essential tool for the adequate mathematical formulation of quantum mechanics is the theory of linear operators acting on *Hilbert spaces*, which is treated in the mathematical branch of *Functional Analysis*. For a general survey and introduction as well as a thorough discussion of the topics mentioned below (*Hilbert space, L^2-space, Hermitian, self-adjoint, etc.*) we refer to the standard textbooks [96], [97], [24] and [80], a specialized account with regard to quantum mechanics is given in the famous treatise by VON NEUMANN [128], for a brief survey see [66]. A detailed exposition with special emphasis on Hamilton operators may be found in [53]. In the following, we collect the key definitions and results which are of importance in our context.

5.2.1. Linear Operators on Hilbert Spaces

Definition 5.1 (Hilbert space)
A vector space $\mathcal{H}$ furnished with a scalar product $\langle \cdot, \cdot \rangle \to \mathbb{C}$ is called Hilbert space, iff it is complete with respect to the norm $\| \cdot \| = \sqrt{\langle \cdot, \cdot \rangle}$ induced by the scalar product. More precisely this means that for every Cauchy sequence $\{f_n\}_{i=1}^{\infty}$, $f_n \in \mathcal{H}$, i.e.

$$\|f_n - f_m\| \to 0 \qquad (m, n \to \infty) \tag{5.3}$$

the limit f exists and is contained in $\mathcal{H}$.

Definition 5.2 (Orthonormal system and orthonormal basis)
Let $\mathcal{H}$ be a Hilbert space. A subset $T \subset \mathcal{H}$ is called orthonormal system, iff $\langle e, f \rangle = \delta_{e,f}$ for all $e, f \in T$. An orthonormal system S is called orthonormal basis, iff

$$S \subset T, \quad T \text{ orthonormal system} \Longrightarrow T = S \tag{5.4}$$

An orthonormal system S is called separable, *iff it is countable and dense in $\mathcal{H}$.*

Theorem 5.3 (Hilbert space $L^2(\mathbb{R}^n)$, Fischer-Riesz)
The set of square-integrable functions

$$L^2(\mathbb{R}^n) = \left\{ f \, : \, \mathbb{R}^n \to \mathbb{C} \text{ measurable and } \left(\int |f(x)|^2 dx \right)^{\frac{1}{2}} < \infty \right\} \tag{5.5}$$

is a Hilbert space equipped with the scalar product

$$\langle f, g \rangle_{L^2(\mathbb{R}^n)} = \int \bar{f} g \, dx \qquad \forall f, g \in L^2(\mathbb{R}^n) \tag{5.6}$$

Proof: see [96]   □

Proposition 5.4 (Banach space $L^\infty(\mathbb{R}^n)$, essential supremum)
The set

$$L^\infty(\mathbb{R}^n) = \{f \ : \ \mathbb{R}^n \to \mathbb{C} \ \text{measurable and } |f| \leq C < \infty \text{ almost everywhere}\} \qquad (5.7)$$

is a Banach space, i.e. it is complete with respect to the norm

$$\|f\|_\infty := \inf \left\{ \sup_{x \in \mathbb{R}^n \backslash \mathcal{N}} |f(x)| \ : \ \mathcal{N} \text{ is a Lebesgue null set} \right\} \qquad (5.8)$$

which is referred to as essential supremum.

Proof: see [80] $\square$

Definition 5.5 (Operators on Hilbert spaces)
A linear map T between the Hilbert spaces $\mathcal{H}_1$ *and* $\mathcal{H}_2$

$$T \ : \ \mathcal{H}_1 \to \mathcal{H}_2 \qquad (5.9)$$

is called bounded, *iff there exists a constant $c > 0$, such that*

$$\|Tx\|_{\mathcal{H}_2} \leq c \cdot \|x\|_{\mathcal{H}_1} \qquad \forall x \in \mathcal{H}_1 \qquad (5.10)$$

The smallest such constant

$$\|T\| := \max_{x \neq 0} \frac{\|Tx\|_{\mathcal{H}_2}}{\|x\|_{\mathcal{H}_1}} = \sup_{\|x\|_{\mathcal{H}_1} = 1} \|Tx\|_{\mathcal{H}_2} \qquad (5.11)$$

is called operator norm *of T. If no bound c can be placed in (5.10), then the operator T is called* unbounded.

Proposition 5.6 (Boundedness and continuity)
A linear operator $T : \mathcal{H}_1 \to \mathcal{H}_2$ is bounded, iff T is continuous.

Proof: see [24] $\square$

Lemma 5.7
Let $A : D(A) \to \mathcal{H}$ be an operator with dense domain $D(A) \subset \mathcal{H}$. Then the following assertions hold

1. $D(A^*) = \{y \in \mathcal{H} \ : \ x \longmapsto \langle Ax, y \rangle \text{ is continuous on } D(A)\}$
   is a linear subspace of $\mathcal{H}$

2. *For any $y \in D(A^*)$ there is a unique $A^*y \in \mathcal{H}$, such that*

$$\langle Ax, y \rangle = \langle x, A^*y \rangle \qquad \forall x \in D(A) \qquad (5.12)$$

   3. $A^* : D(A^*) \to \mathcal{H}$ is linear.

Proof: see [80]   □

Definition 5.8 (Adjoint operator)
Let $A : D(A) \to \mathcal{H}$ be an operator with dense domain $D(A) \subset \mathcal{H}$. Then we call the operator A^* defined in Lemma 5.7 the adjoint of A.

Note, that in contrast to the theory of finite Hermitian matrices (cf. Definition 2.10), the terms *Hermitian* and *self-adjoint* are *not* synonymous for unbounded partial differential operators like the Laplacian defined in (5.21) or the Hamiltonian (5.52) to be discussed in the following section. Unfortunately, this distinction is often missing in the standard textbooks on quantum mechanics. For an operator A to be self-adjoint it is essential, that its domain $D(A)$ and the domain $D(A^*)$ of the adjoint operator A^* coincide, whereas hermiticity in general only implies that $D(A) \subset D(A^*)$. The following definition makes this more precise:

Definition 5.9 (Hermitian, self-adjoint)
   1. An operator $A : D(A) \to \mathcal{H}$ is called Hermitian, iff $D(A)$ is dense in $\mathcal{H}$ and

$$\langle Af, g \rangle = \langle f, Ag \rangle \qquad \forall f, g \in D(A) \tag{5.13}$$

   If A is Hermitian, then $A \subset A^*$, i.e. $D(A) \subset D(A^*)$

   2. An operator $A : D(A) \to \mathcal{H}$ is called self-adjoint, iff A is Hermitian and $A = A^*$, i.e. $D(A) = D(A^*)$.

It is important to stress, that the distinction between Hermitian and self-adjoint is not just a mathematical subtleness, but essential for the proper application of spectral theory, which is only valid for self-adjoint operators (see Theorem 5.12 below).

Definition 5.10 (Projectors)
A bounded operator $P : \mathcal{H} \to \mathcal{H}$ on a Hilbert space $\mathcal{H}$ is called projector, iff

   • P is self-adjoint, i.e. $P = P^*$

   • $P^2 = P$

Definition 5.11 (Spectral family)
A spectral family $(E_\lambda)_{\lambda \in \mathbb{R}}$ is a family of operator-valued functions with the following properties:

   1. E_λ is an orthogonal projection for all $\lambda \in \mathbb{R}$

   2. $E_\lambda E_\mu = E_\mu E_\lambda$ for all $\lambda, \mu \in \mathbb{R}$ with $\lambda < \mu$

3. $\lim\limits_{\mu \to \lambda+} E_\mu x = E_\lambda x$ for all $x \in \mathcal{H}$ and for all $\lambda \in \mathbb{R}$

4. $\lim\limits_{\lambda \to \infty} E_\lambda x = x$ and $\lim\limits_{\lambda \to -\infty} E_\lambda x = 0$ for all $x \in \mathcal{H}$

Theorem 5.12 (Spectral theorem for unbounded self-adjoint operators)
For any self-adjoint operator $A : D(A) \to \mathcal{H}$ with dense domain $D(A) \subset \mathcal{H}$ there exists
a unique spectral family of projection operators E_λ such that

$$\langle \phi, A\psi \rangle = \int_{-\infty}^{+\infty} \lambda \, d\langle \phi, E_\lambda \psi \rangle \qquad \forall \phi \in \mathcal{H}, \psi \in D(A) \tag{5.14}$$

One defines

$$A = \int_{-\infty}^{+\infty} \lambda \, dE_\lambda \tag{5.15}$$

(5.15) is commonly referred to as spectral decompostion.

Proof: [96] □

As opposed to the situation for finite dimensional Hermitian matrices (see Def. 2.2
and Thm. 2.12) or compact operators on *Hilbert spaces*, (see [24]) the spectra of un-
boundedself-adjoint operators need not necessarily consist of eigenvalues and may even
contain continuous intervals on the real line. The following definition of the spectrum
which is equivalent to the case of finite-dimensional matrices (see Def. 2.1 and 2.2) makes
more precise what situations may arise:

Definition 5.13 (Spectrum)
Let A be a linear operator on a Hilbert space $\mathcal{H}$ with domain $D(A)$. The spectrum of
A, $\sigma(A)$ is the set of all points $\lambda \in \mathbb{C}$ for which $A - \lambda I$ is not invertible. We distinguish
between the following cases:

1. The discrete spectrum of A, $\sigma_d(A)$, is the set of all eigenvalues of A with finite
   (algebraic) multiplicity and which are isolated points of $\sigma(A)$

2. The essential spectrum of A is defined as the complement of $\sigma_d(A)$ in $\sigma(A)$:
   $\sigma_{\text{ess}} \equiv \sigma(A) \setminus \sigma_d(A)$.

The distinction between the two cases in Def. 5.13 is necessary because there are different
reasons why $A - \lambda I$ may fail to be invertible. See [96] and [53] for more details.

Remark 5.14 (Characterization of the spectrum by the spectral family)
- $\lambda \in \sigma_d(A) \iff E(\cdot)$ is discontinuous at λ

- $\lambda \in \sigma_{\mathrm{ess}}(A) \Longleftrightarrow E(\cdot)$ is continuous at λ but not constant

□

The Fourier transform, which is introduced in the following theorem, is a very important tool in the theory of partial differential operators and equations:

Theorem 5.15 (Fourier-Plancherel)
The map

$$\mathscr{F} : L^2(\mathbb{R}^n) \;\; \rightarrow \;\; L^2(\mathbb{R}^n)$$
$$f \;\; \mapsto \;\; \widehat{f}$$

defined by

$$\widehat{f}(\xi) = (\mathscr{F}f)(\xi) = \underset{n\to\infty}{\mathrm{l.i.m}} \, (2\pi)^{-\frac{n}{2}} \int\limits_{\|x\|<R_n} e^{-i\langle \xi, x\rangle} \cdot f(x)\, dx \tag{5.16}$$

is a unitary isomorphism on $L^2(\mathbb{R}^n)$, i.e.

$$\|\mathscr{F}f\|_2 = \|f\|_2 \tag{5.17}$$

and its inverse is given by

$$g(x) = \mathscr{F}^{-1}\widehat{f}(x) = \underset{n\to\infty}{\mathrm{l.i.m}} \, (2\pi)^{-\frac{n}{2}} \int\limits_{\|\xi\|<R_n} e^{i\langle \xi, x\rangle} \cdot \widehat{f}(\xi)\, d\xi \tag{5.18}$$

where $R_n \uparrow \infty$ and "l.i.m." (limit in mean) denotes the limit of a sequence of functions in $L^2(\mathbb{R}^n)$. $\mathscr{F}$ is commonly referred to as Fourier transform.

Proof: see [97]   □

Definition 5.16 (Schwartz space)
The Schwartz space (or space of rapidly decreasing functions) $\mathcal{S}$ on $\mathbb{R}^n$ is the function space defined by

$$\mathcal{S}(\mathbb{R}^n) = \left\{ \phi \in C^\infty(\mathbb{R}^n) \mid \forall k \in \mathbb{N}_0, \alpha \in (\mathbb{N}_0)^n \; : \; \sup_{x\in\mathbb{R}^n} |x^k D^\alpha \phi(x)| < \infty \right\} \tag{5.19}$$

where $\alpha = (\alpha_1, \ldots, \alpha_n) \in (\mathbb{N}_0)^n$ is a multi-index *and D^α is a shorthand notation for the differential operator*

$$D^\alpha := \frac{\partial^{\alpha_1}}{\partial x_1^{\alpha_1}} \; \cdots \; \frac{\partial^{\alpha_n}}{\partial x_n^{\alpha_n}} \tag{5.20}$$

The Laplacian is of fundamental importance in quantum mechanics, as it is part of the Hamiltonian (5.52) to be discussed in the following section:

Proposition 5.17 (Laplacian)
The second order partial differential operator $\Delta : \mathcal{S}(\mathbb{R}^n) \to \mathcal{S}(\mathbb{R}^n)$ defined by

$$\Delta := \sum_{i=1}^{n} \frac{\partial^2}{\partial x_i^2} \tag{5.21}$$

is called Laplacian and for any $f \in \mathcal{S}(\mathbb{R}^n)$ it holds

$$\mathscr{F}(\Delta f)(\xi) = \widehat{\Delta f}(\xi) = -|\xi|^2 \cdot \widehat{f}(\xi) \tag{5.22}$$

Proof: Property (5.22) follows by integration in parts and the fact that all boundary terms vanish because $f \in \mathcal{S}(\mathbb{R}^n)$:

$$
\begin{aligned}
\mathscr{F}\left(\frac{\partial^2}{\partial x_i^2} f\right)(\xi) &= (2\pi)^{-\frac{n}{2}} \int_{\mathbb{R}^n} e^{-i\langle \xi, x \rangle} \cdot \left(\frac{\partial^2}{\partial x_i^2} f\right)(x)\, dx \\
&= (-1)^2 (2\pi)^{-\frac{n}{2}} \int_{\mathbb{R}^n} \frac{\partial^2}{\partial x_i^2} e^{-i\langle \xi, x \rangle} \cdot f(x)\, dx \\
&= (-1)^2 (-i)^2 \xi_i^2 (2\pi)^{-\frac{n}{2}} \int_{\mathbb{R}^n} e^{-i\langle \xi, x \rangle} \cdot f(x)\, dx \\
&= -\xi_i^2 \cdot \widehat{f}(\xi)
\end{aligned}
$$

Summation over all x_i finally yields the assertion. $\qquad\square$

The *Schwartz space* $\mathcal{S}(\mathbb{R}^n)$ as a domain for the Laplacian (5.21) is not well-suited for our consideration, since it is not complete with respect to the L^2-norm. Property (5.22) motivates the following definition, which remedies this drawback by introducing a more general class of functions $H^2(\mathbb{R}^n)$ in which $\mathcal{S}(\mathbb{R}^n)$ is densely contained.

Proposition 5.18 (Sobolev space $H^2(\mathbb{R}^n)$)
The space
$$H^2(\mathbb{R}^n) = \left\{ f \in L^2(\mathbb{R}^n) \mid (1 + |\xi|^2)(\mathscr{F}f)(\xi) \in L^2(\mathbb{R}^n) \right\} \tag{5.23}$$
is a Hilbert space furnished with the scalar product

$$\langle f, g \rangle_{H^2(\mathbb{R}^n)} = \int (1 + |\xi|^2)^2 \overline{\widehat{f}(\xi)} \widehat{g}(\xi) d\xi \qquad \forall f, g \in L^2(\mathbb{R}^n) \tag{5.24}$$

and is called Sobolev space of order two.

Proof: see [97], [53] $\qquad\square$

Lemma 5.19 (Citerion for self-adjointness)
Let $A : D(A) \to \mathcal{H}$ be Hermitian. A is self-adjoint, iff $Ran(A \pm i) = \mathcal{H}$.

Proof: see [96]   □

Theorem 5.20 (Self-adjointness of the Laplacian)
The Laplacian $\Delta : D(\Delta) \to L^2(\mathbb{R}^n)$ as defined in (5.21) is self-adjoint on $D(\Delta) = H^2(\mathbb{R}^n)$

Proof: To see the symmetry property (5.13), exploit that the Fourier transform is an isometric isomorphism on $L^2(\mathbb{R}^n)$ and use (5.22), thus

$$\langle \Delta f, g \rangle = \langle \widehat{\Delta f}, \widehat{g} \rangle \;\; = \;\; \int \overline{-|\xi|^2 \widehat{f}} \widehat{g}\, d\xi$$

$$= \int \overline{\widehat{f}}\left(-|\xi|^2 \widehat{g}\right) d\xi = \langle \widehat{f}, \widehat{\Delta g} \rangle = \langle f, \Delta g \rangle \quad \forall f, g \in H^2(\mathbb{R}^n) \tag{5.25}$$

To prove $D(\Delta) = D(\Delta^*)$, we use Lemma 5.19, i.e. we have to show that for each $f \in L^2(\mathbb{R}^n)$ the equation

$$(\Delta \pm i)u = f \tag{5.26}$$

has a solution $u \in H^2(\mathbb{R}^n)$. To this end define

$$\widetilde{u} := \frac{\widehat{f}}{-|\xi|^2 \pm i} \tag{5.27}$$

It is easy to see that $u^* := \mathscr{F}^{-1}\widetilde{u} \in H^2(\mathbb{R}^n)$, because (cf. Prop. 5.18)

$$\left|\, (1 + |\xi|^2)\widetilde{u} \,\right| = \left|(1 + |\xi|^2)\, \frac{\widehat{f}}{-|\xi|^2 \pm i}\right| \leq c|\widehat{f}| \in L^2(\mathbb{R}^n) \tag{5.28}$$

Now a straightforward computation shows that $u = u^*$ is a solution of (5.26) and the proof is complete.   □

The following perturbation theorem is an important tool to prove the self-adjointness of unbounded Hermitian operators arising in quantum mechanics, and it describes what properties an additive perturbation B must have, such that the self-adjointness of an operator A carries over to $A + B$:

Theorem 5.21 (Kato-Rellich)
Suppose that A is self-adjoint, B Hermitian with $D(A) \subset D(B)$ and that there exist positive constants $a < 1$ and b such that

$$\|B\phi\| \leq a\|A\phi\| + b\|\phi\|, \qquad \forall \phi \in D(A) \tag{5.29}$$

Then the operator $A + B$ is self-adjoint on $D(A)$.

Proof: see [97], [53]   □

The following corollary characterizes a class of "admissible" perturbations V for the negative Laplacian $-\Delta$:

Corollary 5.22 (Kato-Rellich potentials)
Let $V \in L^2(\mathbb{R}^n) + L^\infty(\mathbb{R}^n)$ and be real. Then the operator $H \equiv -\Delta + V$, defined on $D(\Delta) = H^2(\mathbb{R}^n)$, is self-adjoint. Functions V with the described property are referred to as Kato-Rellich potentials.

Proof: see [97], [53]   □

The following definition characterizes a special sub class of Kato-Rellich potentials which we refer to as *Kato potentials* (see [53]):

Definition 5.23 (Kato potentials)
A Kato-Rellich potential $V(x)$ is called a Kato potential, if V is real and $V \in L^2(\mathbb{R}^n) + L^\infty(\mathbb{R}^n)_\epsilon$, where the ϵ indicates that for any $\epsilon > 0$ we can decompose $V = V_1 + V_2$ with $V_1 \in L^2(\mathbb{R}^n)$ and $V_2 \in L^\infty(\mathbb{R}^n)$ with $\|V_2\|_\infty \leq \epsilon$.

Let us now come back to the characterization of spectra of self-adjoint unbounded operators and give examples for possible situations:

Example 5.24 (Spectrum of the negative Laplacian)
The spectrum of the self-adjoint operator $-\Delta$ on $H^2(\mathbb{R}^n)$ is purely essential, i.e.

$$\sigma(-\Delta) = \sigma_{\text{ess}}(-\Delta) = [\, 0, \infty\,) \tag{5.30}$$

□

Proof: see [53]   □

The following examples show what impact additional perturbations of the negative Laplacian may have on the spectrum. This is of importance for our considerations in the following sections where we introduce and define the Hamiltonian (5.52) as perturbation of the negative Laplacian by a potential V.

Example 5.25 (Perturbation by a Kato potential)
If V is a Kato potential, then

$$\sigma_{\text{ess}}(-\Delta + V) = \sigma_{\text{ess}}(\Delta) = [\, 0, \infty\,), \tag{5.31}$$

i.e. the essential spectrum of $-\Delta$ is not affected by V.   □

Proof: see [53]   □

Example 5.26 (Perturbation by a Harmonic oscillator potential)
Let $A \in \mathbb{R}^n$ be a positive definite matrix and define

$$K(\lambda) = -\Delta + \lambda^2 \langle x, Ax \rangle \tag{5.32}$$

The operator defined by (5.32) is self-adjoint and its spectrum $\sigma(K(\lambda))$ is purely discrete, i.e. it exclusively consists of eigenvalues.   □

Proof: see [53]   □

5.2.2. Tensor Products of Hilbert Spaces and Operators

Tensor products of *Hilbert spaces* and operators will turn out important later on (see Section 5.8.1, for instance).

Definition 5.27 (Tensor products of Hilbert spaces)
Let $\mathcal{H}_i$ be Hilbert spaces furnished with scalar products $\langle \cdot, \cdot \rangle_{\mathcal{H}_i}$ and orthonormal bases

$$\mathcal{B}_i = \{\phi_{i,j_i}\}_{j_i=1}^{\infty} \qquad (i = 1, \ldots, n) \tag{5.33}$$

Then one can define

1. *formal n-tuples of the basis elements*

$$\otimes_{i=1}^{n} \phi_{i,j_i} := \phi_{1,j_1} \otimes \phi_{2,j_2} \otimes \ldots \otimes \phi_{n,j_n} \tag{5.34}$$

2. *formal n-tuples for arbitrary linear combinations $\psi_i = \sum\limits_{j_i=1} \alpha_{j_i} \phi_i \in \mathcal{H}_i$ by*

$$\otimes_{i=1}^{n} \psi_i := \psi_1 \otimes \psi_2 \otimes \ldots \otimes \psi_n := \sum_{j_1,\ldots,j_n} \alpha_{j_1,\ldots,j_n} \otimes_{i=1}^{n} \phi_{i,j_i} \tag{5.35}$$

3. *the set of all such linear combinations by*

$$\mathcal{H} = \bigotimes_{i=1}^{n} \mathcal{H}_i = \left\{ f \mid f = \sum_{j_1,\ldots,j_n} \alpha_{j_1,\ldots,j_n} \otimes_{i=1}^{n} \phi_{i,j_i}, \quad \sum_{j_1,\ldots,j_n} |\alpha_{j_1,\ldots,j_n}|^2 < \infty \right\} \tag{5.36}$$

By simple verification it follows that $\mathcal{H}$ is a Hilbert space with the orthonormal basis

$$\mathcal{B} = \bigotimes_{i=1}^{n} \mathcal{B}_i = \{\otimes_{i=1}^{n} \phi_{i,j_i}\}_{j_1,\ldots,j_n}^{\infty} \tag{5.37}$$

and furnished with the scalar product

$$\langle \otimes_{i=1}^n \psi_i, \otimes_{i=1}^n \chi_i \rangle_{\mathcal{H}} := \prod_{i=1}^n \langle \psi_i, \chi_i \rangle_{\mathcal{H}_i} \qquad \forall \psi_i, \chi_i \in \mathcal{H}_i \quad i \in \{1, \ldots, n\} \qquad (5.38)$$

It is common to use the following appellations:

- a formal tuple as defined in (5.34) is called *tensor product*.

- a Hilbert space $\mathcal{H}$ as constructed in (5.36) is called product space or *tensor space*.

- a basis $\mathcal{B}$ as constructed in (5.37) is called product basis.

The following corollary is an immediate consequence of the above definition and will be of importance in our following considerations:

Corollary 5.28 (Dimension of a finite dimensional product space)
Let $\mathcal{H}_i$ $(i = 1, \ldots, n)$ be finite dimensional Hilbert spaces, i.e. $\dim \mathcal{H}_i < \infty$ and let $\mathcal{H} = \otimes_{i=1}^n \mathcal{H}_i$ be the resulting product space. Then the dimension of $\mathcal{H}$ is obtained by

$$\dim \mathcal{H} = \prod_{i=1}^n \dim \mathcal{H}_i \qquad (5.39)$$

Proof: follows directly by counting all possible combinations in (5.37). □

The principles from Definition 5.27 can also be used to define tensor products of operators:

Definition 5.29 (Tensor products of operators)
Let $T_i : D(T_i) \to \mathcal{H}_i$ be operators with domains $D(T_i) \subset \mathcal{H}_i$ on the Hilbert spaces $\mathcal{H}_i$ $(i = 1, \ldots, n)$. Then one can define the operator tensor product

$$T = \bigotimes_{i=1}^n T_i : D(T) \to \mathcal{H} \qquad (5.40)$$

with domain $D(T) = \otimes_{i=1}^n D(T_i)$ on the product space $\mathcal{H} = \otimes_{i=1}^n \mathcal{H}_i$ by

$$T\psi = \left(\bigotimes_{i=1}^n T_i \right) (\otimes_{i=1}^n \psi_i) = \otimes_{i=1}^n (T_i \psi_i) \qquad (5.41)$$

where

$$\psi = \otimes_{i=1}^n \psi_i \in D(T) \quad \text{and} \quad \psi_i \in D(T_i) \quad i = 1, \ldots, n \qquad (5.42)$$

The following Corollary is a generalization of the mixed product property (2.75) for Kronecker products in Lemma 2.48:

Corollary 5.30 (Mixed "product" property)
Let S_i, T_i be operators acting on the Hilbert spaces $\mathcal{H}_i$ $(i = 1, \ldots, n)$. Then the following identity holds:

$$\left(\bigotimes_{i=1}^{n} S_i \right) \cdot \left(\bigotimes_{i=1}^{n} T_i \right) = \bigotimes_{i=1}^{n} (S_i T_i) \tag{5.43}$$

For the ease of notation when dealing with matrix representations of operators acting on finite dimensional *Hilbert spaces* we recall a fundamental principle from linear algebra:

Definition 5.31 (Matrix and vector representations)
Let $\mathcal{H}$ be a finite dimensional with an orthonormal basis $\mathcal{B} = \{\phi_i\}_{i=1}^{m}$ and

$$\iota : \mathcal{H} \longrightarrow \mathbb{C}^m$$
$$\psi = \sum_{j=1}^{m} \alpha_j \phi_j \longmapsto \mathbf{c} = (\alpha_1, \ldots, \alpha_m)^T$$

be the canonical isomorphism between $\mathcal{H}$ and $\mathbb{C}^m$ with respect to $\mathcal{B}$. Furthermore, let $\widehat{A} : \mathcal{H} \to \mathcal{H}$ be an operator acting on $\mathcal{H}$.
Following the well-known principles from linear algebra, the columns of the matrix representation $\mathbf{A}$ are the coordinate vectors of the images of the basis vectors ϕ_i under the operator $\widehat{A}$. Therefore, it is sensible to extend the canonical isomorphism ι for the definition of the matrix representation as follows:

$$\mathbf{A} = \iota(\widehat{A}) := [\iota(\widehat{A}\phi_1), \iota(\widehat{A}\phi_2), \ldots, \iota(\widehat{A}\phi_m)] := [\mathbf{A}_1, \mathbf{A}_2, \ldots, \mathbf{A}_m] \in \mathbb{C}^{m \times m} \tag{5.44}$$

These definitions imply that

$$\iota(\widehat{A}\psi) = \iota(\widehat{A})\iota(\psi) = \mathbf{A}\mathbf{c} \tag{5.45}$$

The following corollary shows that tensor products on finite dimensional *Hilbert spaces* isomorphically correspond to Kronecker products on $\mathbb{C}^m$, which will be of importance later on:

Corollary 5.32 (Matrix and vector representation of tensor products)
Let $\mathcal{H}_i$ be finite dimensional Hilbert spaces with orthonormal bases $\mathcal{B}_i = \{\phi_{i,j_i}\}_{j_i=1}^{m_i}$.

1. Let $\psi_i = \sum_{j=1}^{m_i} \alpha_{j_i} \phi_{i,j_i}$ be elements in the Hilbert spaces $\mathcal{H}_i$, $\iota_i : \mathcal{H}_i \to \mathbb{C}^{m_i}$ the canonical isomorphisms on $\mathcal{H}_i$ and $\mathbf{c}_i = \iota_i(\psi_i) \in \mathbb{C}^{m_i}$. Then

$$\iota(\otimes_{i=1}^{n} \psi_i) = \otimes_{i=1}^{n} \iota_i(\psi_i) = \otimes_{i=1}^{n} \mathbf{c}_i \tag{5.46}$$

2. Let $\widehat{A}_i : \mathcal{H}_i \to \mathcal{H}_i$ be linear operators acting on $\mathcal{H}_i$ and $\mathbf{A}_i \in \mathbb{C}^{m_i \times m_i}$ their matrix representations. Then

$$\iota \left\{ \left(\bigotimes_{i=1}^{n} \widehat{A}_i \right) \right\} = \left(\bigotimes_{i=1}^{n} \mathbf{A}_i \right) \tag{5.47}$$

We will be primarily concerned with the following construction of product *Hilbert spaces* (see also [96]):

Definition 5.33 (Tensor product of $L^2(\mathbb{R})$ spaces)
Let $\mathcal{H}_i = L^2(\mathbb{R})$ $(i = 1, \ldots, n)$ and $\mathcal{B}_i = \{f_j(x_i)\}_{j=1}^{\infty}$ be the corresponding orthonormal bases. Then the tensor product of basis elements may be defined as

$$\otimes_{i=1}^{n} f_{j,i}(x_i) = \prod_{i=1}^{n} f_{j,i}(x_i) \tag{5.48}$$

i.e. as a simple point-wise multiplication of the basis function with respect to different variables x_i. According to Definition 5.27 this construction leads to a product space $\mathcal{H}$ which can be isomorphically identified with $L^2(\mathbb{R}^n)$, i.e.

$$\mathcal{H} = \bigotimes_{i=1}^{n} L^2(\mathbb{R}) \cong L^2(\mathbb{R}^n) \tag{5.49}$$

and the orthogonal basis

$$\mathcal{B} = \bigotimes_{i=1}^{n} \mathcal{B}_i = \left\{ \prod_{i=1}^{n} f_{j_i}(x_i) \right\}_{j_1, \ldots, j_n}^{\infty} \tag{5.50}$$

5.3. Schrödinger Equation for One-Particle Systems

Let us again turn our attention to the computation of the discrete energy levels E_i of a physical system. One of the fundamental principles in quantum mechanics is the fact that the discrete energy levels E_m of a given physical system at atomic level are the real eigenvalues of the Hamiltonian $\widehat{H}$, which is formalized in the following operator eigenvalue problem:

$$\widehat{H}\psi_m = E_m\psi_m \tag{5.51}$$

(5.51) is known as time-independent *Schrödinger equation*. The eigenfunctions $\psi_m \in L^2(\mathbb{R}^3)$ also have a physical meaning, because $|\psi(X,Y,Z)|^2$ may be interpreted as probability density functions. A concrete example will be given later on in Chapter 6 (see Fig. 6.3) when the *Double Renner effect* is discussed. The partial differential operator

$$\widehat{H} \ : \ D(\widehat{H}) \to L^2(\mathbb{R}^3)$$
$$\widehat{H} \ = \ \widehat{T} + V(X,Y,Z) = -\frac{\hbar^2}{2m}\Delta + V(X,Y,Z) \tag{5.52}$$

is densely defined and self-adjoint on an appropriately chosen subset $D(\widehat{H})$ (see discussion in Example 5.34 below) of the $L^2(\mathbb{R}^3)$. $\widehat{H}$ is well-known as *Hamilton operator* or *Hamiltonian*. The self-adjointness of $\widehat{H}$ implies that all its eigenvalues are real (the proof for the corresponding Theorem 2.11 on Hermitian matrices carries over almost literally). $\widehat{H}$ decomposes into

- the kinetic energy operator $\widehat{T}$

- the potential energy operator $V(X, Y, Z)$

The definition of $\widehat{H}$ in (5.52) only covers the case that one particle with mass m and the Cartesian coordinates (X, Y, Z) in a space fixed Cartesian coordiante system ("laboratory system") is considered. In the following section an extension of this definition with respect to molecules is derived. The discussion in the previous section has shown that, strictly speaking, one has to determine a domain $D(\widehat{H})$ on which $\widehat{H}$ is self-adjoint in order to be able to apply the spectral theory (Theorem 5.12 and the related characterizations of the spectrum in Section 5.2.1). The following example demonstrates for the simple case of the hydrogen atom what steps – at least in principle – have to be carried out for a proper theoretical treatment:

Example 5.34 (Hydrogen atom)
The hydrogen atom is known to consist of one proton and one electron, and thus, provides an example of a particularly simple molecule. Using the reduced mass

$$\mu = \frac{m_e m_p}{m_e + m_p} \tag{5.53}$$

it can be treated as a one-particle system as per (5.52). Neglecting spin-orbit coupling and relativistic effects due to the motion of the electron, the Hamiltonian is given by

$$\widehat{H} = -\frac{\hbar^2}{2\mu}\Delta + \frac{e^2}{4\pi\epsilon_0}\frac{1}{R} \tag{5.54}$$

where

$$R := \sqrt{X^2 + Y^2 + Z^2} \tag{5.55}$$

See Table 5.1 for a description of the arising physical constants.

- *Self-adjointness:*
  First of all, we need to prove that $\widehat{H}$ is self-adjoint on a proper domain $D(\widehat{H})$. Using the results from the previous section, it is possible to specify such a domain explicitly. In Theorem 5.20 it was shown that the Laplacian $-\Delta$ is self-adjoint on the domain

  $$D(-\Delta) = H^2(\mathbb{R}^3) \subset L^2(\mathbb{R}^3) \tag{5.56}$$

  where $H^2(\mathbb{R}^3)$ is the Sobolev space of order two (Def. 5.18). To show that the Hamiltonian (5.54) is self-adjoint on the same domain, one can apply the perturbation results from the previous section (the Kato-Rellich theorem 5.21 and its Corollary 5.22). For the concrete case of the hydrogen atom the potential is $V(R) = 1/R$ (Coulomb potential, $R = \|x\|$) and it is easy to see that it can be decomposed as

  $$V(R) = 1/R = \chi_B 1/R + (1 - \chi_B)1/R \in L^2(\mathbb{R}^3) + L^\infty(\mathbb{R}^3) \tag{5.57}$$

where χ_B is the characteristic function of the unit ball $B = \{x \in \mathbb{R}^3 \mid \|x\| \leq 1\}$ in $\mathbb{R}^3$. Hence, $V(R)$ is a Kato-Rellich potential, and from Corollary 5.22 it follows immediately that $\widehat{H}$ is also self-adjoint on the domain of the Laplacian, i.e.

$$D(\widehat{H}) = H^2(\mathbb{R}^3) \subset L^2(\mathbb{R}^3) \tag{5.58}$$

- Spectrum of the Hamiltonian and solutions of the Schrödinger equation:
  Defining χ_ϵ as the characteristic function of the set $\{x \mid \|x\| \leq (c\epsilon)^{-1}\}$ and decomposing the potential as

$$V(x) = c\chi_\epsilon(x)\|x\|^{-1} + c(1 - \chi_\epsilon(x))\,\|x\|^{-1} = V_1(x) + V_2(x) \tag{5.59}$$

we have $V_1 \in L^2(\mathbb{R}^n)$ and

$$\sup_{x \in \mathbb{R}^n} |\,c(1 - \chi_\epsilon(x))\,\|x\|^{-1}\,| \leq \epsilon, \tag{5.60}$$

i.e. $V(R)$ is actually a Kato potential in the sense of Def. 5.23 which implies that the essential spectrum of the Hamiltonian for the hydrogen atom (5.54) is a non-empty set and given by (as per Example 5.25):

$$\sigma_{\text{ess}}(\widehat{H}) = [\,0, \infty\,) \tag{5.61}$$

This shows that the spectrum of the Hamiltonian is not purely discrete in case of the hydrogen atom.
Let us now turn to the characterization of the discrete spectrum σ_d: The hydrogen atom is one of the few cases, for which it is possible to solve the Schrödinger equation explicitly, i.e. it is possible to give closed analytic expressions for both, energy levels E_m (the discrete eigenvalues of (5.54)) and the corresponding eigenfunctions ψ_m. To do so, one can express the Hamiltonian with respect to spherical coordinates (by means of the chain rule of differential calculus) and exploit the radial symmetry of the potential V. This allows to pursue a separation approach and split the equation into three ordinary differential equations (each with respect to a spherical coordinate), which can be treated independently. The total wavefunction ψ as solution of the Schrödinger equation is then the product of the solutions of the partial problems, and it can be shown that the discrete energy levels are

$$E_m = -\frac{\mu}{\hbar^2}\left(\frac{e^2}{4\pi\epsilon_0}\right)\frac{1}{m^2} \tag{5.62}$$

so that the total spectrum is obtained as

$$\sigma(\widehat{H}) = \sigma_d(\widehat{H}) \cup \sigma_{\text{ess}}(\widehat{H}) = \{E_m \mid m \in \mathbb{N}\} \cup [\,0, \infty\,) \tag{5.63}$$

For full details on the related computation see [109].

For more elaborate molecules consisting of several nuclei and electrons it will be in general much more complicated to show the self-adjointness of the related molecular Hamiltonian $\widehat{H}$ and to determine a proper domain $D(\widehat{H})$. Further interesting examples along with a detailed discussion on this issue may be found [97] and [53]. In the following, we will, for pragmatic reasons, no more examine the self-adjointness of the Hamiltonians that we will be concerned with, because the involved potentials are in general not available as analytic expressions which makes the analysis too complicated. The major difficulty, however, is to solve the Schrödinger equation, i.e. to determine the discrete spectrum σ_d of the Hamiltonian. The computation for the hydrogen atom leading to (5.62) is already a rather complicated matter (see [109]), and for a general molecule consisting of more than just one electron and one proton it is in general not feasible to state solutions in terms of closed analytic expressions for eigenvalues and eigenfunctions. Consequently, one is forced to rely on numerical techniques and approximations, and this is what the description in the following sections will be dealing with.

5.4. Molecular Hamiltonian

The Hamiltonian defined in (5.52) only describes the simple case that the system under consideration involves one particle. By contrast, the systems we are concerned with are molecules, and these are known to be composed of more than one particle, more precisely speaking

- N nuclei (consisting of protons and neutrons)

- n electrons

Thus, to derive the Hamiltonian of a general molecule, we have to take into account the masses of the N nuclei M_η ($\eta = 1, \ldots, N$) (obtained by summing up the masses of the protons and neutrons according to the isotope for the nucleus under consideration, cf. Table 5.1) and their coordinates

$$R_\eta = (X_\eta, Y_\eta, Z_\eta), \qquad \eta = 1, \ldots, N \tag{5.64}$$

in the Cartesian system. Analogously, the coordinates of the n electrons are given by

$$r_i = (x_i, y_i, z_i), \qquad i = 1, \ldots, n \tag{5.65}$$

The mass of an electron is given by m_e (cf. Table 5.1). For ease of notation we write $\mathbf{R}$ for all nuclear coordinates R_η and $\mathbf{r}$ for all electronic coordinates r_i.

In the following, we neglect the contribution made by the spin of the electrons and the nuclei. Furthermore, we can assume the molecule to be isolated in space, i.e. there is no interaction with other molecules (molecules in gases are sufficiently well separated from each other) such that the Hamiltonian for a general molecule can be written as

$$\widehat{H} = \widehat{T}_n + \widehat{T}_e + V_{\text{Coulomb}}(\mathbf{R}, \mathbf{r}) \tag{5.66}$$

where

$$\widehat{T}_n = -\frac{\hbar^2}{2} \sum_{\eta=1}^{N} \frac{1}{M_\eta} \frac{\partial^2}{\partial R_\eta^2} = \frac{\hbar^2}{2} \sum_{\eta=1}^{N} \frac{1}{M_\eta} \left[\frac{\partial^2}{\partial X_\eta^2} + \frac{\partial^2}{\partial Y_\eta^2} + \frac{\partial^2}{\partial Z_\eta^2} \right] \tag{5.67}$$

is the kinectic energy operator related to the motion of the nuclei and

$$\widehat{T}_e = -\frac{\hbar^2}{2m_e} \sum_{i=1}^{n} \frac{\partial^2}{\partial r_i^2} = \frac{\hbar^2}{2m_e} \sum_{i=1}^{n} \left[\frac{\partial^2}{\partial x_i^2} + \frac{\partial^2}{\partial y_i^2} + \frac{\partial^2}{\partial z_i^2} \right] \tag{5.68}$$

the kinetic energy operator related to the motion of the electrons. For the potential energy $V(\mathbf{R}, \mathbf{r})$ we have to take into account all coulomb interactions between all particles, i.e.

- interaction between the electrons and nuclei (total electron-nucleus Coulombic attraction in the system)

- interaction between the electrons (total electron-electron Coulombic repulsion)

- interaction between the nuclei (total nucleus-nucleus Coulombic repulsion)

which leads to the following expression for the potential:

$$V_{\text{Coulomb}}(\mathbf{R}, \mathbf{r}) = \sum_{\eta < \eta'} \frac{C_\eta C_{\eta'} e^2}{\|R_\eta - R_{\eta'}\|_2} + \sum_{i < i'} \frac{e^2}{\|r_i - r_{i'}\|_2} - \sum_{\eta=1}^{N} \sum_{i=1}^{n} \frac{C_\eta e^2}{\|R_\eta - r_i\|_2} \tag{5.69}$$

The fact that the Hamiltonian now depends on $n + N$ particles makes the solution of the Schrödinger equation (5.51) complicated and rather often even untractable. For this reason, one tries to find approximations that reduce the complexity of the problem without losing too much accuracy. A common approach for this purpose is discussed in the following section.

5.5. Born-Oppenheimer Approximation

The approach presented in the following, the so-called *Born-Oppenheimer approximation*, (briefly: BO approximation) relies on the fact that the nuclei of a molecule are 10^4 to 10^5 times heavier than the electrons, and consequently, the nuclei move around much more slowly than the electrons do. This motivates to consider the nuclei to be fixed in space for several (in principle infinitely many) geometries and to treat the motion of the electrons and the nuclei seperately. To do so, we consider the previously derived Schrödinger equation for the spatial motion of the particles of a molecule,

$$\widehat{H}\psi_{\mathrm{ne}}(\mathbf{R}, \mathbf{r}) = [\widehat{T}_n + \widehat{T}_e + V_{\mathrm{Coulomb}}(\mathbf{R}, \mathbf{r})]\psi_{\mathrm{ne}}(\mathbf{R}, \mathbf{r}) = E_{\mathrm{ne}}\psi_{\mathrm{ne}}(\mathbf{R}, \mathbf{r}) \qquad (5.70)$$

where the indices **ne** indicate, that energy contributions from the motion of both nuclei and electrons are taken into account. We now write the sought-after eigenfunction ψ_{ne} as a product of two factors

$$\psi_{\mathrm{ne}}(\mathbf{R}, \mathbf{r}) = \psi_{\mathrm{nuc}}(\mathbf{R})\psi_{\mathrm{elec}}(\mathbf{R}, \mathbf{r}) \qquad (5.71)$$

where the wavefunction of rotation and vibration $\psi_{\mathrm{nuc}}(\mathbf{R})$ only depends on the nuclear coordinates $\mathbf{R}$, whereas the wavefunction of the electronic motion $\psi_{\mathrm{elec}}(\mathbf{R}, \mathbf{r})$ depends on the nuclear *and* electronic coordinates. The latter wavefunction is determined from the so-called *electronic Schrödinger equation*

$$[\widehat{T}_e + V_{\mathrm{Coulomb}}(\mathbf{R}^{(0)}, \mathbf{r_e})]\psi_{\mathrm{elec}}(\mathbf{R}^{(0)}, \mathbf{r}) = V(\mathbf{R}^{(0)})\psi_{\mathrm{elec}}(\mathbf{R}^{(0)}, \mathbf{r}) \qquad (5.72)$$

which is solved for fixed nuclear coordinates $\mathbf{R} = \mathbf{R}^{(0)}$, such that the eigenvalues $V(\mathbf{R}^{(0)})$ depend parametrically on the nuclear coordinates. The classical *ab initio theory* is exclusively concerned with the solution of (5.72) and represents a wide field of research of its own account (see [62], [51] for an introduction and more details). Solving (5.72) for arbitrary nuclear geometries $\mathbf{R}^{(0)}$ leads to the so called *Born-Oppenheimer potential function* and is often referred to as *Potential Energy Surface* (briefly: PES).

Inserting (5.71) and (5.72) into (5.70) we obtain

$$[\widehat{T}_n + V(\mathbf{R})]\psi_{\mathrm{nuc}}(\mathbf{R})\psi_{\mathrm{elec}}(\mathbf{R}, \mathbf{r}) = E_{\mathrm{ne}}\psi_{\mathrm{nuc}}(\mathbf{R})\psi_{\mathrm{elec}}(\mathbf{R}, \mathbf{r}) \qquad (5.73)$$

We now introduce the following approximation in (5.73):

$$\widehat{T}_n[\psi_{\mathrm{nuc}}(\mathbf{R})\psi_{\mathrm{elec}}(\mathbf{R}, \mathbf{r})] \approx \psi_{\mathrm{elec}}(\mathbf{R}, \mathbf{r})[\widehat{T}_n\psi_{\mathrm{nuc}}(\mathbf{R})] \qquad (5.74)$$

In other words, the effect of the differential operators in $\widehat{T}_n$ on the electronic wave funtion is neglected, which is the computational consequence of the above introductory considerations to motivate the BO approximation. For a more in-depth discussion and a detailed justification of the approximation made in (5.74) see [21] and [109].

We can now plug (5.74) into (5.73) and arrive at the following simplified Schrödinger equation

$$\boxed{[\widehat{T}_n + V(\mathbf{R})]\psi_{\mathrm{nuc}}(\mathbf{R}) = E_{\mathrm{ne}}\psi_{\mathrm{nuc}}(\mathbf{R})} \qquad (5.75)$$

which does not depend on the electronic coordinates $\mathbf{r}$ any longer. The effect of the electronic motion is now incorporated in the potential energy surface $V(\mathbf{R})$

Remark 5.35

- *Of course, it is not possible to compute solutions of the electronic Schrödinger equation (5.72) for infinitely many fixed molecular coordinates $\mathbf{R}^{(0)}$. In practice, one has to rely on sensible approximations, in which one determines $V(\mathbf{R}_j^{(0)})$ for a couple of selected geometries $\mathbf{R}_j^{(0)}$. These discrete points $(\mathbf{R}_j^{(0)}, V(\mathbf{R}_j^{(0)}))$ are then employed to construct an approximate PES (see Fig. 5.5). This can be done by using Taylor expansions, interpolation, splines or further analytical fitting techniques. We cannot go into detail here and refer to the standard literature for a comprehensive account on the matter.*

Figure 5.5.: Construction of a Potential Energy Surface (PES)

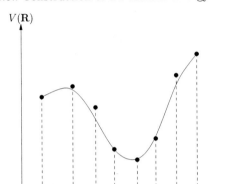

- *In this thesis we are exclusively concerned with the solution of the nuclear Schrödinger equation (5.75), i.e. we have precomputed high-quality PES at hand for the molecules we are dealing with.*

- *The BO approximation outlined above is not always appropriate. This is the case when two PES related to different electronic states are close to each other in energy, or if they are partially degenerate. Then one can no more neglect their interaction and has to take into account both of them. The general strategy, however, remains the same, i.e. in order to solve the general molecular Schrödinger equation a 2-step-approach is pursued, i.e.*

  1. *solve the electronic Schrödinger equation (5.72) for several fixed nuclear geometries to construct an approximate PES $V(\mathbf{R})$ (ab initio step)*

  2. *Employ the PES $V(\mathbf{R})$ obtained in step 1 (or combine several PES, if necessary) to solve the nuclear Schrödinger equation (5.75)*

□

5.6. Nuclear Motion and Coordinate Systems

As can be seen from the derivation of the nuclear Schrödinger equation (5.75), one has to take into account the kinetic energy of the nuclei, whose contribution is incorporated in the kinetic energy operator $\widehat{T}_n$ (cf. (5.67)) and describes the motion of the nuclei in space. Consequently, molecules are not rigid entities, their nuclei perform

- vibrational motion (stretching and bending)
  (the positions of the nuclei in space relative to each other are changing)

- rotational motion
  (the molecule rotates about axes running through the center-of-mass of the nuclei)

- translation
  (the molecule performs a translation in space, i.e. the center-of-mass of its nuclei is moving in a certain direction with constant velocity)

The following description is geared to [61] and gives a brief outline on the choice of coordinate systems, as far as it is of importance for the derivation and computational complexity of the eigenvalue problem. For a full account see [21]. The space fixed coordinate system we have made use of so far ("laboratory system") is not always appropriate for the adequate description of the vibrational and rotational motion of the nuclei, because it neither provides any information on how much the molecule deviates from its equilibrum structure, nor does it allow for conclusions on how the molecule rotates in space. To remedy this drawback, one introduces an additional Cartesian coordinate system, a so-called molecule-fixed system, in a first step. A formal introduction of both systems is given in the following definition:

Definition 5.36 (Space fixed and molecule fixed coordinate system)
We consider a molecule with N nuclei and masses M_η ($\eta = 1, \ldots, N$).

1. *The position of a nucleus in the* space *fixed coordinate system ("laboratory system"), is given by its Cartesian coordiantes (in capital letters)*

$$R_\eta = (X_\eta, Y_\eta, Z_\eta) \tag{5.76}$$

The laboratory system is called space fixed, because it is attached to the observer and does not follow the translational and rotational motion of the molecule.

2. *By contrast, the position of a nucleus in a molecule fixed coordinate system given by its Cartesian coordinates (in small letters)*

$$r_\eta = (x_\eta, y_\eta, z_\eta) \tag{5.77}$$

is described relative to a system having its origin in the center of mass of the N nuclei of the molecule

$$R_0 = (X_0, Y_0, Z_0) \tag{5.78}$$

and is given by its space fixed coordinates

$$X_0 = \frac{1}{M} \sum_{\eta=1}^{N} M_\eta X_\eta, \quad Y_0 = \frac{1}{M} \sum_{\eta=1}^{N} M_\eta Y_\eta, \quad Z_0 = \frac{1}{M} \sum_{\eta=1}^{N} M_\eta Z_\eta \qquad (5.79)$$

where

$$M = \sum_{\eta=1}^{N} M_\eta \qquad (5.80)$$

The molecule fixed system follows the rotational and translational motion of a molecule in space (it is "attached" to the molecule).

Before we can show how to transform between both coordinate systems, we need the following definition:

Definition 5.37 (Euler angles)
To describe the orientation of a molecule fixed system $\mathcal{M} = (x, y, z)$ relative to a space fixed system $\mathcal{S} = (X, Y, Z)$, it is common to use the so-called Euler angles (cf. Fig. 5.6). Without loss of generality we may assume that both, $\mathcal{M}$ and $\mathcal{S}$, have the same origin $\mathbf{O}$ The Euler angles

$$E = (\theta, \phi, \chi) \qquad (5.81)$$

are defined by the following steps that have to be carried out in order to make $\mathcal{S}$ coincide with $\mathcal{M}$:

- *$\mathcal{S}$ is rotated about the Z-axis by the angle ϕ, such that the Y-axis coincides with the so-called line of nodes ON (which is defined as the intersection between the xy-plane and the XY-plane).*

- *The rotated coordinate system is now rotated about the ON-axis by the angle θ, which transfers the Z-axis into the z-axis.*

- *Finally, the rotated coordinate system is rotated about the z-axis by the angle χ, which carries over the ON-axis to the y-axis*

We are now in a position to give an explicit formula for the transformation between space fixed and molecule fixed coordinate system

Proposition 5.38 (Transformation space fixed $\leftrightarrow$ molecule fixed system)
Let the space fixed coordinate system $\mathcal{S}$ and the molecule fixed coordinate system $\mathcal{M}$ be defined according to Definition 5.36. Then a coordinate $\mathbf{r}_\eta$ in $\mathcal{M}$ is obtained from the corresponding coordinate $\mathbf{R}_\eta$ in $\mathcal{S}$ by

$$\mathbf{r}_\eta = \mathbf{S}(\theta, \phi, \chi) (\mathbf{R}_\eta - \mathbf{R}_0) \qquad (5.82)$$

Figure 5.6.: Euler angles

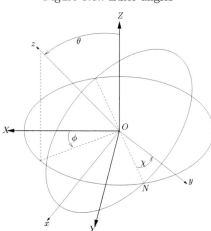

where θ, ϕ, χ are the Euler angles explained above in Definition 5.37 and $\mathbf{S}(\theta, \phi, \chi)$ the rotation matrix defined by

$\mathbf{S}(\theta, \phi, \chi) =$

$$
\begin{bmatrix}
\cos\theta\cos\phi\cos\chi - \sin\phi\sin\chi & -\cos\theta\cos\phi\sin\chi - \sin\phi\cos\chi & \sin\theta\cos\phi \\
\cos\theta\sin\phi\cos\chi + \cos\phi\sin\chi & -\cos\theta\sin\phi\sin\chi + \cos\phi\cos\chi & \sin\theta\sin\phi \\
-\sin\theta\cos\chi & \sin\theta\cos\chi & \cos\theta
\end{bmatrix} \quad (5.83)
$$

Proof: simple verification   □

Remark 5.39 (Rotational, vibrational and translational coordinates)

- *Obviously, the Euler angles* $E = (\theta, \phi, \chi)$ *describe the rotational motion of the molecule under consideration, and therefore, they are also referred to as* rotational coordinates.

- *The center-of-mass* $R_0 = (X_0, Y_0, Z_0)$ *describes the translation of the molecule, and thus, can be regarded as* translational coordinate.

- *The molecule fixed coordinate system* $\mathcal{M}$ *still does not reflect the vibrational behavior of the molecule well, but it can be employed as a starting point for a more suitable characterization. Independent of the concrete choice of the coordinate system(s), $3N$ coordinates are required to determine the positions of the N nuclei in space. As we have already consumed 3 coordinates for the center-of-mass R_0 defined in Def. 5.36, and 3 further coordinates for the Euler angles in Def. 5.37, there are only $3N - 6$ degrees of freedom left for the definition of the vibrational* modes. *Denoting the current position of a nucleus with index η by its molecule fixed coordinate $\mathbf{r}_\eta$ and its position in the equilibrum geometry of the molecule by $\mathbf{a}_\eta$, it seems to be natural to define the vibration as current displacement d_η, i.e.*

$$\mathbf{d}_\eta = \mathbf{r}_\eta - \mathbf{a}_\eta \quad \eta = 1, \dots, N \tag{5.84}$$

Obviously, 6 of the N coordinates defined in (5.84) are redundant. This in turn implies, that one can define 6 equations connecting all $3N$ components. The resulting formulae are well-known as Eckart conditions and have the nice property that the coupling between rotational and vibrational motion is minimized. We do not discuss the derivation here and refer to [61] and [21].

- *Note that the coordinates defined in (5.84) are not the only possible way to describe the vibrational motion of the nuclei in a molecule and that there are other popular choices that may be more appropriate. One example for such a set of vibrational coordinates is provided by the so-called Jacobi coordinates, which are particularly well-suited for the description of the bending motion within a triatomic molecule. A definition is given below and we will make use of it later on, when the Double Renner effect is discussed.*

$\square$

Figure 5.7.: Jacobi coordinates for a triatomic molecule

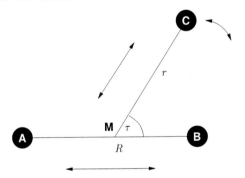

Definition 5.40 (Jacobi coordinates for triatomic molecules)
Let $\mathbf{ABC}$ be a generic triatomic molecule consisting of the the nuclei $\mathbf{A}$, $\mathbf{B}$ and $\mathbf{C}$ along with the masses m_A, m_B and m_C. Let R be the bond-length of the $\mathbf{AB}$ moiety and M be the center of mass with respect to the nuclei $\mathbf{A}$ and $\mathbf{B}$. Furthermore, let r be the length of the distance $\mathbf{MC}$ and the angle τ be defined as $\angle(\mathbf{MB}, \mathbf{MC})$ (cf. Fig. 5.7). Then the vibrational motion of $\mathbf{ABC}$ can be described by the Jacobi coordinates (R, r, τ), where R and r are referred to as stretching coordinates and τ as bending coordinate. Note that this definiton is consistent with the allowed degrees of freedom for the number of vibrational coordinates, since $N = 3$ and consequently $3N - 6 = 3$ coordinates are admissible.

We can now exploit the definition of space fixed and molecule fixed coordinates in Def.
5.36 to show that the translational motion of a molecule can be neglected in our considerations:

Remark 5.41 (Separation of the translational motion)
*Translation moves all nuclei in the same spatial direction with constant velocity, i.e. the
space fixed center-of-mass coordinate $\mathbf{R}_0$ is submitted to change, whereas the position
of the nuclei relative to each other is not affected, which means that their molecule
fixed coordinates $\mathbf{r}_\eta$ are constant. The definition of the potential V in (5.69) only
involves Euclidean distances of particles. In other words, V is invariant with respect to
translation and rotation, and thus, only depends on the molecule fixed coordinates of
the particles (electrons and nuclei), such that we can write it as a function $V(x, y, z)$.
The Hamiltonian of the molecule now can be re-arranged as*

$$
\begin{aligned}
\widehat{H} &= \widehat{T} + V(x, y, z) \\
&= \widehat{T}_0 + \widehat{T}_{rel} + V(x, y, z) \\
&= \widehat{H}_{\text{trans}} + \widehat{H}_{\text{rel}}
\end{aligned}
\tag{5.85}
$$

where $\widehat{H}_{\text{trans}} = \widehat{T}_0$ is the Hamiltonian for the translational motion and

$$
\widehat{H}_{\text{rel}} = \widehat{T}_{rel} + V(x, y, z)
\tag{5.86}
$$

*denotes the Hamiltonian related to the realtive motion of the nuclei. Obviously, $\widehat{H}_{\text{trans}}$
only depends on the space fixed coordinates X_0, Y_0, Z_0, whereas $\widehat{H}_{\text{rel}}$ only depends on the
molecule fixed coordinates x, y, z. By the separation of the coordinates we have achieved
that the Schrödinger equation can be split into two eigenvalue problems*

$$
\widehat{H}_{\text{trans}} \psi_{\text{trans}}(X_0, Y_0, Z_0) = E_{\text{trans}} \psi_{\text{trans}}(X_0, Y_0, Z_0)
\tag{5.87}
$$

and

$$
\widehat{H}_{\text{rel}} \psi_{\text{rel}}(x, y, z) = E_{\text{rel}} \psi_{\text{rel}}(x, y, z)
\tag{5.88}
$$

that can be solved independently. The total Schrödinger equation

$$
\widehat{H}_{\text{trans-rel}} \psi_{\text{trans-rel}}(X_0, Y_0, Z_0, x, y, z) = E_{\text{trans-rel}} \psi_{\text{trans-rel}}(X_0, Y_0, Z_0, x, y, z)
\tag{5.89}
$$

then gives the eigenvalues

$$
E_{\text{trans-rel}} = E_{\text{trans}} + E_{\text{rel}}
\tag{5.90}
$$

and the corresponding wave functions

$$
\psi_{\text{trans-rel}}(X_0, Y_0, Z_0, x, y, z) = \psi_{\text{trans}}(X_0, Y_0, Z_0) \psi_{\text{rel}}(x, y, z)
\tag{5.91}
$$

*Spectroscopic transitions (induced by electromagnetic radiation) as described in Figure
5.4 and Formula (5.1) cannot have any impact on the translation state, i.e. such a
transition combines two wave functions*

$$
\psi_{\text{trans}}(X_0, Y_0, Z_0) \psi_{\text{rel}}''(x, y, z) \longrightarrow \psi_{\text{trans}}(X_0, Y_0, Z_0) \psi_{\text{rel}}'(x, y, z)
\tag{5.92}
$$

with the same translation wave function, but two different wave functions $\psi''_{\text{rel}}(x, y, z)$ and $\psi'_{\text{rel}}(x, y, z)$ for the relative motion. Consequently, the related energy difference amounts to

$$\Delta E = E'_{\text{rel}} - E''_{\text{rel}} \tag{5.93}$$

and the contribution of the translation energy E_{trans} cancels out. Consequently, we can restrict ourselves to the solution of the Schrödinger equation for the relative motion (5.88) in the following, as spectroscopists are primarily interested in differences between energy levels. □

Summary 5.42
Let us now briefly summarize the essential points from this section:
The motion of the nuclei of a molecule may be described by means of appropriately chosen coordinates, a survey of which is given in Table 5.2.

Table 5.2.: Coordinates for the different kinds of nuclear motion

Type of motion	Description by	# coordinates
translation	center-of-mass $R_0 = (X_0, Y_0, Z_0)$ (see Def. 5.36)	3
rotation	Euler angles $E = (\theta, \phi, \chi)$ (see Def. 5.37)	3
vibration	several possibilities (cf. Rem. 5.39 and Def. 5.40)	$3N - 6$

Furthermore, the discussion in Remark 5.41 has shown that one can neglect the translational motion, such that the Schrödinger equation for the relative motion (5.88) only depends on $3N - 6 + 3 = 3N - 3$ vibrational and rotational coordinates. □

5.7. Variational Approach and Matrix Eigenvalue Problem

Let $\mathcal{B} = \{\phi_i\}_{i=0}^{\infty}$ be an orthonormal basis of $L^2(\mathbb{R}^n)$. In the following we approximate the molecular Hamiltonian $\widehat{H}$ by orthogonal projection onto a finite dimensional subspace

$$\mathcal{K} = \text{span}\{\phi_1, \ldots, \phi_M\} \tag{5.94}$$

spanned by the first M basis vectors of $\mathcal{B}$ in order to solve the Schrödinger equation

$$\widehat{H}\psi_m(q_1, q_2, \ldots, q_n) = E_m \psi_m(q_1, q_2, \ldots, q_n) \tag{5.95}$$

numerically, where we use a set of arbitrary coordinates q_i ($i = 1, \ldots, n$) and the shorthand notation $dV = dq_1 dq_2 \ldots dq_n$ for the differential volume element in integrals. Provided that the basis functions are chosen appropriately, this leads to a real-symmetric matrix representation $\mathbf{H} \in \mathbb{R}^{M \times M}$ which can be diagonalized using one of the eigen-solvers discussed in Chapters 3 and 4. Obviously, this approach is an extension of the *Rayleigh-Ritz procedure* (Alg. 3.12) to the more general situation that the vector space under consideration is not finite-dimensional and we can proceed in almost complete analogy: We are looking for approximate eigenpairs $(\widetilde{E}, \widetilde{\psi})$ where $\widetilde{\psi} \in \mathcal{K}$. Then, impos-ing the *Galërkin condition*

$$\widehat{H}\widetilde{\psi} - \widetilde{E}\widetilde{\psi} \perp \mathcal{K} \tag{5.96}$$

on the residual implies that

$$\langle \chi, \widehat{H}\widetilde{\psi} - \widetilde{E}\widetilde{\psi} \rangle = 0 \qquad \forall \chi \in \mathcal{K} \tag{5.97}$$

The arising inner products are scalar products on $L^2(\mathbb{R}^n)$ in the sense of (5.6). Expressing $\widetilde{\psi}$ as linear combination of the basis vectors

$$\widetilde{\psi} = \sum_{j=1}^{M} \alpha_j \phi_j \tag{5.98}$$

it follows from (5.97) that

$$\left\langle \phi_i, \widehat{H} \sum_{j=1}^{M} \alpha_j \phi_j \right\rangle - \widetilde{E} \left\langle \phi_i, \sum_{j=1}^{M} \alpha_j \phi_j \right\rangle = 0 \qquad \forall i = 1, \ldots, M$$

$$\sum_{j=1}^{M} \alpha_j \langle \phi_i, \widehat{H}\phi_j \rangle = \widetilde{E}\alpha_i \qquad \forall i = 1, \ldots, M \tag{5.99}$$

We now define $\mathbf{H} \in \mathbb{C}^{M \times M}$ by its *matrix elements*

$$\mathbf{H}_{ij} = \langle \phi_i, \widehat{H}\phi_j \rangle = \int_{\mathbb{R}^n} \phi_i^* \widehat{H}\phi_j \, dV \qquad \forall i, j = 1, \ldots, M \tag{5.100}$$

and the coordinate vector $\widetilde{\mathbf{c}}$ as

$$\widetilde{\mathbf{c}} = \iota(\widetilde{\psi}) = (\alpha_1, \ldots, \alpha_M)^T \tag{5.101}$$

where $\iota : \mathcal{K} \to \mathbb{C}^M$ is the canonical isomorphism. Then (5.99) can be re-written as the desired matrix eigenvalue problem:

$$\boxed{\mathbf{H}\widetilde{\mathbf{c}} = \widetilde{E}\widetilde{\mathbf{c}}} \tag{5.102}$$

Definition 5.43 (Finite basis representation (FBR))

The matrix $\mathbf{H} \in \mathbb{C}^{M \times M}$ *is called Hamiltonian matrix or Finite Basis Representation (FBR) of the molecular Hamiltonian* $\widehat{H}$, *which acknowledges that* $\mathbf{H}$ *is an approximation of* $\widehat{H}$ *obtained by using finite basis expansions of* ψ_m *in (5.98). It is important to note that by a proper choice of the basis functions it can always be achieved that* $\mathbf{H}$ *has only real matrix coefficients. The computational scheme suggested above to determine eigenpairs of the molecular Hamiltonian* $\widehat{H}$ *is commonly referred to as FBR approach or variational approach and the terms introduced in Section 3.3.1.1 carry over as follows:*

- $\mathbf{H} \in \mathbb{R}^{M \times M}$ is the interaction matrix of $\widehat{H}$ with respect to $\mathcal{K}$

- $\widetilde{E}_m$ $(m = 1, \ldots, M)$ are the Ritz values of $\widehat{H}$ with respect to $\mathcal{K}$.

- $\widetilde{\psi}_m$ $(m = 1, \ldots, M)$ are the Ritz functions of $\widehat{H}$ with respect to $\mathcal{K}$ (see (5.98))

- $(\widetilde{E}_m, \widetilde{\psi}_m)$ $(m = 1, \ldots, M)$ are the Ritz pairs of $\widehat{H}$ with respect to $\mathcal{K}$.

Finally, the following lemma provides a compact formulation of the above considerations which turns out useful for the formal derivation of FBRs with respect to product bases in the following section. Furthermore, it clarifies the analogy between the interaction matrix (3.39) defined in Section 3.3.1.1 and the Hamiltonian matrix $\mathbf{H}$ in (5.100) above and shows that the projector $\mathcal{P}$ is the suitable generalization to the orthogonal matrix V holding the basis vectors.

Lemma 5.44
Let the subspace $\mathcal{K} \subset L^2(\mathbb{R}^n)$ be defined as in (5.94) and define the cut-off projection operator with respect to a fixed number $M \in \mathbb{N}$

$$\mathcal{P} : L^2(\mathbb{R}^n) \quad \rightarrow \quad \mathcal{K}$$
$$\psi = \sum_{i=1}^{\infty} \alpha_i \phi_i \quad \mapsto \quad \widetilde{\psi} = \sum_{i=1}^{M} \alpha_i \phi_i$$

Then the interaction operator

$$\widehat{M} = \mathcal{P}^* \widehat{H} \mathcal{P} : \mathcal{K} \rightarrow \mathcal{K} \tag{5.103}$$

is the projection of $\widehat{H}$ onto $\mathcal{K}$ and the Hamiltonian matrix defined by (5.100) can be expressed as

$$\mathbf{H} = \iota(\widehat{M}) \tag{5.104}$$

where ι is the extension of the canonical isomorphism between $\mathcal{K}$ and $\mathbb{C}^M$ as per (5.44) in Def. 5.31.

HYLLERAAS and UNDHEIM [57] as well as MACDONALD [78] were the first to realize the variational character of the approach in the context of quantum chemical computations at the beginning of the 30's of the last century and for this reason the following theorem is well-known under the names *MacDonald's theorem* or *Hylleraas-Undheim-MacDonald interleaving theorem*. It is a generalization of the interleaving theorems for finite dimensional Hermitian matrices discussed in Section 2.2.2:

Theorem 5.45 (Hylleraas, Undheim, MacDonald)
Let the eigenvalues $E_i \in \mathbb{R}$ of the molecular Hamiltonian $\widehat{H}$ and the eigenvalues $\widetilde{E}_i^{(N)} \in \mathbb{R}$ of the Hamiltonian matrix $\mathbf{H}^{(N)} \in \mathbb{R}^{N \times N}$ with respect to a finite basis $\{\phi_i\}_{i=1}^{N}$ be ordered by ascending magnitude. Then the following statements hold

1. Let $H^{(N)} \in \mathbb{R}^{M \times M}$, $H^{(L)} \in \mathbb{R}^{L \times L}$ be Hamiltonian matrices w.r.t. to basis sizes $L, M \in \mathbb{N}$, where $L > M$. Then the kth eigenvalue $\widetilde{E}^{(M)}$ of $\mathbf{H}^{(M)}$ is an upper bound to the kth eigenvalue $\widetilde{E}^{(L)}$ of $\mathbf{H}^{(L)}$, i.e.

$$\widetilde{E}_k^{(M)} \geq \widetilde{E}_k^{(L)} \qquad k = 1, \ldots, M \tag{5.105}$$

2. the kth eigenvalue of $\mathbf{H}^{(M)}$ is an upper bound to the corresponding exact kth eigenvalue of $\widehat{H}$, i.e.

$$\widetilde{E}_k^{(M)} \geq E_k \qquad k = 1, \ldots, M \tag{5.106}$$

Proof: The first part of the theorem, where the spectra of finite dimensional matrices are compared, is a direct consequence of the inclusion principle (Lemma 2.18) for Hermitian matrices, since $\mathbf{H}^{(M)}$ can be obtained from $\mathbf{H}^{(L)}$ by deleting the last $L - M$ rows and columns. For a mathematically satisfactory proof of the second assertion see [111], where the variational characterisations for Hermitian matrices presented in Section 2.2.2 are generalized to unbounded self-adjoint operators acting on *Hilbert spaces* $\mathcal{H}$. $\square$

Remark 5.46

- *The variational approach discussed above provides a viable means for the numerical solution of the Schrödinger equation. The computation of the matrix elements defined in (5.100) involves integration over $\mathbb{R}^n$, which is in general only feasible by means of numerical techniques and requires enormous computational effort. The arising costs scale rather unfavorably as $\mathcal{O}(n^2)$ which becomes perceivable especially for large problem sizes n. However, the overall costs are dominated by the subsequent numerical eigenvalue computation, for which the costs are $\mathcal{O}(n^3)$, if one of the direct solvers discussed in Section 3.2 is employed. This gives additional motivation for our investiagtions on iterative projection methods in this context.*

- *MacDonald's theorem 5.45 suggests that the sequence of the upper bounds $\widetilde{E}_k^{(N)}$ converges monotonously from above to the corresponding exact eigenvalue E_k of $\widehat{H}$ as N tends to infinity*

$$\widetilde{E}_k^{(N)} \downarrow E_k \qquad N \to \infty \tag{5.107}$$

   A mathematically stringent convergence analysis for the case that the variational approach is applied to the solution of the electronic Schrödinger equation (5.72) may be found in [68] where further interesting references on the topic are given. For a systematic and general discussion on the convergence theory for spectral approximations of operators, see the monograph by CHATELIN [23].

- *MacDonald's theorem 5.45 states that the approximations $\widetilde{E}_m$ are upper bounds to the exact eigenvalues E_m. Unfortunately, knowing that $\widetilde{E}_k^{(N)}$ is an upper bound to E_k in general does not imply anything on the quality of the approximation $\widetilde{E}_k^{(N)}$. A common strategy in practical computations is thus to choose a reasonably large basis with dimension N and to assess how much some successive approximations*

$\widetilde{E}_k^{(N+1)}, \widetilde{E}_k^{(N+2)}, \ldots$ *deviate from* $\widetilde{E}_k^{(N)}$. *If the difference is only small or* $\widetilde{E}_k^{(M)}$ *seems to be constant for* $M > N$, *then* $\widetilde{E}_k^{(N)}$ *can be assumed to be converged to* E_k *which is called E-convergence in the literature (convergence of the energy levels* E_m). *Analogously, one speaks of* ψ-*convergence when monitoring the differences of wavefunctions* $\widetilde{\psi}_k^{(M)}$ *with respect to the* L^2-*norm. Of course, this is only a heuristic guideline, and no convergence criterion in a strictly mathematical sense, but experience has shown that it works rather reliably.*

- *To obtain reliable inclusions of energy levels one also requires* lower bounds *on the related eigenvalues. Again, this is a general issue in the theory of self-adjoint operators. A theoretical exposition as well as practical hints for a possible realization may be found in [10] and the references therein.*

- *The variational FBR approach described in this section is not the only way to obtain a matrix eigenvalue problem. There are also grid-based methods, so called* Discrete Variable Representation *(DVR) methods discussed in the literature, which are formally related to the FBR approach by a transformation* **T** *involving Gaussian quadrature points and weights of the (polynomial) basis functions. In general, the variational character in a strict sense, as it is known from the FBR approach, is lost for grid-based methods. However, these methods share the fortunate aspect that the potential energy matrix is diagonal which makes the Hamiltonian matrix very sparse. We cannot go into detail here and refer to [18], [19] and [20] for a further discussion.*

□

5.8. Product vs. Contracted Basis and Direct vs. Iterative Eigensolver

In this section we combine the essential insights of the preceding sections in order to discuss in more detail how the orthonormal basis $\{\phi\}_{i=1}^{\infty}$ of the $\mathcal{H} = L^2(\mathbb{R}^n)$ required in the previously derived variational approach is actually constructed and what impact this has on the complexity of the arising symmetric eigenvalue problems. Eventually, we will outline four principle strategies for their solution. For the sake of generality, the discussion will be held on an abstract level, as we are primarily interested in the structure and the computational complexity. For related surveys with more emphasis on the technical details related to theoretical spectroscopy and quantum chemistry along with concrete examples of applications see [20] (Chapter 9, JONATHAN TENNYSON, *Variational calculations of rotation-vibration spectra*), and the papers by CARRINGTON and his co-workers [19], [104], [129] and [130].

5.8.1. Product Basis

In Section 5.6 we have extensively discussed the choice of coordinates for the appropriate description of the nuclear motion. For the sake of generality, let us first assume, that the motion is described by n coordinates $q_1, \ldots, q_n$. Using the so-called *Podolsky trick* (see [21], [61]) it is possible to express the nuclear kinetic energy operator (KEO) $\widehat{T}_n$ in (5.75) with respect to other than Cartesian coordinates. Depending on the concrete choice this leads to different types of Hamiltonians (cf. [59], [60] and [20], for instance). The general situation is that the kinectic energy operator $\widehat{T}_n$ may be expressed as a sum of operator tensor products (cf. Def. 5.29) such that one obtains the following representation

$$\widehat{H} = \widehat{T} + V = \sum_{j=1}^{k} \bigotimes_{i_j=1}^{n} T_{i_j} + V(q_1, \ldots, q_n) \tag{5.108}$$

where

$$T_{i_j}(q_{i_j}) : \mathcal{H}_{i_j} \to \mathcal{H}_{i_j} \qquad i_j = 1, \ldots, n \tag{5.109}$$

denotes an operator acting on the $\mathcal{H}_{i_j} = L^2(\mathbb{R})$ related to the coordinate q_{i_j}. The potential V depends on the same coordinates, but it may not always be possible to expand it as a sum of tensor products. This is primarily depending on how the potential energy surface (PES) discussed in Section 5.5 is actually constructed. If, for instance, a polynomial expansion (e.g. Taylor approximation or a force field expression) is employed as numerical approximation technique, it is possible to expand V accordingly. The common strategy pursued in variational approaches (cf. [20], [19]) is to construct a product basis

$$\mathcal{B} = \bigotimes_{j=1}^{n} \mathcal{B}_j = \mathcal{B}_1 \otimes \mathcal{B}_2 \otimes \ldots \otimes \mathcal{B}_n \tag{5.110}$$

of $\mathcal{H} = L^2(\mathbb{R}^n)$ as per Def. 5.33, where each basis $\mathcal{B}_i$ is related to one vibrational resp. rotational coordinate. In general, one attempts to employ bases $\mathcal{B}_i$ whose functions give an appropriate description of the molecular motion related to the coordinate q_i. Typical choices are the eigenfunctions of the harmonic oscillator (see [21] for a general description) or the eigenfunctions of the Morse oscillator [60] for bases related to stretching coordinates q_i. A frequent choice for basis functions related to the rotational coordinates are the spherical harmonics as eigenfunctions of the angular momentum operator (see [21],[133] for more details). To obtain the finite dimensional Hamiltonian matrix $\mathbf{H}$ one now has to cut off each basis $\mathcal{B}_i$ in (5.110) after $M_i \in \mathbb{N}$ elements which results in the truncated product basis of (5.110)

$$\mathcal{B}^{(M)} = \bigotimes_{j=1}^{n} \mathcal{B}_j^{(M_j)} = \mathcal{B}_1^{(M_1)} \otimes \mathcal{B}_2^{(M_2)} \otimes \ldots \otimes \mathcal{B}_n^{(M_n)} \tag{5.111}$$

and the correspnding cut-off projections

$$\mathcal{P}_i : \mathcal{H}_i \rightarrow \mathcal{K}_i \qquad i = 1, \ldots, n \tag{5.112}$$

onto the subspaces

$$\mathcal{K}_i = \mathrm{span}\{\mathcal{B}_i^{(M_i)}\} \qquad i = 1, \ldots, n \tag{5.113}$$

Consequently, the total projection is obtained by the tensor product

$$\mathcal{P} := \bigotimes_{i=1}^{n} \mathcal{P}_i \; : \mathcal{H} \rightarrow \mathcal{K} \tag{5.114}$$

where

$$\mathcal{K} := \bigotimes_{i=1}^{n} \mathcal{K}_i = \mathrm{span}\{\mathcal{B}^{(M)}\} \tag{5.115}$$.

From Corollary 5.28 it is known that the problem size is M determined by

$$M = \dim \mathcal{K} = \dim \left\{ \bigotimes_{i=1}^{n} \mathcal{K}_i \right\} = \prod_{i=1}^{n} \dim \mathcal{K}_i = \prod_{i=1}^{n} M_i \tag{5.116}$$

Application of Lemma 5.44 to the operator $\widehat{H}$ defined in (5.108) using the projection $\mathcal{P}$ (5.114) and successive exploitation of the tensor product properties (see Corollary 5.30 and Lemma 5.32) then yields the finite basis representation $\mathbf{H}$ as a sum of matrix Kronecker products for the kinetic energy matrix $\mathbf{T}$ and the potential matrix $\mathbf{V}$:

$$\mathbf{H} = \mathbf{T} + \mathbf{V} = \sum_{j=1}^{k} \bigotimes_{i_j=1}^{n} \mathbf{T}_{i_j} + \mathbf{V} \tag{5.117}$$

where

$$\mathbf{H}, \mathbf{T}, \mathbf{V} \in \mathbb{R}^{M \times M} \quad \text{and} \quad \mathbf{T}_{i_j} \in \mathbb{R}^{M_{i_j} \times M_{i_j}}, \quad i_j = 1, \ldots, n \tag{5.118}$$

Relation (5.116) reveals the general and fundamental problem with the variational approach: To obtain reasonably accurate eigenvalue approximations, one has to choose the product basis $B^{(M)}$ sufficiently large, which in turn is accomplished by choosing the sizes M_i of factor bases $\mathcal{B}_i^{(M_i)}$ appropriately. The concrete choice and weighting of these limiting numbers is problem dependent, but in general all bases have to be represented adequately. A fortunate aspect is that rather often the Hamiltonian matrix (5.117) becomes block diagonal with respect to suitably chosen bases related to the 3 rotational coordinates, i.e. each block can be considered seperately and the complexity of the arising eigenvalue problems thus effectively depends on the $n = 3N - 6$ vibrational coordinates. In the following we will take this situation for granted, since it also applies for the *Double Renner Hamiltonian* we will be discussing in detail in the subsequent chapters. However, this is still bad enough, as the following example impressively illustrates: Given an N-atom molecule, assume moderate limiting numbers $M_i = 5$, $M_i = 7$

or $M_i = 10$ for all $3N - 6$ vibrational bases. Table 5.3 shows what this implies for the arising matrix dimensions.

The problem sizes rapidly increase for small molecules and even small changes in the limiting parameters M_i can make the dimension of the Hamiltonian matrix $\mathbf{H}$ explode. In general, the arising eigenvalue problems become intractable for the direct eigensolvers presented in Section 3.2 for $N > 4$ because of the tremendous storage requirements. Iterative projection methods offer a way out of this difficulty, and from the discussion in Section 3.3 it is known that the user-supplied matrix-vector multiplication is a key ingredient for these algorithms. There are two possible scenarios in our context:

1. The potential matrix $\mathbf{V}$ is "factorizable", i.e. it may be written as a sum of Kronecker products analogous to the matrix $\mathbf{T}$ in (5.117) arising from the kinetic energy operator $\widehat{T}$. Then iterative projection methods are particularly well-suited, because the product basis structure can be exploited for the computation of matrix-vector products. Let ℓ be the number of terms in $\mathbf{V}$ and $g = k + \ell$. Then the Hamiltonian matrix may be expressed as

$$\mathbf{H} = \sum_{j=1}^{g} \bigotimes_{i_j=1}^{n} \mathbf{H}_{i_j} \tag{5.119}$$

This has very favorable consequences for the storage requirements and the computational costs, because a matrix-vector product $\mathbf{H}\mathbf{x}$ may be evaluated by exploiting the mixed-product property for Kronecker products (see Lemma 2.48, property 5), i.e.

$$\mathbf{H}\mathbf{x} = \left(\sum_{j=1}^{g} \bigotimes_{i_j=1}^{n} H_{i_j} \right) \cdot \otimes_{i=1}^{n} \mathbf{x}_i = \sum_{j=1}^{g} \otimes_{i_j=1}^{n} (\mathbf{H}_{i_j} \mathbf{x}_{i_j}) \tag{5.120}$$

where $\mathbf{x}$ is also expressed as a tensor product.

$$\mathbf{x} = \otimes_{i=1}^{n} \mathbf{x}_i \tag{5.121}$$

Let us again assume all factor bases $\mathcal{B}_i^{(M_i)}$ $(i = 1, \ldots, n)$ to be of equal dimension m, such that $\dim \mathbf{H} = M = m^n$. Then $n \cdot m^{n+1} = (m/\log m) M \log M$ scalar

Table 5.3.: Dimensions of FBR Hamiltonian matrices

nuclei N	$n = 3N - 6$	$\dim(\mathbf{H})$, $M_i = 5$	$\dim(\mathbf{H})$, $M_i = 7$	$\dim(\mathbf{H})$, $M_i = 10$
3	3	125	343	1000
4	6	15625	117649	1000000
5	9	1953125	40353607	10000000000
6	12	244140625	13841287201	10000000000000

multiplications are required, i.e. the computational costs amount to $\mathcal{O}(M \log M)$. Furthermore, it is only necessary to store $g \cdot n$ lower dimensional matrices $\mathbf{H}_{i_j} \in \mathbb{R}^{m \times m}$ and n vectors $\mathbf{x}_i \in \mathbb{R}^m$, such the storage costs for the matrix entries are $\mathcal{O}(\sqrt[n]{M}^2)$. In many cases some of the factors $\mathbf{H}_{i_j}$ are identity matrices I_m, such that the computation further simplifies. For a detailed discussion on the complexity of the matrix-vector multiplication see [18].

2. If the potential matrix $\mathbf{V}$ is *not* factorizable, it is no more possible to take advantage of the product basis structure directly, but one can exploit the sparsity of $\mathbf{H}$, which in general exhibits a very regular structure due to the terms arising from the Kronecker products in $\mathbf{T}$ and often reduces to a great deal memory consumption and computational costs.

5.8.2. Contracted Basis

The examples in Table 5.3 show that the size of the Hamiltonian matrix $\mathbf{H}$ is often too large for practical computations, even for molecules consisting of a small number of nuclei. Hence, a common strategy is to construct a smaller matrix

$$\mathbf{V}^* \mathbf{H} \mathbf{V} = \widetilde{\mathbf{H}} \in \mathbb{R}^{k \times k}, \qquad k \ll M \qquad (5.122)$$

obtained by an additional subsequent application of the *Rayleigh-Ritz procedure* (Alg. 3.12). From the Poincaré separation theorem (Corollary 2.19) it is known that the eigenvalues of $\widetilde{\mathbf{H}}$ are upper bounds to the eigenvalues of $\mathbf{H}$, and consequently, also upper bounds to the exact eigenvalues of the Hamiltonian $\widehat{H}$, such that the variational character of the approach is maintained. The matrix $\widetilde{\mathbf{H}}$ is represented with respect to a *contracted basis* $\mathcal{C}$ formed by the columns of $\mathbf{V}$ and the *Rayleigh-Ritz procedure* is often referred to as a *contraction scheme* in this context. To carry out the projection, a suitable orthogonal matrix $\mathbf{V} \in \mathbb{R}^{M \times k}$ has to be constructed and for a good quality it is important that the subspace $\mathcal{K}$ spanned by the columns of $\mathbf{V}$ and the related eigenspace $\mathcal{X}$ spanned by the first k eigenvectors of $\mathbf{H}$ make a small angle. Fortunately, the Hamiltonian matrix $\mathbf{H}$ is often dominant with respect to relatively large square blocks on its diagonal, i.e. the Frobenius norms are large as compared to the norms of the off-diagonal blocks. Thus, a general recipe for the construction of the projection matrix $\mathbf{V}$ is to compute eigensystems of these blocks $\mathbf{B}_i$ and to use selected subsets of the eigenvectors stored in the transformation matrices $\mathbf{T}_i$ to build up the projection matrix $\mathbf{V}$. A concrete example along with a detailed description of the construction (cf. Alg. 7.5) and application (cf. Alg. 7.6) of such a projection $\mathbf{V}$ will be given later on in Section 7.3 in the context of the *Double Renner effect*. A general problem with this approach is, however, how to control the quality of the eigenapproximations: To play it safe, one actually has to compare the results from the contracted and the related product basis computation. On the one hand, this is often not feasible due to the problem size of the product basis problem and on the other hand it obviates our efforts to reduce the computatinal costs. In [87] this is discussed in detail for the concrete situation of triatomic molecules exhibiting the

Double Renner effect and a couple of numerical examples is given as a guideline on how to construct the projection. Note that there are further, more sophisticated contraction schemes discussed in the literature. For more details see [19], [129] and [130]. Owing to their small size contracted matrices $\widetilde{\mathbf{H}}$ may be treated by direct solvers and this often leads to reasonable results, provided that the quality of the constructed projection is sufficiently good. Iterative projection methods are applicable as well, but the situation is less favorable, because the product basis structure is destroyed, such that it is no more possible to express the matrix-vector multiplication as a sum of Kronecker products and in general only the sparsity on a block level is conserved.

5.8.3. Dichotomies and General Approaches

In Chapter 3 we have already pointed out the dichotomy between direct solvers and iterative projection methods for eigenvalue problems. The above discussion shows that in the context of eigenvalue problems in theoretical spectroscopy this dichotomy is enriched by an additional "dimension". According to the terminology used by BRAMLEY and CARRINGTON [19] one can distinguish the following four general strategies:

1. *direct-product approach*
   apply one of the direct solvers presented in Section 3.2 to the product-basis Hamiltonian matrix $\mathbf{H}$ (5.117)

2. *direct-contracted approach*
   apply a direct solver to the contracted-basis Hamiltonian matrix $\widetilde{\mathbf{H}}$ (5.122)

3. *iterative-product approach*
   apply one of the iterative projection methods discussed in Section 3.3 and Chapter 4 to the product-basis Hamiltonian matrix $\mathbf{H}$ (5.117)

4. *iterative-contracted approach*
   apply an iterative projection method to the contracted-basis Hamiltonian matrix $\widetilde{\mathbf{H}}$ (5.122)

Summary 5.47
We have seen, that the direct-contracted calculation is in general the least time-consuming approach, but suffers from the drawback that it is often not clear how good the computed eigenvalue approximations actually are. Direct-product computations are more accurate, but often not feasible, because the problem size n is governed by the size of the finite product basis, which impedes explicit storage of the Hamiltonian matrix $\mathbf{H}$ (cf. Table 5.3). Furthermore, the huge matrix sizes make the time complexity of $\mathcal{O}(n^3)$ a severe drawback. Therefore, the iterative-product approach offers a valuable alternative and is of great use as a means of verification, because one is often only intrested in a relatively small fraction of the spectrum (some of the lowest eigenvalues) and one can take advantage of the product basis structure or the sparsity structure in matrix-vector

multiplications. Iterative-product approaches will be in the center of our interest when we will discuss the Double Renner effect. It also may happen that the dimension of a contracted problem is so large that iterative-contracted approaches become competetive with their direct-contracted counterparts. Hence, depending on the type of problem and the chosen projection, each of the four possibilities has its justification. A related and extensive discussion on the pros and cons of the four approaches may be found in [19].
□

5.9. General Framework for the Computation of Energy Levels

In this chapter we have so far collected the essential components for the numerical solution of the time-independent Schrödinger equation and we have seen that the Born-Oppenheimer approximation discussed in Section 5.5 allows to separate the contribution of nuclei and electrons which simplifies the problem considerably. Furthermore, one can ignore the contribution of translational motion, such that the complexity of the nuclear Schrödinger equation depends on $3N - 3$ vibrational and rotational coordinates. For the sake of lucidity it is now appropriate to give a brief summary. To compute the discrete energy levels E_i of a molecule consisting of N nuclei from the simplified nuclear Schrödinger equation, the following ingredients are required:

- A potential energy surface (PES) (see Section 5.5 and Fig. 5.5) for the molecule under consideration obtained by means of *ab initio calculations* and/or experimental data

- A proper choice of $3N - 3$ coordinates q_i for the appropriate description of the rotational and vibrational motion of the nuclei. The molecular Hamiltonian then has to be expresssed with respect to these coordinates using the chain rule or the so-called *Podolsky trick*.

- A choice of n bases $\mathcal{B}_i$ representing the rotational/vibrational motion in each coordinate q_i

Fortunately, the computational complexity often only depends on the $n = 3N - 6$ vibrational coordinates, because in many cases the arising Hamiltonian matrix is block diagonal with respect to the rotational basis functions:

1. Choose finite bases $\mathcal{B}_i^{(M_i)}$ by cutting off $\mathcal{B}_i$ after $M_i \in \mathbb{N}$ elements resulting in a finite product basis $\mathcal{B}^{(M)} = \bigotimes_{i=1}^n \mathcal{B}_i^{(M_i)}$ of dimension $M = \prod_{i=1}^n M_i$.

2. Compute the matrix elements for the various terms $\mathbf{T}_{i_j}$ and the potential V in the Hamiltonian (5.108). This is done by using the relation (5.100) for a Hamiltonian

matrix element $\mathbf{H}_{ij}$ and exploiting the definition of the operator tensor product (Def. 5.29) and the scalar product (5.38) on the product $\mathcal{H} = L^2(\mathbb{R}^n)$ (cf. Def. 5.33). Depending on the choice of basis functions, some of these terms may be evaluated by means of explict analytic formulae. However, in general, one has to resort to numerical techniques. For the matrix elements $\mathbf{V}_{ij}$ related to the potential the situation is even worse, because rather often the potential function V cannot be written as sum of tensor product expressions. This implies that simultaneous numerical quadrature in all n coordinates is required for the computation of one potential matrix element. Assuming K (typically 10-30) quadrature points in each coordinate this leads to K^n evaluations and makes the computation of the matrix elements an extremely demanding and time-consuming problem.

3. Collect the matrix blocks computed in the preceding step to construct the final Hamiltonian matrix $\mathbf{H}$.

4. Compute approximate eigenpairs by using one of the four recently presented general approaches (application of direct/iterative solver to product/contracted-basis representation, see Section 5.8.3).

6. The Double Renner Effect for Triatomic Molecules

In this chapter we come to the problem, which is actually in the center of our interest, the computation of energy levels of triatomic molecules with the Renner effect property. A full theoretical account on this matter is given in the PhD thesis by ODAKA [86], where all ingredients (choice of coordinates, Hamiltonian, choice of basis functions, computation of matrix elements, numerical integration and diagonalization of the Hamiltonian matrix) required for the variational calculation of energy levels along with a software (FORTRAN 90 [85] code DR) are discussed in detail. In the following we will focus on those aspects that are of importance to understand how the Hamiltonian matrix is constructed and describe its structure and its properties.

6.1. Breakdown of the Born-Oppenheimer Approximation

In the preceding chapter we have seen that the Born-Oppenheimer approximation plays a key role in the variational computation of energy levels, because it allows one to treat the motion of the electrons and the nuclei separately and greatly reduces the complexity of the problem. The contribution of the electronic motion is represented by the potential energy surface (PES), $V(R)$ which is obtained in the *ab initio step*, i.e. by solving the electronic Schrödinger equation (5.72) for several fixed nuclear geometries. A molecule has infinitely many electronic states, for each of which – at least in principle – one could determine a PES. When two such potential energy surfaces become close to each other in energy or degenerate at certain molecular geometries, it is no more possible to neglect the interaction of electronic motion and vibration of the nuclei, and consequently, the Born-Oppenheimer approximation fails. RENNER [98] was the first to give an example of such a possible break-down. He realized, that if the electronic energy in a triatomic molecule is doubly degenerate at linear geometries, it necessarily splits into two separate components when the molecule bends. The two resulting electronic states are close in energy and the Born-Oppenheimer approximation fails. If one linear molecular geometry is accessible at which the degeneracy arises, this is called *Renner effect*. In case of two accessible linear geometries this is referred to as *Double Renner effect*. An example of such a molecule is provided by the **MgNC/MgCN** molecule. The sketch below illustrates the double degeneracy at linear geometries and the splitting upon bending. Besides, it shows how electronic energies change on isomerization from an **MgNC** molecule ($\tau = 0$) to an **MgCN** molecule ($\tau = \pi$), where the Jacobi angle τ (cf. Def. 5.40 and Fig. 6.2 below) describes the bending motion of the **Mg** nucleus.

Figure 6.1.: Degeneracy at linear geometries

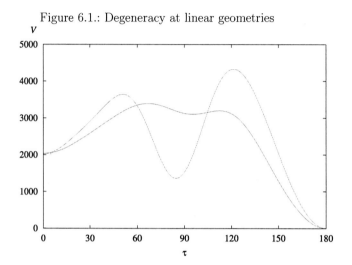

Obviously, one has to modify the Born-Oppenheimer approach outlined in Section 5.5 in order to be able to compute the discrete energy levels using a variational approach. The suitable remedy is to combine the two potential energy surfaces, which we will refer to as "upper" $V^{(+)}$ and "lower" $V^{(-)}$ surface (see [86] for details on how this is accomplished). The drawback with this approach is that the resulting PES has more than one local minimum and therefore it is not appropriate to approximate it by means of Taylor expansions. This in turn means that the potential matrix may not be expanded as a sum of tensor products, such that one can only exploit the sparsity of the Hamiltonian matrix for the design of matrix-vector multiplications (see the related discussion in Section 5.8.1). The Renner effect also explains the technical term *rovibronic*, which is a portmanteau made up of *ro*-tational, *vib*-rational and elect-*ronic*: *Rovibronic* energy levels arise from the coupling of electronic motion and vibrational motion of the nuclei (as the description of the Renner effect shows) as well as the rotation of the molecule.

It is appropriate to distinguish between two classes of molecules (we will examine one example for each class),

1. **ABC** type molecules consisting of three different nuclei **A**, **B** and **C**.
   Example: **MgNC** resp. **MgCN** molecule (see above)

2. **ABB** type molecules consisting of two identical nuclei **B**
   Example: **HOO** molecule

This distinction allows to take advantage of molecular symmetry porperties and reduces the computational effort for **ABB** type molecules (see [86]), as will become evident later on.

In order to compute rovibronic energy levels we have to solve the Schrödinger equation and to this end we have to follow the general procedure outlined in Section 5.9.

6.2. The Double Renner Hamiltonian

According to the list of ingredients in Section 5.9 one requires a potential energy surface, which is obtained by computing two PES obtained in an *ab initio step* (see above discussion). Then an appropriate coordinate system for the description of the nuclear motion has to be chosen. The *Jacobi coordinates* (see Def. 5.40) turn out useful as vibrational coordinate system, because they adequately reflect the bending nature, e.g. in the motion of the **Mg** nucleus (see Fig. 6.2). In order to describe the rotation of the molecule we further need a space-fixed Cartesian xyz-coordinate system having its origin in the center of mass **O** of the three involved nuclei. Both, molecule and space fixed coordinate systems, are illustrated below for the case of an **MgCN** molecule, where the z-axis points into the plane:

Figure 6.2.: **MgCN** molecule: coordinate systems

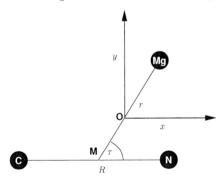

Using the *Podolsky trick* ([21], [61]) one can now express the molecular Hamiltonian $\widehat{H}_{DR}$ with respect to these coordinates (the subscript DR stands for *Double Renner*):

$$\hat{H}_{DR} = \frac{\hbar^2}{2\mu_R R^2}\{\hat{N}_x^2 + \hat{N}_y^2 + \cot^2\tau(\hat{N}_z - \hat{L}_z)^2 + \hat{N}_\tau^2 + \hat{N}_x\hat{N}_\tau + \hat{N}_\tau\hat{N}_x$$
$$+ \cot\tau(\hat{N}_y(\hat{N}_z - \hat{L}_z) + (\hat{N}_z - \hat{L}_z)\hat{N}_y)\}$$
$$+ \frac{\hbar^2}{2\mu_r r^2}\left\{\frac{1}{\sin^2\tau}(\hat{N}_z - \hat{L}_z)^2 + \hat{N}_\tau^2\right\}$$
$$- \frac{\hbar^2}{8}\left\{\frac{1}{\mu_R R^2} + \frac{1}{\mu_r r^2}\right\}\left\{1 + \frac{1}{\sin^2\tau}\right\}$$
$$+ \frac{1}{2\mu_R}\hat{P}_R^2 + \frac{1}{2\mu_r}\hat{P}_r^2 + \hat{H}_{SO} + V(R, r, \tau) \qquad (6.1)$$

The following list gives a brief explanation of some symbols and expressions (see [86] for details):

- μ_r and μ_R are the reduced masses in a *Jacobi coordinate* system:

$$\mu_R = \frac{m_B m_C}{m_B + m_C}$$
$$\mu_r = \frac{m_A \cdot (m_B + m_C)}{m_A + m_B + m_C}$$

(Here we label the atoms of a molecule by A, B and C).

- $\hat{L}_z$ is the projection of the angular momentum operator $\hat{L}$ onto the z-axis.

- $\hat{N}_x$, $\hat{N}_y$, $\hat{N}_z$ are the projections of the angular momentum operator $\hat{N}$ onto the x-, y- and z-axes respectively.

- $\hat{N}_\tau$, $\hat{P}_r$, $\hat{P}_R$ are kinetic energy operators.

- $\hat{H}_e$ is the part of the Hamiltonian related to the potential

- $\hat{H}_{SO}$ is the spin-orbit coupling term

Figure 6.3.: **HOO** molecule: 395th energy state for $J = 1/2$ and $\Gamma_{rve} = B_2$

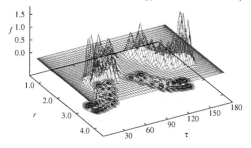

Before we proceed, let us briefly come back to the physical interpretation of the eigenfunctions ψ of the Hamiltonian. In Section 5.3 it has already been mentioned that $|\psi|^2$ may be interpreted as a probability density function. Figure 6.3 shows a three-dimensional plot (for a fixed stretching coordinate R) of $|\psi|^2$, where ψ is an eigenfuction of the *Double Renner Hamiltonian* (6.1) related to an **HOO** molecule. The peaks of the plot indicate the most likely molecular geometries.

6.3. Choice of the Basis Set

For the sake of simplicity we introduce the so called *bra-ket* notation, which is widely used in the field of quantum mechanics and quantum chemistry:

Notation 6.1 (Bra-ket)
Let V, W be a vector spaces and V^*, W^* their dual spaces. Then

- *$|\,v\rangle$ (ket) denotes an element $v \in V$*

- *$\langle v\,|$ (bra) denotes an element $v^* \in V^*$*

- *$|v\rangle|w\rangle$ (ket-ket) denotes the tensor product $v \otimes w$ for elements $v \in V$ and $w \in W$*

- *$\langle w\,|\,v\rangle$ (bra-ket) denotes the scalar product w^*v for elements $v, w \in V$*

This notation has the advantage that it is coordinate-free and allows to include additional parameters, e.g. quantum numbers. The ket $|N_R, \Gamma_R\rangle$, for instance, denotes the set of all basis functions depending on the quantum number $N_R = 0, 1, \ldots$ $\qquad\square$

The symmetry properties of a given molecule are described by its *molecular symmetry group*, which has a prominent role to play in theoretical spectroscopy. Taking advantage of it can simplify calculations and theoretical considerations to a great deal. The theoretical background of group theory is beyond the scope of this thesis (a full account may be found in [21]), we only require the following notation for our purposes:

Definition 6.2 (Molecular symmetry)
The molecular symmetry is labeled by

- *$\Gamma_{rve} \in \{A_1, A_2, B_1, B_2\}$ for **ABB** type molecules*

- *$\Gamma_{rve} \in \{A', A''\}$ for **ABC** type molecules*

In case of the *Double Renner effect* one makes use of a product-like basis, which differs from the product basis described in the discussion of the general case (see Section 5.8.1) in that one basis element is now defined by a sum of tensor products:

$$\Psi_{rve}^{J,M_J,S,\Gamma_{rve}} = \sum_{N=|J-S|}^{J+S} \sum_{K=0}^{N} \sum_{\Gamma_{rve},N_r,N_R,\eta,v_2^\eta} c_{\eta,N,K,v_2^\eta,N_r,N_R,\Gamma_{vib}^\eta}^{J,M_J,S,\Gamma_{rve}}$$

$$\times \ |N_R,\Gamma_R\rangle |N_r,\Gamma_r\rangle |v_2^\eta,K,\Gamma_{v_2}^\eta\rangle |\eta;N,J,S,K,M_J,p\rangle \qquad (6.2)$$

The summation is nested and runs over several quantum numbers and one can see that the total basis set is composed of four particular bases:

- $|N_R,\Gamma_R\rangle$ is the basis set describing the stretching motion along the R-bond labeled by the quantum number N_R.

- $|N_r,\Gamma_r\rangle$ is the basis set related to the r-bond stretching and labeled by the quantum number N_r.

- The basis set $|v_2^\eta,K,\Gamma_{v_2}^\eta\rangle$ represents the bending motion along the Jacobi angle τ together with K-type rotational angular momentum. The related quantum numberis v_2^η where the parameter η can take two different symbolic values, $\eta = a$ and $\eta = b$ and splits the bending basis in two different parts.

- $|\eta,N,J,S,K,M_J,p\rangle$ describes the electronic motion, the effects of electron spin and the rotation of the molecule. J is referred to as *rotational quantum number*.

- $c_{\eta,N,K,v_2^\eta,N_r,N_R,\Gamma_{vib}^\eta}^{J,M_J,S,\Gamma_{rve}}$ are the coefficients which allow to construct any basis function by just taking the corresponding linear combination.

To obtain a finite-dimensional basis one now has to choose appropriate limiting numbers for cutting off of the factor bases. As we are only interested in the basis sizes and in order to be consistent with the terminology employed in [87] we introduce the following notation

Notation 6.3 (Quantum numbers, cut-off numbers and basis sizes)
The superscript (max) denotes the cut-off numbers, i.e. the maximum value of a quantum number in a finite basis set. Since the quantum numbers start at zero one obtains the actual size of the finite basis (lim) by adding 1:

$$N_R = 0,1,\ldots,N_R^{(max)} \qquad\qquad N_R^{(lim)} = N_R^{(max)} + 1, \qquad (6.3)$$

$$N_r = 0,1,\ldots,N_r^{(max)} \qquad\qquad N_r^{(lim)} = N_r^{(max)} + 1, \qquad (6.4)$$

$$v_2^a = 0,1\ldots,(v_2^a)^{(max)} \qquad\qquad (v_2^a)^{(lim)} = (v_2^a)^{(max)} + 1, \qquad (6.5)$$

$$v_2^b = 0,1\ldots,(v_2^b)^{(max)} \qquad\qquad (v_2^b)^{(lim)} = (v_2^b)^{(max)} + 1, \qquad (6.6)$$

$\square$

6.4. Construction and Structure of the Hamiltonian Matrix

One of the advantageous properties of the basis set (6.2) lies in the following basic

Fact 6.4
The Hamiltonian matrix **H** *is block diagonal with respect to the quantum numbers* J, M_J, S *and* Γ_{rve}, *which are therefore often referred to as good quantum numbers. We can omit* M_J *in the following, since the matrix elements do not depend on it. Consequently, the Hamiltonian matrix* **H** *splits into blocks* $\mathbf{H}^{(J,S,\Gamma_{rve})}$, *which are labeled by* J, S *and* Γ_{rve} *and can be treated independently. There are*

- *4 such diagonal matrix blocks in case of an* **ABB** *type molecule*

- *2 such diagonal matrix blocks in case of an* **ABC** *type molecule*

per J-*quantum number, i.e. each matrix block is associated with one irreducible representation in the corresponding molecular symmetry group (see Def. 6.2).* □

As a consequence, we can restrict ourselves to the analysis of a matrix eigenvalue problem related to the block $\mathbf{H}^{(J,S,\Gamma_{rve})}$. Using the finite basis set defined by (6.2), the fixed triple (J, S, Γ_{rve}) and the cut-off numbers (see Def. 6.3) the elements of the Hamiltonian matrix block $\mathbf{H}^{(J,S,\Gamma_{rve})}$ now may be determined by means of the variational approach presented in Section 5.7:

$$\mathbf{H_{ij}}^{(J,S,\Gamma_{rve})} = \left\langle \Psi_i^{J,M_J,S,\Gamma_{rve}} | \hat{H}_{DR} | \Psi_j^{J,M_J,S,\Gamma_{rve}} \right\rangle \tag{6.7}$$

In the following we will describe the structure, properties and the construction of $\mathbf{H}^{(J,S,\Gamma_{rve})}$ in more detail:

6.4.1. Hierarchies and Partitioning into Blocks

The dependencies of the N and K quantum numbers in the nested summation of the total basis set (6.2) impose a partitioning of the Hamiltonian matrix $\mathbf{H}^{(J,S,\Gamma_{rve})}$ into related matrix blocks, which we will describe in the following. For the ease of presentation we introduce the following shorthand notation

Definition 6.5
For pairs of K *quantum numbers* (K', K'') *and* N *quantum numbers* (N', N'') *we define:*

$$\Delta N = |N' - N''| \tag{6.8}$$
$$\Delta K = |K' - K''| \tag{6.9}$$

To describe the structure of a Hamiltonian matrix block, we now define:

Definition 6.6 (J-blocks, N-blocks, K-blocks and addresses)

- *The Hamiltonian matrix block* $\mathbf{H}^{(J,S,\Gamma_{rve})}$ *is called J-block*

- *A sub-block* $\mathbf{H}^{(J,S,\Gamma_{rve})}_{N',N''}$ *related to the* N *quantum number is called N-block and is uniquely determined by its address* N', N''

- *A sub-block* $\mathbf{H}^{(J,S,\Gamma_{rve})}_{(N',K'),(N'',K'')}$ *related to the* K *quantum number is called K-block and is uniquely determined by its address* $(N', K'), (N'', K'')$.

- *An N-block is called diagonal N-block, iff* $\Delta N = 0$ *and is uniquely determined by its address* $N = N' = N''$, *i.e. the block is notated as* $\mathbf{H}^{(J,S,\Gamma_{rve})}_{N}$

- *A K-block is called diagonal K-block, iff* $\Delta K = 0$ *and* $\Delta N = 0$ *and is uniquely determined by its address* N, K *where* $N = N' = N''$ *and* $K = K' = K''$. *If there is no danger of confusion with off-diagonal N blocks we denote it by* $\mathbf{H}^{(J,S,\Gamma_{rve})}_{N,K}$ *for the sake of brevity.*

A full specification in terms of absolute coordinates and dimensions will be given in Corollary 6.9 in the following section.

The following example clarifies the meaning of the definition:

Table 6.1.: Block partitioning of $\mathbf{H}^{(J,S,\Gamma_{rve})}$ for $J = 5/2$ and $S = 1/2$

		$N'' = 2$			$N'' = 3$			
		$K'' = 0$	$K'' = 1$	$K'' = 2$	$K'' = 0$	$K'' = 1$	$K'' = 2$	$K'' = 3$
$N' = 2$	$K' = 0$							
	$K' = 1$							
	$K' = 2$							
$N' = 3$	$K' = 0$							
	$K' = 1$		*					
	$K' = 2$							
	$K' = 3$							

Example 6.7
From the dependence of J and S in (6.2) it follows that $N = 2, 3$ and $K = 0, \ldots, N$. Table 6.1 shows the resulting partitioning into N- and K-blocks. According to the terminology introduced in Def. 6.6 the Hamiltonian matrix can be partitioned into 4 N-blocks and 49 K-blocks. There are 2 diagonal N-blocks and 7 diagonal K-blocks. The K-block marked by the asterisk has the address $(3, 1), (2, 1)$. □

6.4.2. Block and Problem Sizes

Obviously, we can restrict ourselves to analyze the dimensions of diagonal K-blocks $\mathbf{H}_{N,K}^{(J,S,\Gamma_{rve})}$, which depend on the sizes of the finite vibrational basis sets $N_R^{(lim)}$, $N_r^{(lim)}$, $(v_2^a)^{(lim)}$, $(v_2^b)^{(lim)}$. The difficulty is, that depending on the N and K quantum numbers, different parts of the bending basis (which is composed of $|a\rangle$- and $|b\rangle$ functions) are included such that the dimensions of the blocks can take different values. More precisely speaking, one has to take into account the influence of the *vibrational symmetry* Γ_{vib} (not to be confused with Γ_{rve} from Def. 6.2, see [86] and [21] for more details):

Definition 6.8 (Vibrational symmetry)
The vibrational symmetry of a molecule is labeled by

- $\Gamma_{vib} \in \{A_1, B_2\}$ *for* **ABB** *type molecules*

- $\Gamma_{vib} \in \{A'\}$ *for* **ABC** *type molecules.*

The following table gives a list of the different possibilities (three for an **ABC** type molecule and six for an **ABB** type molecule) and shows how the corresponding sizes of the bending basis are determined from $(v_2^a)^{(lim)}$ and $(v_2^b)^{(lim)}$:

Table 6.2.: Size of the bending basis depending on K, Γ_{vib} and η

	K	η	Γ_{vib}	size of bending basis	
ABC	0	a	A'	$(v_2^a)^{(lim)}$	
	0	b	A'	$(v_2^b)^{(lim)}$	
	$\neq 0$	a,b	A'	$(v_2^{a,b})^{(lim)}$	$= (v_2^a)^{(lim)} + (v_2^b)^{(lim)}$
ABB	0	a	A_1	$(v_2^{a,A_1})^{(lim)}$	$= (v_2^a)^{(lim)}/2$
	0	b	A_1	$(v_2^{b,A_1})^{(lim)}$	$= (v_2^b)^{(lim)}/2$
	$\neq 0$	a,b	A_1	$(v_2^{A_1})^{(lim)}$	$= (v_2^{a,A_1})^{(lim)} + (v_2^{b,A_1})^{(lim)}$
	0	a	B_2	$(v_2^{a,B_2})^{(lim)}$	$= (v_2^a)^{(lim)} - (v_2^{a,A_1})^{(lim)}$
	0	b	B_2	$(v_2^{b,B_2})^{(lim)}$	$= (v_2^b)^{(lim)} - (v_2^{b,A_1})^{(lim)}$
	$\neq 0$	a,b	B_2	$(v_2^{B_2})^{(lim)}$	$= (v_2^{a,B_2})^{(lim)} + (v_2^{b,B_2})^{(lim)}$

Which of the above combinations (η and Γ_{vib}) actually arises, depends on the quantum numbers Γ_{rve}, N and K and is explained in the Tables 6.3 and 6.4 below:

Table 6.3.: Combinations for an **ABC** type molecule

Γ_{rve}	$K = 0$		$K \neq 0$
	N even	N odd	
A'	$\eta = a$	$\eta = b$	$\eta = a, b$
A''	$\eta = b$	$\eta = a$	$\eta = a, b$

Table 6.4.: Combinations for an **ABB** molecule

Γ_{rve}	$K = 0$		K even	K odd
	N even	N odd		
A_1	$\eta = a$	$\eta = b$		
	$\Gamma_{vib} = A_1$		$\Gamma_{vib} = A_1$	$\Gamma_{vib} = B_2$
A_2	$\eta = b$	$\eta = a$		
	$\Gamma_{vib} = B_2$		$\Gamma_{vib} = B_2$	$\Gamma_{vib} = A_1$
B_1	$\eta = b$	$\eta = a$		
	$\Gamma_{vib} = A_1$		$\Gamma_{vib} = A_1$	$\Gamma_{vib} = B_2$
B_2	$\eta = a$	$\eta = b$		
	$\Gamma_{vib} = B_2$		$\Gamma_{vib} = B_2$	$\Gamma_{vib} = A_1$

We denote the different possibilities for the K-block dimensions in case of an **ABC** type molecule by

$$d^a = N_R^{(lim)} \cdot N_r^{(lim)} \cdot (v_2^a)^{(lim)} \tag{6.10}$$

$$d^b = N_R^{(lim)} \cdot N_r^{(lim)} \cdot (v_2^b)^{(lim)} \tag{6.11}$$

$$d = d^a + d^b \tag{6.12}$$

and for an **ABB** type molecule by

$$d^{a,A_1} = N_R^{(lim)} \cdot N_r^{(lim)} \cdot (v_2^{a,A_1})^{(lim)} \tag{6.13}$$

$$d^{b,A_1} = N_R^{(lim)} \cdot N_r^{(lim)} \cdot (v_2^{b,A_1})^{(lim)} \tag{6.14}$$

$$d^{A_1} = d^{a,A_1} + d^{b,A_1} \tag{6.15}$$

$$d^{a,B_2} = N_R^{(lim)} \cdot N_r^{(lim)} \cdot (v_2^{a,B_2})^{(lim)} \tag{6.16}$$

$$d^{b,B_2} = N_R^{(lim)} \cdot N_r^{(lim)} \cdot (v_2^{b,B_2})^{(lim)} \tag{6.17}$$

$$d^{B_2} = d^{a,B_2} + d^{b,B_2} \tag{6.18}$$

Algorithm 6.1 determines the size of a diagonal K-block using the above definitions and the Tables 6.2, 6.3 and 6.4:

Algorithm 6.1: Dimension of a diagonal K-block ($\Delta N = 0$, $\Delta K = 0$)

---

1   **function** $d = \dim_{\mathrm{K}}(\Gamma_{rve},\ N,\ K,\ N_R^{(lim)},\ N_r^{(lim)},\ (v_2^a)^{(lim)},\ (v_2^b)^{(lim)},\ \mathrm{ABC}\,)$

2   compute d, d^a, d^b, d^{a,A_1}, d^{b,A_1}, d^{A_1}, d^{a,B_2}, d^{b,B_2}, d^{B_2} ; /* ←cf. formulae (6.10) - (6.18) */

3   **if** *(ABC=true)* **then** ←————————————————————————**ABC** type molecule?

4   **if** *(K = 0)* **then**

5   **if** *(N* mod $2 = 0$*)* **then** ←————————————————————————*N* even?

6   **if** *($\Gamma_{rve} = A'$)* **then**

7 | **return** d^a

8   **else**

9 | **return** d^b

10   **end if**

11   **else**

12   **if** *($\Gamma_{rve} = A''$)* **then**

13 | **return** d^a

14   **else**

15 | **return** d^b

16   **end if**

17   **end if**

18   **else**

19 | **return** d

20   **end if**

21   **else**

22   **if** *($\Gamma_{rve} = A_1$ or $\Gamma_{rve} = B_2$)* **then**

23   **if** *(K = 0)* **then**

24   **if** *(N* mod $2 = 0$*)* **then**

25 | **return** d^{a,A_1}

26   **else**

27 | **return** d^{b,A_1}

28   **end if**

29   **else**

30   **if** *(K* mod $2 = 0$*)* **then**

31 | **return** d^{A_1}

32   **else**

33 | **return** d^{B_2}

34   **end if**

35   **end if**

36   **else**

37   **if** *(K = 0)* **then**

38   **if** *(N* mod $2 = 0$*)* **then**

39 | **return** d^{b,B_2}

40   **else**

41 | **return** d^{a,B_2}

42   **end if**

43   **else**

44   **if** *(K* mod $2 = 0$*)* **then**

45 | **return** d^{B_2}

46   **else**

47 | **return** d^{A_1}

48   **end if**

49   **end if**

50   **end if**

51 **end if**

This in turn enables us to determine the dimension of a diagonal N-block (Alg. 6.2) and the whole J-Block (Alg. 6.3):

Algorithm 6.2: Dimension of a diagonal N-block ($\Delta N = 0$)

1 **function** $d = \dim_N(\Gamma_{rve}, N, N_R^{(lim)}, N_r^{(lim)}, (v_2^a)^{(lim)}, (v_2^b)^{(lim)}, \text{ABC})$
2 $\quad d = 0$
3 $\quad$ **for** $K = 0, \ldots, N$ **do**
4 $\qquad | \quad d = d + \dim_K(\Gamma_{rve}, N, K, N_R^{(lim)}, N_r^{(lim)}, (v_2^a)^{(lim)}, (v_2^b)^{(lim)}, \text{ABC})$
5 $\quad$ **end for**
6 $\quad$ **return** d

Algorithm 6.3: Dimension of the Hamiltonian J-block $\mathbf{H}^{(J,S,\Gamma_{rve})}$

1 **function** $d = \dim_J(\Gamma_{rve}, S, J, N_R^{(lim)}, N_r^{(lim)}, (v_2^a)^{(lim)}, (v_2^b)^{(lim)}, \text{ABC})$
2 $\quad N_{min} = |J - S|, N_{max} = J + S$
3 $\quad d = 0$
4 $\quad$ **for** $N = N_{min}, \ldots, N_{max}$ **do**
5 $\qquad | \quad d = d + \dim_N(\Gamma_{rve}, N, N_R^{(lim)}, N_r^{(lim)}, (v_2^a)^{(lim)}, (v_2^b)^{(lim)}, \text{ABC})$
6 $\quad$ **end for**
7 $\quad$ **return** d

Besides, we can determine the starting position of a diagonal K-block with address (K, N), which will turn out useful later on, when we discuss algorithms for the matrix-vector multiplication:

Algorithm 6.4: Starting coordinate of a diagonal K-block within a J-block

1 **function** $i = \text{pos}_K(\Gamma_{rve}, S, J, N, K, N_R^{(lim)}, N_r^{(lim)}, (v_2^a)^{(lim)}, (v_2^b)^{(lim)}, \text{ABC})$
2 $\quad N_{min} = |J - S|$
3 $\quad i = 1$
4 $\quad$ **for** $N' = N_{min}, \ldots, N - 1$ **do**
5 $\qquad$ **for** $K' = 0, \ldots, N'$ **do**
6 $\qquad\quad | \quad i = i + \dim_K(\Gamma_{rve}, N', K', N_R^{(lim)}, N_r^{(lim)}, (v_2^a)^{(lim)}, (v_2^b)^{(lim)}, \text{ABC})$
7 $\qquad$ **end for**
8 $\quad$ **end for**
9 $\quad$ **for** $K' = 0, \ldots, K - 1$ **do**
10 $\qquad | \quad i = i + \dim_K(\Gamma_{rve}, N, K', N_R^{(lim)}, N_r^{(lim)}, (v_2^a)^{(lim)}, (v_2^b)^{(lim)}, \text{ABC})$
11 $\quad$ **end for**
12 $\quad$ **return** i

Algorithm 6.5: Starting coordinate of a diagonal N-block within a J-block

1 **function** $i = \text{pos}_N(\Gamma_{rve}, S, J, N, N_R^{(lim)}, N_r^{(lim)}, (v_2^a)^{(lim)}, (v_2^b)^{(lim)}, \text{ABC})$
2 $\quad i = \text{pos}_K(\Gamma_{rve}, S, J, N, 0, N_R^{(lim)}, N_r^{(lim)}, (v_2^a)^{(lim)}, (v_2^b)^{(lim)}, \text{ABC})$
3 $\quad$ **return** i

Using these auxiliary procedures one can finally give a precise specification of the matrix blocks introduced in Definition 6.6 in terms of absolute coordinates and dimensions:

Corollary 6.9 (Dimensions and positions of K- and N-blocks)
Any K-block with the address $(N', K'), (N'', K'')$ within a Hamiltonian J-block $\mathbf{H}^{(J,S,\Gamma_{rve})}$ is fully determined by

$$\mathbf{H}^{(J,S,\Gamma_{rve})}_{(N',K'),(N'',K'')} = \mathbf{H}^{(J,S,\Gamma_{rve})}[i : i + d' - 1, j : j + d'' - 1] \in \mathbb{R}^{d' \times d''} \qquad (6.19)$$

where

$$
\begin{aligned}
d' &= \dim_K(\,\Gamma_{rve},\, N',\, K',\, N_R^{(lim)},\, N_r^{(lim)},\, (v_2^a)^{(lim)},\, (v_2^b)^{(lim)},\, \text{ABC}\,) & (6.20) \\
d'' &= \dim_K(\,\Gamma_{rve},\, N'',\, K'',\, N_R^{(lim)},\, N_r^{(lim)},\, (v_2^a)^{(lim)},\, (v_2^b)^{(lim)},\, \text{ABC}\,) & (6.21) \\
i &= \text{pos}_K(\,S,\, J,\, \Gamma_{rve},\, N',\, K',\, N_R^{(lim)},\, N_r^{(lim)},\, (v_2^a)^{(lim)},\, (v_2^b)^{(lim)},\, \text{ABC}\,) & (6.22) \\
j &= \text{pos}_K(\,S,\, J,\, \Gamma_{rve},\, N'',\, K'',\, N_R^{(lim)},\, N_r^{(lim)},\, (v_2^a)^{(lim)},\, (v_2^b)^{(lim)},\, \text{ABC}\,) & (6.23)
\end{aligned}
$$

Using Algorithms 6.2 and 6.5 one obtains analogous assertions on N-blocks.

The knowledge of the dimensions of the diagonal K-blocks fully determines the dimensions of all K-blocks, the N-blocks and the whole Hamiltonian matrix block $\mathbf{H}^{(J,S,\Gamma_{rve})}$. In the following we demonstrate the consequences on the problem sizes for the basis sets defined in Table 6.5, which we will employ in our numerical experiments.

Table 6.5.: Big and small basis sets

Type	Molecule	Size	$N_R^{(lim)}$	$N_r^{(lim)}$	$(v_2^a)^{(lim)}$	$(v_2^b)^{(lim)}$
ABC	MgNC	big	6	16	31	31
ABC	MgNC	small	4	9	16	16
ABB	HOO	big	16	8	24	16

Table 6.6 shows the dimensions of the K-blocks and the whole J-block with respect to the big basis for the **HOO** molecule ($J = 5/2$, $S = 1/2$ and the four different symmetries), Table 6.7 gives a corresponding survey for the **MgNC** molecule (with respect to both, big and small basis).

Fig. 6.4 shows that the problem size (i.e. the dimension of a J-block) grows linearly with the J quantum number. Furthermore one can recognize the great gap between the problem sizes for the "small" and "big" basis set, which is due to the unfavorable product basis behavior discussed in Section 5.8.1.

Table 6.6.: J-block and K-block dimensions for $J = 5/2$, **HOO** molecule (big basis)

N	K	$\Gamma_{rve} = A_1$	$\Gamma_{rve} = A_2$	$\Gamma_{rve} = B_1$	$\Gamma_{rve} = B_2$
	0	1536	1024	1024	1536
2	1	2560	2560	2560	2560
	2	2560	2560	2560	2560
	0	1024	1536	1536	1024
3	1	2560	2560	2560	2560
	2	2560	2560	2560	2560
	3	2560	2560	2560	2560
total		15360	15360	15360	15360

Table 6.7.: J-block and K-block dimensions for $J = 5/2$, **MgNC** molecule

N	K	small basis		big basis	
		$\Gamma_{rve} = A'$	$\Gamma_{rve} = A''$	$\Gamma_{rve} = A'$	$\Gamma_{rve} = A''$
	0	1536	576	2976	2976
2	1	2560	1152	2976	5952
	2	2560	1152	5952	5952
	0	576	576	2976	2976
3	1	1152	1152	5952	5952
	2	1152	1152	5952	5952
	3	1152	1152	5952	5952
total		6912	6912	35712	35712

Figure 6.4.: Linear growth of J-block dimension (**MgCN** molecule)

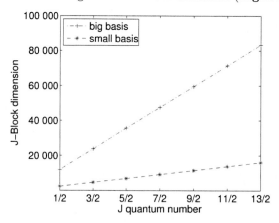

6.4.3. Hamiltonian Matrix Blocks

Now that it is known how to determine the block sizes, we need to describe how the block elements are actually computed. The following discussion is also of central importance for the construction of suitable preconditioners for the *Jacobi-Davidson method*, because it heavily relies on the construction principles for a Hamiltonian J-Block (see following chapter). Using the relation $\hat{N}^2 = \hat{N}_x^2 + \hat{N}_y^2 + \hat{N}_z^2$ one can derive the following decomposition of the Hamiltonian, (6.1) which is more appropriate for the construction of the Hamiltonian matrix blocks:

$$\hat{H} = \hat{H}_u + \hat{H}_p \tag{6.24}$$

$\hat{H}_u$ is the part of the Hamiltonian that produces diagonal K-blocks, which we will refer to as *basic information* and $\hat{H}_p$ contains perturbation terms that produce contributions to diagonal and off-diagonal K-blocks:

$$\hat{H}_u = \underbrace{\hat{H}_e}_{\text{potential}} + \underbrace{\hat{H}_{Pr} + \hat{H}_{PR}}_{\text{stretching motion}} + \underbrace{\hat{H}_b + \hat{H}_{ba} + \hat{H}_{bb}}_{\text{bending motion}} \tag{6.25}$$

$$\hat{H}_p = \underbrace{\hat{H}_{nk} + \hat{H}_{dk} + \hat{H}_{SO}}_{\text{perturbation terms}} \tag{6.26}$$

The corresponding terms are:

$$\hat{H}_{Pr} = \frac{1}{2\mu_r}\hat{P}_r^2 \tag{6.27}$$

$$\hat{H}_{PR} = \frac{1}{2\mu_R}\hat{P}_R^2 \tag{6.28}$$

$$\hat{H}_b = \left\{\frac{\hbar^2}{2\mu_R R^2} + \frac{\hbar^2}{2\mu_r r^2}\right\}\left\{\hat{N}_\tau^2 - \frac{1}{4}\left(1 + \frac{1}{\sin^2(\tau)}\right)\right\} \tag{6.29}$$

$$\hat{H}_{ba} = \left\{\frac{\hbar^2}{2\mu_R R^2}\cot^2(\tau)\right\}(\hat{N}_z^2 + \hat{L}_z^2 - 2\hat{N}_z\hat{L}_z) \tag{6.30}$$

$$\hat{H}_{bb} = \left\{\frac{\hbar^2}{2\mu_r r^2}\frac{1}{\sin^2(\tau)}\right\}(\hat{N}_z^2 + \hat{L}_z^2 - 2\hat{N}_z\hat{L}_z) \tag{6.31}$$

$$\hat{H}_{nk} = \frac{\hbar^2}{2\mu_R R^2}(\hat{N}^2 - \hat{N}_z^2) \tag{6.32}$$

$$\hat{H}_{dk} = \frac{\hbar^2}{2\mu_R R^2}\left\{\hat{N}_x\hat{N}_\tau + \hat{N}_\tau\hat{N}_x \right.$$

$$\left. + \frac{\hat{N}_y\hat{N}_z + \hat{N}_z\hat{N}_y - (\hat{N}_y\hat{L}_z + \hat{L}_z\hat{N}_y)}{\tan(\tau)}\right\} \tag{6.33}$$

$$\tag{6.34}$$

One can now define the multiplication operators

$$\hat{M}_{r^{-2}} = \frac{\hbar^2}{2\mu_r r^2} \tag{6.35}$$

$$\hat{M}_{R^{-2}} = \frac{\hbar^2}{2\mu_R R^2} \tag{6.36}$$

which are part of $\hat{H}_{nk}$ and $\hat{H}_{dk}$, $\hat{H}_b$, $\hat{H}_{ba}$ and $\hat{H}_{bb}$ and rewrite these operators as:

$$\hat{H}_b = \left\{ \hat{M}_{R^{-2}} + \hat{M}_{r^{-2}} \right\} \hat{H}'_b \tag{6.37}$$

$$\hat{H}_{ba} = \hat{M}_{R^{-2}} \hat{H}'_{ba} \tag{6.38}$$

$$\hat{H}_{bb} = \hat{M}_{r^{-2}} \hat{H}'_{bb} \tag{6.39}$$

$$\hat{H}_{dk} = \hat{M}_{R^{-2}} \hat{H}'_{dk} \tag{6.40}$$

$$\hat{H}_{nk} = \hat{M}_{R^{-2}} \hat{H}'_{nk} \tag{6.41}$$

This economizes the construction of matrix elements because $M_{r^{-2}}$ and $M_{R^{-2}}$ have matrix represenations of their own, which only need to be computed once (by means of Gauß-Laguerre integration, see [86]) and can be re-used repeatedly for the computation of matrix representations of the operators defined in (6.38)-(6.41). Note that the chaining of operators with respect to different coordinates is a tensor product of operators.

6.4.3.1. Basic diagonal K-blocks

The advantage of the decomposition of the *Double Renner Hamiltonian* in (6.24) lies in the following central fact (see [86] for details), which will also be of key importance in the derivation of suitable preconditioners for the *Jacobi-Davidson method* later on:

Fact 6.10 (Basic diagonal K-blocks)
The matrix representation $\mathbf{H}_u$ of $\widehat{H}_u$ (6.25) is block diagonal in K and N. The diagonal blocks $(\mathbf{H}_u)_{N,K}^{(S,J,\Gamma_{rve})}$ are referred to as basic K-blocks. *Since their elements only depend on the quantum numbers K, Γ_{vib} and η, it is appropriate to denote one such block by $\mathbf{B}^{K,\Gamma_{vib},\eta}$. Thus, it holds*

$$(\mathbf{H}_u)_{N,K}^{(S,J,\Gamma_{rve})} = \mathbf{B}^{K,\Gamma_{vib},\eta} \tag{6.42}$$

where Γ_{vib} and η are determined from the rules in the Tables 6.4 and 6.3. □

This directly leads to the following corollary:

Corollary 6.11 (Multiple occurences of basic K-blocks)
For spin multiplicity $S = 1/2$ (which is the case for both, the $\mathbf{MgCN}$ and the $\mathbf{HOO}$ molecule), basic K-blocks w.r.t. $K \leq N_{min} = |J - S|$ and $K \neq 0$ appear twice in $\mathbf{H}_u$. Analogous assertions can be made for spin multiplicities other than $S = 1/2$.

Example 6.12 $(J = 5/2$ and $S = 1/2)$
For $J = 5/2$ and $S = 1/2$ we have $N_{min} = 2$, i.e. the basic K-blocks with respect to $K = 1$ and $K = 2$ appear twice in $\mathbf{H}^{(1/2, 5/2, \Gamma_{rve})}$ (see also Table 6.1). □

As explained in Section 5.8.1 the kinetic energy operator (KEO) can be expressed as a sum of tensor products (5.108), which we now exploit to derive the matrix representation for $\hat{H}_u$ (see (6.25)). To this end one first determines the matrix representations for operators related to one particular vibrational coordinate (and the corresponding quantum number), which are listed in Table 6.8, where $v_2^{(lim)}$ is the appropriate size of the bending basis (according to the rules in the Tables 6.2, 6.4 and 6.3). In analogy to the discussion of the general case in Section 5.8.1, where the operator tensor product (5.108) carries over to the Kronecker product (5.117), the finite basis representation (FBR) of $\hat{H}_u$ (6.25) with respect to the basis set defined in Notation 6.3 is now obtained as the following expression by adding the terms listed in Table 6.9:

$$
\begin{aligned}
\mathbf{B}^{K,\Gamma_{vib},\eta} &= \mathbf{H}_e + \mathbf{H}_{PR} + \mathbf{H}_{Pr} + \mathbf{H}_{ba} + \mathbf{H}_{bb} + \mathbf{H}_b \\
&= \mathbf{H}_e + \mathbf{H}'_{PR} \otimes \mathbf{I}_{N_r^{(lim)}} \otimes \mathbf{I}_{v_2^{(lim)}} + \mathbf{I}_{N_R^{(lim)}} \otimes \mathbf{H}'_{Pr} \otimes \mathbf{I}_{v_2^{(lim)}} \\
&\quad + \mathbf{M}_{R^{-2}} \otimes \mathbf{I}_{N_r^{(lim)}} \otimes \mathbf{H}'_{ba} + \mathbf{I}_{N_R^{(lim)}} \otimes \mathbf{M}_{r^{-2}} \otimes \mathbf{H}'_{bb} \\
&\quad + \mathbf{M}_{R^{-2}} \otimes \mathbf{I}_{N_r^{(lim)}} \otimes \mathbf{H}'_b + \mathbf{I}_{N_R^{(lim)}} \otimes \mathbf{M}_{r^{-2}} \otimes \mathbf{H}'_b
\end{aligned}
\tag{6.43}
$$

Unfortunately, due to the rather complex structure of the potential energy surface (PES) the matrix related matrix term $\mathbf{H}_e$ cannot be written as a Kronecker product, which makes it impossible to exploit the product basis structure in the design of a routine for matrix-vector multiplication (see Scenarios 1 and 2 in Section 5.8.1). Furthermore, the computation of its matrix elements is a time-consuming affair, because their computation involves simultaneous numerical integration with respect to the three vibrational *Jacobi coordinates* (R, r, τ) (see [86] for details). Since the the entries of the potential matrix $\mathbf{H}_e$ are in general non-zero, the basic diagonal K-blocks are dense.

Remark 6.13 (Dependence of the matrix components on K, Γ_{vib} and η)
Table 6.8 shows that the dependence of the basic diagonal K-blocks $\mathbf{B}^{K,\Gamma_{vib},\eta}$ on the quantum numbers K, Γ_{vib} and η is due to the operators related to the bending motion ($\hat{H}_b$, $\hat{H}_{ba}$ and $\hat{H}_{bb}$) and the potential $\hat{H}_e$, whereas the operators related to the stretching motion (H_{PR}, H_{Pr}, $M_{R^{-2}}$ and $M_{r^{-2}}$) exclusively depend on the stretching quantum numbers N_R and N_r. If it is necessary to point out these dependencies, we denote the arising matrices with the corresponding superscripts, e.g. $\mathbf{H}_b^{K,\Gamma_{vib}}$ and $\mathbf{H}_e^{K,\Gamma_{vib},\eta}$ (analogous to the notation for the basic diagonal K-blocks). □

Table 6.8.: Matrix components for the construction of a basic K-block $\mathbf{B}^{K,\Gamma_{vib},\eta}$

Operator	Formula	Matrix	Dimension	Dependence					
				K	Γ_{vib}	η	N_R	N_r	v_2^η
$\hat{H}'_b$	(6.37)	$\mathbf{H}'_b$	$v_2^{(lim)} \times v_2^{(lim)}$	✓	✓				✓
$\hat{H}'_{ba}$	(6.38)	$\mathbf{H}'_{ba}$	$v_2^{(lim)} \times v_2^{(lim)}$	✓	✓				✓
$\hat{H}'_{bb}$	(6.39)	$\mathbf{H}'_{bb}$	$v_2^{(lim)} \times v_2^{(lim)}$	✓	✓				✓
$\hat{H}'_{PR}$	(6.28)	$\mathbf{H}'_{PR}$	$N_R^{(lim)} \times N_R^{(lim)}$				✓		
$\hat{H}'_{Pr}$	(6.27)	$\mathbf{H}'_{Pr}$	$N_r^{(lim)} \times N_r^{(lim)}$					✓	
$\hat{M}_{R-2}$	(6.36)	$\mathbf{M}_{R-2}$	$N_R^{(lim)} \times N_R^{(lim)}$				✓		
$\hat{M}_{r-2}$	(6.35)	$\mathbf{M}_{r-2}$	$N_r^{(lim)} \times N_r^{(lim)}$					✓	
$\hat{H}_e$		$\mathbf{H}_e$	$(N_R^{(lim)} \cdot N_r^{(lim)} \cdot v_2^{(lim)})^2$	✓	✓	✓	✓	✓	✓

Table 6.9.: Kronecker product representations of the terms in $\hat{H}_u$

Operator	Formula	Matrix	Kronecker Product					
$\hat{H}_b$	(6.37)	$\mathbf{H}_b$	$\mathbf{M}_{R-2}$	$\otimes$	$\mathbf{I}_{N_r^{(lim)}}$	$\otimes$	$\mathbf{H}'_b$	
			$+$ $\mathbf{I}_{N_R^{(lim)}}$	$\otimes$	$\mathbf{M}_{r-2}$	$\otimes$	$\mathbf{H}'_b$	
$\hat{H}_{ba}$	(6.38)	$\mathbf{H}_{ba}$	$\mathbf{M}_{R-2}$	$\otimes$	$\mathbf{I}_{N_r^{(lim)}}$	$\otimes$	$\mathbf{H}'_{ba}$	
$\hat{H}_{bb}$	(6.39)	$\mathbf{H}_{bb}$	$\mathbf{I}_{N_R^{(lim)}}$	$\otimes$	$\mathbf{M}_{r-2}$	$\otimes$	$\mathbf{H}'_{bb}$	
$\hat{H}_{Pr}$	(6.27)	$\mathbf{H}_{Pr}$	$\mathbf{I}_{N_R^{(lim)}}$	$\otimes$	$\mathbf{H}'_{Pr}$	$\otimes$	$\mathbf{I}_{v_2^{(lim)}}$	
$\hat{H}_{PR}$	(6.28)	$\mathbf{H}_{PR}$	$\mathbf{H}'_{PR}$	$\otimes$	$\mathbf{I}_{N_r^{(lim)}}$	$\otimes$	$\mathbf{I}_{v_2^{(lim)}}$	

6.4.3.2. Diagonal and Off-Diagonal Perturbation K-blocks

Let us now discuss the matrix elements related to the perturbation part $\hat{H}_p$ of the Hamiltonian. Table 6.10 summarizes the block sparsity rules for the occurences of the matrix blocks arising from $\hat{H}_p$ defined in (6.26). We see that the operators $\hat{H}_{so}$ and $\hat{H}_{nk}$ do not only produce off-diagonal matrix blocks, but also involve perturbations to the diagonal basic K-blocks, whereas $\hat{H}_{dk}$ exclusively has off-diagonal matrix representations. Table 6.12 shows what these rules imply for our standard example ($S = 1/2$ and $J = 5/2$). Analogous to the way one proceeds for the basic diagonal K-blocks one can exploit for the computation of the matrix blocks that the multiplication operator $\hat{M}_{R-2}$ (6.36) is part of $\hat{H}_{nk}$ and $\hat{H}_{nk}$. This is reflected in the Kronecker product expressions in Table 6.11 (see [86] for derivation and details), which induce regular sparsity patterns for the off-diagonal matrix blocks originating from $\hat{H}_{so}$ and $\hat{H}_{dk}$ because of the involved identity matrices. We will discuss in more detail later on how to take advantage of the sparsity for the design of an efficient matrix-vector multiplication algorithm and for compact storage.

Table 6.10.: Block sparsity rules for perturbation K-blocks

Operator	Occurences of matrix blocks	
$\hat{H}_{nk}$	$\Delta K = 0$	$\Delta N = 0$
$\hat{H}_{dk}$	$\Delta K = 1$	$\Delta N = 0$
$\hat{H}_{so}$	$\Delta K = 0$	$\Delta N = 0, 1$

Table 6.11.: Kronecker product representations of the terms in $\hat{H}_p$

Operator	Formula	Matrix	Kronecker Product	
$\hat{H}_{dk}$	(6.40)	$\mathbf{H}_{dk}$		$\mathbf{M}_{R-2} \otimes \mathbf{I}_{N_r^{(lim)}} \otimes \mathbf{H}'_{dk}$
$\hat{H}_{nk}$	(6.41)	$\mathbf{H}_{nk}$	$f(N, K)$	$\cdot\, \mathbf{M}_{R-2} \otimes \mathbf{I}_{N_r^{(lim)}} \otimes \mathbf{I}_{v_2^{(lim)}}$
$\hat{H}_{so}$		$\mathbf{H}_{so}$	$g(J, S, N', N'', K')$	$\cdot\, \mathbf{I}_{N_R^{(lim)}} \otimes \mathbf{I}_{N_r^{(lim)}} \otimes \mathbf{H}'_{so}$

The functions f and g in Table 6.11 are defined by

$$f(N, K) = \hbar\{\, N(N+1) - K^2 \,\} \tag{6.44}$$

$$g(J, S, N', N'', K') = (-1)^{N'+N''+S+J-K'}$$
$$\sqrt{(2S+1)S(S+1)(2N'+1)(2N''+1)}$$
$$\times \begin{pmatrix} N' & 1 & N'' \\ -K' & 0 & K'' \end{pmatrix} \begin{Bmatrix} N' & S & J \\ S & N'' & 1 \end{Bmatrix} \tag{6.45}$$

where the expression in the parentheses is a $3j$-symbol (cf. [133]) and the expression in curly braces is a $6j$-symbol (cf. [133]).

Remark 6.14 (Dependencies of the perturbation terms)
The matrix components $\mathbf{H}'_{so}$ and $\mathbf{H}'_{dk}$ have the following dependencies (see [86]):

- $\mathbf{H}'_{so}$ *depends on K, Γ_{vib} and the bending quantum number v_2^η*
  A comparison with the terms $\mathbf{H}'_b$, $\mathbf{H}'_{ba}$ and $\mathbf{H}'_{bb}$ in Table 6.9 reveals that $\mathbf{H}'_{so}$ has the same dependencies, which is exploited in the construction of the Hamiltonian matrix $\mathbf{H}^{(J,S,\Gamma_{rve})}$ in the program DR (see [86]).

- $\mathbf{H}'_{dk}$ *depends on N, K', K'' and v_2^η*

$\square$

Table 6.12.: Occurences of perturbation blocks in $\mathbf{H}^{(S,J,\Gamma_{rve})}$ $(S = 1/2,\ J = 5/2)$

| | | N = 2 | | | N = 3 | | | |
		K = 0	K = 1	K = 2	K = 0	K = 1	K = 2	K = 3
N = 2	K = 0	so,nk	dk		so			
	K = 1	dk	so,nk	dk		so		
	K = 2		dk	so,nk			so	
N = 3	K = 0	so			so,nk	dk		
	K = 1		so		dk	so,nk	dk	
	K = 2			so		dk	so,nk	dk
	K = 3						dk	so,nk

6.4.4. Construction of the Hamiltonian Matrix

We now collect the explanations of the preceding sections and summarize the construction of a Hamiltonian J-block $\mathbf{H}^{(J,S,\Gamma_{rve})}$ in the form of two algorithms. Before we do so, we make the following useful definition:

Definition 6.15 (DIAG-, DK- and SO-blocks)
We call

- *the diagonal K-blocks of the final Hamiltonian matrix $\mathbf{H}^{(J,S,\Gamma_{rve})}$ DIAG blocks. These are the sum of basic diagonal K-blocks and diagonal perturbation blocks originating from the operators $\hat{H}_{so}$ and $\hat{H}_{nk}$.*

- *the matrix blocks originating from the operator $\hat{H}_{dk}$ DK-blocks.*

- *the off-diagonal K-blocks originating from the operator $\hat{H}_{so}$ SO-blocks. Note that the diagonal SO-blocks (cf. Table 6.12) are already comprised in the DIAG-blocks.*

The block sparsity of the Hamiltonian matrix $\mathbf{H}^{(J,S,\Gamma_{rve})}$ *(i.e. the occurrences of these blocks) is determined by the rules in Table 6.13.*

We have already seen that the basic diagonal K-blocks are the fundamental elements in the construction of Hamiltonian matrices. If, for instance, one is interested in the construction of Hamiltonian J-blocks $\mathbf{H}^{(J,S,\Gamma_{rve})}$ with respect to all molecular symmetries Γ_{rve} and all J quantum numbers less or equal than a predefined bound J_{max}, then one can take advantage of the fact they turn up repeatedly for equal K quantum numbers greater than zero (cf. Corollary 6.11). The program DR by ODAKA [86] economizes the J-block construction by computing all possible combinations of K, Γ_{vib} and η for the diagonal K-blocks (see Table 6.2) related to $K \leq J_{max} + S = N_{max}$ once and storing them on disk. Whenever a specific $\mathbf{B}^{K,\Gamma_{vib},\eta}$ is required in the course of the construction, it can be loaded into memory. Furthermore, it pays off to store the matrices $\mathbf{H}_{PR}$, $\mathbf{H}_{Pr}$, $\mathbf{M}_{R^{-2}}$ and $\mathbf{M}_{r^{-2}}$. These ideas are summarized in Algorithm 6.6, where ABC is a boolean parameter that indicates whether or not the molecule under consideration is **ABC** type. Note that the dimensions of the arising matrices are either determined by Table 6.8 or by Algorithm 6.1.

Algorithm 6.6: Construction of a basic diagonal K-block

1   **procedure Construct-basic-K**(S, J_{max}, $N_R^{(lim)}$, $N_r^{(lim)}$, $(v_2^a)^{(lim)}$, $(v_2^b)^{(lim)}$, ABC)

2   construct matrix representations for operators that only depend on the stretching quantum numbers N_R or N_r (see Table 6.8):

3   (a) $\mathbf{H}_{PR}$

4   (b) $\mathbf{H}_{Pr}$

5   (c) $\mathbf{M}_{R^{-2}}$

6   (d) $\mathbf{M}_{r^{-2}}$

7   $N_{max} = J_{max} + S$

8   **for** $K = 0, \ldots, N_{max}$ **do**

9   **for all** Γ_{vib} **do**

10   construct matrix representations for operators that depend on K and Γ_{vib}:

11   (a) $(\mathbf{H}'_b)^{K,\Gamma_{vib}}$

12   (b) $(\mathbf{H}'_{ba})^{K,\Gamma_{vib}}$

13   (c) $(\mathbf{H}'_{bb})^{K,\Gamma_{vib}}$

14   (d) $(\mathbf{H}'_{so})^{K,\Gamma_{vib}}$

15   **for** $\eta = a, b$ **do**

16   construct matrix representations for operators depending on K, Γ_{vib}, η:

17   (a) $\mathbf{H}_e^{K,\Gamma_{vib},\eta}$ (potential matrix)

18   (b) $\mathbf{B}^{K,\Gamma_{vib},\eta}$ as per (6.43)   (diagonal basic K-block)

19   **end for**

20   **end for**

21 **end for**

Finally, Algorithm 6.7 summarizes how *one* Hamiltonian J-block $\mathbf{H}^{(J,S,\Gamma_{rve})}$ related to a fixed spin multiplicity S, a fixed molecular symmetry Γ_{rve} and a fixed rotational quantum number J is constructed. Essentially, it adds perturbation terms to the basic diagonal K-blocks resulting in the final DIAG-blocks and constructs the off-diagonal DK- and SO-blocks. Note that the absolute block positions and dimensions are implicitly determined by the block addresses according to Corollary 6.9.

Algorithm 6.7: Construction of the Hamiltonian J-block $\mathbf{H}^{(J,S,\Gamma_{rve})}$

---

1 **procedure Construct-J-Block**(Γ_{rve}, S, J, $N_R^{(lim)}$, $N_r^{(lim)}$, $(v_2^a)^{(lim)}$, $(v_2^b)^{(lim)}$, ABC)

2 **if** *not yet computed* **then**

3 $\quad$ **Construct-basic-K**(S, J, $N_R^{(lim)}$, $N_r^{(lim)}$, $(v_2^a)^{(lim)}$, $(v_2^b)^{(lim)}$, ABC)

4 **end if**

5 $N_{min} = |J - S|$, $N_{max} = J + S$

6 **for** $N' = N_{min}, \ldots, N_{max}$ **do**

7 $\quad$ **for** $K' = 0, \ldots, N'$ **do**

8 $\quad\quad$ **for** $N'' = N_{min}, \ldots, N_{max}$ **do**

9 $\quad\quad\quad$ **for** $K'' = 0, \ldots, N''$ **do**

10 $\quad\quad\quad\quad$ $\Delta N = |N' - N''|$, $\Delta K = |K' - K''|$

$\quad\quad\quad\quad$ /* DIAG-Block ? */

11 $\quad\quad\quad\quad$ **if** $(\Delta N = 0$ **and** $\Delta K = 0)$ **then**

12 $\quad\quad\quad\quad\quad$ determine Γ_{vib} and η as a function of N, K, Γ_{rve}

$\quad\quad\quad\quad\quad$ from Tables 6.3 resp. 6.4

13 $\quad\quad\quad\quad\quad$ $(\mathbf{H}_{so})_{(N',K'),(N'',K'')} = g(J, S, N', N'', K') \cdot I_{N_R^{(lim)}} \otimes I_{N_r^{(lim)}} \otimes (\mathbf{H}'_{SO})^{K,\Gamma_{vib}}$

14 $\quad\quad\quad\quad\quad$ $(\mathbf{H}_{nk})_{(N',K'),(N'',K'')} = f(N,K) \cdot \mathbf{M}_{R-2} \otimes I_{N_r^{(lim)}} \otimes I_{v_2^{(lim)}}$

15 $\quad\quad\quad\quad\quad$ $\mathbf{H}_{(N',K'),(N'',K'')} =$
$\quad\quad\quad\quad\quad\quad\quad \mathbf{B}^{K,\Gamma_{vib},\eta} + (\mathbf{H}_{nk})_{(N',K'),(N'',K'')} + (\mathbf{H}_{so})_{(N',K'),(N'',K'')}$

16 $\quad\quad\quad\quad$ **end if**

$\quad\quad\quad\quad$ /* DK-Block ? */

17 $\quad\quad\quad\quad$ **if** $(\Delta N = 0$ **and** $\Delta K = 1)$ **then**

18 $\quad\quad\quad\quad\quad$ $\mathbf{H}_{(N',K'),(N'',K'')} = \mathbf{M}_{R-2} \otimes I_{N_r^{(lim)}} \otimes (\mathbf{H}'_{dk})^{N',K',K'',v_2^\eta}$

19 $\quad\quad\quad\quad$ **end if**

$\quad\quad\quad\quad$ /* SO-Block ? */

20 $\quad\quad\quad\quad$ **if** $(\Delta N = 1$ **and** $\Delta K = 0)$ **then**

21 $\quad\quad\quad\quad\quad$ $\mathbf{H}_{(N',K'),(N'',K'')} = g(J, S, N', N'', K') \cdot I_{N_R^{(lim)}} \otimes I_{N_r^{(lim)}} \otimes (\mathbf{H}'_{SO})^{K,\Gamma_{vib}}$

22 $\quad\quad\quad\quad$ **end if**

23 $\quad\quad\quad$ **end for**

24 $\quad\quad$ **end for**

25 $\quad$ **end for**

26 **end for**

Figure 6.5 shows the typical sparsity pattern of a Hamiltonian matrix block for our standard example ($S = 1/2$, $J = 5/2$). The sparsity of the off-diagonal DK- and SO-blocks will be analyzed in depth later on.

Figure 6.5.: Sparsity pattern of the Hamiltonian matrix $\mathbf{H}^{(J,S,\Gamma_{rve})}$ ($S = 1/2$, $J = 5/2$)

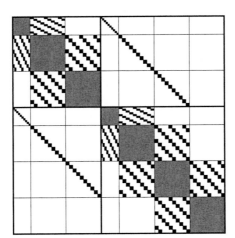

Table 6.13.: Block sparsity rules

Term	Rule for K	Rule for N
DIAG-Block	$\Delta K = 0$	$\Delta N = 0$
DK-Block	$\Delta K = 1$	$\Delta N = 0$
SO-Block	$\Delta K = 0$	$\Delta N = 1$

Remark 6.16 (Construction of J-blocks in the program DR)
As already explained above the program DR ([86]) does not only construct one specific J-block, but all Hamiltonian matrices $\mathbf{H}^{(J,S,\Gamma_{rve})}$ for $J_{min} \leq J \leq J_{max}$ and all symmetries Γ_{vib}. This is described in Algorithm 6.8: □

Algorithm 6.8: Construction of all Hamiltonian J-blocks $\mathbf{H}^{(J,S,\Gamma_{rve})}$

---

1 **procedure All-J-Blocks**(S, J_{min}, J_{max}, $N_R^{(lim)}$, $N_r^{(lim)}$, $(v_2^a)^{(lim)}$, $(v_2^b)^{(lim)}$, ABC)
2 **Construct-basic-K**(S, J_{max}, $N_R^{(lim)}$, $N_r^{(lim)}$, $(v_2^a)^{(lim)}$, $(v_2^b)^{(lim)}$, ABC)
3 **for** $J = J_{min}, \ldots, J_{max}$ **do**
4   **for all** Γ_{rve} **do**
5   **Construct-J-Block**(Γ_{rve}, S, J, $N_R^{(lim)}$, $N_r^{(lim)}$, $(v_2^a)^{(lim)}$, $(v_2^b)^{(lim)}$, ABC)
6   **end for**
7 **end for**

Part III.

Application to the Problem

7. Eigensolvers for the Computation of Rovibronic Energy Levels

In this chapter we come back to the main concern of this thesis, the computation of eigenvalues of the Hamiltonian J-Blocks $\mathbf{H}^{(J,S,\Gamma_{rve})}$, which correspond to the rovibronic energy levels of the molecule under consideration. We will especially examine how to apply the JDQR variants for computing several eigenpairs (see Section 4.3). To do so, we have to provide

- a problem specific procedure for matrix-vector multiplication

- suitable and efficient preconditioners

The knowledge of these algorithmic ingredients allows to briefly highlight the use of alternative iterative projection methods, such as the IRLM (Alg. 3.16), *Davidson's method* (Alg. 3.17) and *Olsen's method* (see Section 4.2.2.3). Furthermore, we will discuss the basis contraction scheme for the *Double Renner Hamiltonian* discussed in [86] and the four general computational strategies presented in Section 5.8.3, the main focus being on the product basis problems. In this context we will contrast the most efficient direct eigensolver (combination of two-stage tridiagonalization and *RRR method* for tridiagonal matrices) with the preconditioned standard JDQR method. Finally, we comment on the shared memory parallelization of our JDQR implementation using OpenMP™ ([3],[22]).

7.1. Matrix Properties and Specification of the Eigenproblem

Before we explain in more detail the algorithmic ingredients (matrix-vector multiplication, contraction scheme and preconditioners) let us briefly collect the most important properties of a Hamiltonian matrix block $\mathbf{H}^{(J,S,\Gamma_{rve})}$:

Fact 7.1 (Properties of $\mathbf{H}^{(J,S,\Gamma_{rve})}$)
- $\mathbf{H}^{(J,S,\Gamma_{rve})}$ *is positive definite.*
  This is a consequence of its construction in the program DR (see [86]), where a physically motivated zero point energy is introduced such that all eigenvalues are positive.

- $\mathbf{H}^{(J,S,\Gamma_{rve})}$ *is "block diagonally dominant" with respect to its diagonal K-blocks $\mathbf{H}_{N,K}^{(J,S,\Gamma_{rve})}$, in the sense that the Frobenius norm of a diagonal K-block is greater* *than the sum of the Frobenius norms of the off-diagonal K-blocks in the same block row. This can be recognized from the Tables 7.1, 7.2 and 7.3, where the Frobenius norms of the arising DIAG-, DK- and SO-blocks for $J = 1/2, \ldots, J = 13/2$ are listed (**MgCN** molecule, big basis).*

- *The diagonal K-blocks $\mathbf{H}_{N,K}^{(J,S,\Gamma_{rve})}$ result from very tiny perturbations of the basic K-blocks $\mathbf{B}^{K,\Gamma_{vib},\eta}$ (cf. Line 15 in Algorithm 6.7).*
  The additional NK- and SO-terms only slightly change the Frobenius norm of the basic K-blocks (see Table 7.1). This property will turn out important for the design of the contraction scheme and for the construction of effective preconditioners.

- *$\mathbf{H}^{(J,S,\Gamma_{rve})}$ exhibits a block sparsity pattern according to the rules in Table 6.13. However, the matrix in total is neither sparse nor dense in the classical sense, i.e. the diagonal K-blocks $\mathbf{H}_{N,K}^{(J,S,\Gamma_{rve})}$ are dense whereas the off-diagonal DK- and SO-blocks are sparse (cf. Fig. 6.5). Their sparsity structure will be analyzed in Section 7.2.1.*

□

The following definition is useful for the specification of the eigenvalue problems we are interested in:

Definition 7.2 (Hartree and cm^{-1})
In quantum chemical calculations it is common to use the atomic unit of energy

$$E_h = \frac{m_e e^4}{\hbar^2} \tag{7.1}$$

which is referred to as hartree (see Table 5.1 for a survey of some constants of nature). To obtain the results in this unit one divides both sides of the Schrödinger equation by E_h. In spectroscopical considerations it is sometimes more appropriate to express energies in terms of wavenumbers, *the unit being cm^{-1}. The conversion factor between energies in hartree and the corresponding wavenumbers in cm^{-1} is*

$$conv = 4.556333827 \cdot 10^{-6} \frac{hartree}{cm^{-1}} \tag{7.2}$$

The matrices we are dealing with are produced by the software DR ([86]) and based on hartree. Consequently, the unit of the eigenvalues obtained in our computations is hartree as well and to obtain the corresponding values in wavenumbers one has to divide by the conversion factor (7.2).

Table 7.1.: Frobenius norms of the DIAG-blocks

J	N	$K=0$	$K=1$	$K=2$	$K=3$	$K=4$	$K=5$	$K=6$	$K=7$
1/2	0	128.05578							
	1	134.04131	186.19294						
3/2	1	134.04131	186.19306						
	2	128.05624	186.19342	171.54983					
5/2	2	128.05624	186.19348	171.54995					
	3	134.04210	186.19408	171.55051	157.66387				
7/2	3	134.04209	186.19413	171.55059	157.66399				
	4	128.05731	186.19496	171.55139	157.66476	144.52929			
9/2	4	128.05731	186.19500	171.55145	157.66485	144.52940			
	5	134.04349	186.19606	171.55248	157.66585	144.53039	132.19880		
11/2	5	134.04349	186.19608	171.55253	157.66593	144.53048	132.19891		
	6	128.05899	186.19737	171.55379	157.66717	144.53171	132.20013	120.68041	
13/2	6	128.05899	186.19739	171.55384	157.66723	144.53179	132.20022	120.68051	
	7	134.04552	186.19890	171.55532	157.66869	144.53324	132.20167	120.68198	109.92012
K Basic		134.04115	186.19291	171.54969	157.66361	144.52892	132.19831	120.67980	109.91938

Table 7.2.: Frobenius norms of the DK-blocks

J	N	$K=0$	$K=1$	$K=2$	$K=3$	$K=4$	$K=5$	$K=6$	$K=7$
1/2	0								
	1		0.02011						
3/2	1		0.02012						
	2		0.03250	0.02828					
5/2	2		0.03250	0.02828					
	3		0.04928	0.04472	0.03577				
7/2	3		0.04928	0.04472	0.03577				
	4		0.05934	0.05999	0.05464	0.04256			
9/2	4		0.05933	0.05999	0.05464	0.04256			
	5		0.07791	0.07482	0.07154	0.06384	0.04895		
11/2	5		0.07791	0.07482	0.07154	0.06384	0.04895		
	6		0.08599	0.08943	0.08762	0.08242	0.07261	0.05508	
13/2	6		0.08599	0.08943	0.08762	0.08242	0.07261	0.05508	
	7		0.10645	0.10391	0.10326	0.09982	0.09288	0.08107	0.06102

Table 7.3.: Frobenius norms of the SO-blocks

J	N	$K=0$	$K=1$	$K=2$	$K=3$	$K=4$	$K=5$	$K=6$	$K=7$
1/2	0								
	1	0.00445							
3/2	1								
	2	0.00445	0.00544						
5/2	2								
	3	0.00445	0.00592	0.00469					
7/2	3								
	4	0.00445	0.00608	0.00545	0.00417				
9/2	4								
	5	0.00445	0.00615	0.00577	0.00505	0.00379			
11/2	5								
	6	0.00445	0.00619	0.00594	0.00546	0.00471	0.00350		
13/2	6								
	7	0.00445	0.00622	0.00603	0.00570	0.00519	0.00443	0.00326	

The following definition specifies what range of the spectrum one is typically interested in:

Definition 7.3 (Specification of the eigenvalue problem)
We consider the eigendecomposition

$$\mathbf{H}^{(J,S,\Gamma_{rve})} = \mathbf{V}^* \mathbf{\Lambda} \mathbf{V} \tag{7.3}$$

As usually we assume the eigenvalues λ_i in $\mathbf{\Lambda} = diag(\lambda_1, \ldots, \lambda_n)$ to be ordered by ascending magnitude. The partial eigensystem of interest is specified by an energy limit E_{max} (typical values for E_{max} in practical computations may be 2500 cm^{-1}, 5000 cm^{-1} etc.), i.e. we are looking for eigenpairs (λ_i, v_i) where $\lambda_i \leq E_{max}$.

How large one actually chooses the energy limit E_{max} in Def. 7.3 depends on the purpose of the calculation and the size of the product basis. The eigenvalues of the Hamiltonian matrix $\mathbf{H}^{(J,S,\Gamma_{rve})}$, which are known to be upper bounds to the true eigenvalues of the Hamiltonian $\widehat{H}$ (see Thm. 5.45), are only good approximations below a certain energy limit, which is difficult to predict in practical situations. To make sure that the eigenapproximations near E_{max} are appropriate one actually has to run a reference computation with a larger basis. Figure 7.1 demonstrates the potential danger of choosing the basis too small (as usually we use the bases defined in Table 6.5).

In general, for fixed sizes of the vibrational bases the number of eigenvalues below a fixed energy limit E_{max} increases with the rotational quantum number J as can be recognized from Table 7.4 and can amount to several thousands for larger values of J. We will see that – in principle – the preconditioned standard JDQR variant (Alg. 4.7) is capable to cope with this relatively large number of sought-after eigenpairs. However, in order to better describe the complexity of the problem depending on J and for reasons that will become evident later on we will often restrict ourselves to partial eigensystems with a fixed number k of eigenpairs, say $k = 200$.

Table 7.4.: Number of eigenvalues below $E_{max} = 5000$ cm^{-1} (**MgCN**, big basis)

J	Problem size n	Storage	# of $\lambda_i \leq E_{max}$
1/2	11904	1.06 G	192
3/2	23808	4.22 G	374
5/2	35712	9.50 G	531
7/2	47616	16.89 G	668
9/2	59520	26.39 G	786
11/2	71424	38.00 G	n.a.

Figure 7.1.: Spectra of $\mathbf{H}^{(1/2,\,1/2,\,A')}$ for big and small basis

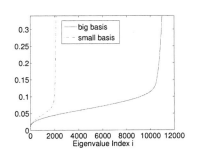

7.2. Matrix-Vector Multiplication and Storage Scheme

We have already pointed out that, unfortunately, it is not possible to directly take advantage of the product basis structure , because the potential matrix cannot be written as a sum of Kronecker products, which impedes the application of the ideas described in Secenario 1 in Section 5.8.1. Thus, it only remains to exploit the sparsity of the Hamiltonian matrix block and a first straightforward but rather naïve approach exploits the block sparsity rules in Table 6.13 which results in Algorithm 7.1. However, this procedure does not take advantage of the sparsity of the DK- and SO-blocks and working with the addresses $(N', K'), (N'', K'')$ (see Def. 6.6) to access the K-blocks is not well-suited for the implementation in a computer code. For this reason we derive a storage scheme for the Hamiltonian matrix blocks that allows to access the K-blocks efficiently and which exploits the sparsity of the off-diagonal matrix blocks.

7.2.1. Sparsity of the Off-Diagonal Hamiltonian Matrix Blocks

Unlike the DIAG blocks an off-diagonal DK- or SO-block $\mathbf{H}_{(N',K'),(N'',K'')} \in \mathbb{R}^{d' \times d''}$ need not necessarily be a square matrix. To facilitate notation we define the index sets

$$I' = \{1, \ldots, d'\} \tag{7.4}$$
$$I'' = \{1, \ldots, d''\} \tag{7.5}$$

and the auxiliary variables

$$v_2^{(lim)'} = d' / (N_R^{(lim)} \cdot N_r^{(lim)}) \tag{7.6}$$
$$v_2^{(lim)''} = d'' / (N_R^{(lim)} \cdot N_r^{(lim)}) \tag{7.7}$$

each of which can take three different values in case of an **ABC** type molecule ((6.10) - (6.12)) and six different values in case of an **ABB** molecule ((6.13) - (6.18)), respectively.

Algorithm 7.1: Matrix-vector multiplication exploiting block sparsity (naïve)

```
 1  function y = mult( H, x, Γ_rve, S, J, N_R^(lim), N_r^(lim), (v_2^a)^(lim), (v_2^b)^(lim), ABC )
 2  y = 0
 3  N_min = |J − S|, N_max = J + S
 4  for N' = N_min, . . . , N_max do
 5    for K' = 0, . . . , N' do
 6      i = pos_K( Γ_rve, S, J, N', K', N_R^(lim), N_r^(lim), (v_2^a)^(lim), (v_2^b)^(lim), ABC )
 7      k = dim_K( Γ_rve, N', K', N_R^(lim), N_r^(lim), (v_2^a)^(lim), (v_2^b)^(lim), ABC )
 8      for N'' = N_min, . . . , N_max do
 9        for K'' = 0, . . . , N'' do
10          j = pos_K( Γ_rve, S, J, N'', K'', N_R^(lim), N_r^(lim), (v_2^a)^(lim), (v_2^b)^(lim), ABC )
11          l = dim_K( Γ_rve, N'', K'', N_R^(lim), N_r^(lim), (v_2^a)^(lim), (v_2^b)^(lim), ABC )
12          ΔN = | N' − N'' |, ΔK = | K' − K'' |
            /* DIAG-block ?                                                      */
13          if ( ΔN = 0 and ΔK = 0 ) then
14            DIAG := H_(N',K'),(N'',K'')
15            y[i : i + k − 1] += DIAG · x[j : j + l − 1]
16          end if
            /* DK-block ?                                                        */
17          if ( ΔN = 0 and ΔK = 1 ) then
18            if ( K' > K'' ) then
19              DK := H_(N',K'),(N'',K'')
20              y[i : i + k − 1] += DK · x[j : j + l − 1]
21            else
22              DK := H_(N'',K''),(N',K')
23              y[i : i + l − 1] += DK^T · x[j : j + k − 1]
24            end if
25          end if
            /* SO-block ?                                                        */
26          if ( ΔN = 1 and ΔK = 0 ) then
27            if ( N' > N'' ) then
28              SO := H_(N',K'),(N'',K'')
29              y[i : i + k − 1] += SO · x[j : j + l − 1]
30            else
31              SO := H_(N'',K''),(N',K')
32              y[i : i + l − 1] += SO^T · x[j : j + k − 1]
33            end if
34          end if
35        end for
36      end for
37    end for
38  end for
39  return y
```

The off-diagonal DK- and SO-blocks exhibit a regular block sparsity pattern, which is induced by identity matrices $\mathbf{I}$ arising as factors in the corresponding Kronecker product representations (see Table 6.11). As a consequence, both DK- and SO-blocks have tiny sub-blocks of the dimension $(v_2^{(lim)'} \times v_2^{(lim)''})$, which we call v_2-*blocks* in the following. Fortunately, to reduce memory consumption, it is not necessary to resort to conventional storage schemes for sparse matrices such as CSR ("Compressed Sparse Row") or CSC ("Compressed Sparse Column") (cf. [101], [43]) involving quite a lot of overhead due to the need to hold index information in additional arrays. Instead, we will compress the matrix block under consideration into a dense (and much smaller) rectangular matrix by collecting the v_2-blocks and derive suitable index mappings that (back-)transform an index pair $(\tilde{i}, \tilde{j})$ of a compactly stored matrix block to its proper position (i, j) in the original matrix block. These mappings σ ((7.27), (7.28)) and τ ((7.17), (7.18)) seem somewhat cumbersome at first glance, but the underlying principle is straightforward and easy to understand as we will illustrate graphically later on in Figs. 7.2 and 7.3. Before we discuss how to proceed in detail we give some useful definitions and statements about the sparsity of matrices (cf. [43], for instance):

Definition 7.4
For a rectangular matrix $\mathbf{A} \in \mathbb{C}^{m \times n}$ we define:

a) the index set associated with the non-zero entries of $\mathbf{A}$

$$struct(\mathbf{A}) = \{(i, j) \in \{1, \ldots, m\} \times \{1, \ldots, n\} \, | \mathbf{A}_{i,j} \neq 0\} \qquad (7.8)$$

b) the number of non-zero entries of $\mathbf{A}$

$$nnz(\mathbf{A}) = | \, struct(\mathbf{A}) \, | \qquad (7.9)$$

c) the sparsity ratio (density) of $\mathbf{A}$

$$sr(\mathbf{A}) = nnz(\mathbf{A})/(m \cdot n) \qquad (7.10)$$

The following lemma shows how the number of non-zero elements of a Kronecker product (cf. Def. 2.47) is related to the nnz of its Kronecker factors:

Lemma 7.5
Let $\mathbf{A}_i \in \mathbb{K}^{m_i \times n_i}$ $(i = 1, \ldots, k)$. Then the number of non-zero entries of the Kronecker product

$$\mathbf{C} = \bigotimes_{i=1}^{k} \mathbf{A_i} \qquad (7.11)$$

can be expressed as:

$$nnz(\mathbf{C}) = nnz \left(\bigotimes_{i=1}^{k} \mathbf{A_i} \right) = \prod_{i=1}^{k} nnz(\mathbf{A_i}) \qquad (7.12)$$

Proof: The statement of the lemma is a direct consequence of the properties and the definition of the Kronecker product. We will conduct the proof by induction over k:

$k = 2$: (induction basis)

We know that for $k = 2$, $\mathbf{A} := \mathbf{A}_1$, $\mathbf{B} := \mathbf{A}_2$ the result $\mathbf{C} = \mathbf{A} \otimes \mathbf{B}$ is a $m_1 \times n_1$ block matrix whose (i, j) block is the $m_2 \times n_2$ matrix $a_{ij}\mathbf{B}$. It is plain to see that there are $nnz(\mathbf{A})$ such non-zero blocks. For $a_{ij} \neq 0$ obviously $nnz(a_{ij}\mathbf{B}) = nnz(\mathbf{B})$. Thus,

$$nnz(\mathbf{A}_1 \otimes \mathbf{A}_2) = nnz(\mathbf{A}_1) \cdot nnz(\mathbf{A}_2)$$

$k - 1 \to k$ (induction step)

Making use of the associativity of the Kronecker product (Lemma 2.48) we obtain

$$\bigotimes_{i=1}^{k} \mathbf{A_i} = \left(\bigotimes_{i=1}^{k-1} \mathbf{A_i} \right) \otimes \mathbf{A_k}$$

and, thus,

$$
\begin{aligned}
nnz\left(\bigotimes_{i=1}^{k} \mathbf{A_i} \right) &= nnz\left\{ \left(\bigotimes_{i=1}^{k-1} \mathbf{A_i} \right) \otimes \mathbf{A_k} \right\} \\
&= nnz\left(\bigotimes_{i=1}^{k-1} \mathbf{A_i} \right) \cdot nnz(\mathbf{A_k}) \\
&= \prod_{i=1}^{k-1} nnz(\mathbf{A_i}) \cdot nnz(\mathbf{A_k}) \\
&= \prod_{i=1}^{k} nnz(\mathbf{A_i})
\end{aligned}
$$

$\square$

For the ease of presentation we introduce the following

Notation 7.6

- $\mathbf{F}_{d',d''} \in \mathbb{R}^{d' \times d''}$ denotes a dense matrix whose entries are exclusively ones.

- The pattern matrix $\mathbf{SA} \in \mathbb{R}^{d' \times d''}$ of a matrix $A \in \mathbb{R}^{d' \times d''}$ is defined as

$$\mathbf{SA}_{ij} = \begin{cases} 1 & : \quad \mathbf{A}_{ij} \neq 0 \\ 0 & : \quad \mathbf{A}_{ij} = 0 \end{cases}$$

$\square$

7.2.1.1. Sparsity and Compact Storage of the SO-Blocks

According to Table 6.11 the sparsity pattern of an SO-block ($\Delta N = 1$, $\Delta K = 0$) **JSO** $\in \mathbb{R}^{d' \times d''}$ can be described by the pattern matrix **SSO** $\in \mathbb{R}^{d' \times d''}$ (under the assumption that the matrix $\mathbf{H}'_{SO}$ is dense, which we can take for granted):

$$
\begin{aligned}
\mathbf{SSO} &= \mathbf{I}_{N_R^{(lim)}} \otimes \mathbf{I}_{N_r^{(lim)}} \otimes \mathbf{F}_{v_2^{(lim)'},v_2^{(lim)''}} \\
&= \mathbf{I}_{N_R^{(lim)} \cdot N_r^{(lim)}} \otimes \mathbf{F}_{v_2^{(lim)'},v_2^{(lim)''}}
\end{aligned}
\tag{7.13}
$$

This implies

$$
\begin{aligned}
nnz(\mathbf{SSO}) &= nnz\big(\mathbf{I}_{N_R^{(lim)} \cdot N_r^{(lim)}} \otimes \mathbf{F}_{v_2^{(lim)'},v_2^{(lim)''}}\big) \\
&= nnz(\mathbf{I}_{N_R^{(lim)} \cdot N_r^{(lim)}}) \cdot nnz(\mathbf{F}_{v_2^{(lim)'},v_2^{(lim)''}}) \\
&= N_R^{(lim)} \cdot N_r^{(lim)} \cdot (v_2^{(lim)'} \cdot v_2^{(lim)''})
\end{aligned}
\tag{7.14}
$$

and

$$
\begin{aligned}
sr(\mathbf{SSO}) &= nnz(\mathbf{SSO}) \,/\, \big(\,(N_R^{(lim)})^2 \cdot (N_r^{(lim)})^2 \cdot (v_2^{(lim)'} \cdot v_2^{(lim)''})\,\big) \\
&= 1 \,/\, (N_R^{(lim)} \cdot N_r^{(lim)})
\end{aligned}
\tag{7.15}
$$

Hence, the density of an SO-block is determined by the parameters $N_R^{(lim)}$ and $N_r^{(lim)}$.

Figure 7.2.: Sparsity and compact storage of an SO-block $\quad (N_R^{(lim)} = 7,\ N_r^{(lim)} = 5)$

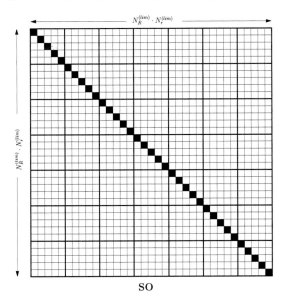

From (7.13) and the definition of a Kronecker product one can easily conclude that an SO-block is block-diagonal, consisting of $N_R^{(lim)} \cdot N_r^{(lim)}$ blocks of dimension $(v_2^{(lim)'} \cdot v_2^{(lim)''})$. This directly gives rise to the compact storage in the matrix **CSO** by stacking the blocks on the diagonal of **SO** in one block column as illustrated in Figure 7.2. More precisely, this is realised by the following bijection τ, which enables us to identify an index pair $(\tilde{i}, \tilde{j})$ of the matrix **CSO** with the proper position (i, j) in $struct(\mathbf{SO})$:

$$\tau : I' \times \{1, \ldots, v_2^{(lim)''}\} \longrightarrow struct(\mathbf{SSO}) \subset I' \times I''$$
$$(\tilde{i}, \tilde{j}) \longmapsto (\tau_1(\tilde{i}, \tilde{j}), \tau_2(\tilde{i}, \tilde{j})) =: (i, j) \tag{7.16}$$

$$i = \tau_1(\tilde{i}, \tilde{j}) = \tilde{i} \tag{7.17}$$
$$j = \tau_2(\tilde{i}, \tilde{j}) = \left\lfloor \tilde{i}/v_2^{(lim)'} \right\rfloor \cdot v_2^{(lim)''} + \tilde{j} \tag{7.18}$$

The inverse map τ^{-1} is given by:

$$\tau^{-1} : struct(\mathbf{SSO}) \subset I' \times I'' \longrightarrow I' \times \{1, \ldots, v_2^{(lim)''}\}$$
$$(i, j) \longmapsto ((\tau^{-1})_1(i, j), (\tau^{-1})_2(i, j)) =: (\tilde{i}, \tilde{j}) \tag{7.19}$$

$$\tilde{i} = (\tau^{-1})_1(i, j) = i \tag{7.20}$$
$$\tilde{j} = (\tau^{-1})_2(i, j) = j \bmod v_2^{(lim)''} \tag{7.21}$$

7.2.1.2. Sparsity and Compact Storage of the DK-Blocks

As per Table 6.11 the sparsity pattern of a DK-block $(\Delta N = 0, \Delta K = 1)$ $\mathbf{DK} \in \mathbb{R}^{d' \times d''}$ can be described by the pattern matrix $\mathbf{SDK} \in \mathbb{R}^{d' \times d''}$ (assuming that $\mathbf{H}'_{dk}$ is dense):

$$\mathbf{SDK} = \mathbf{F}_{N_R^{(lim)}} \otimes \mathbf{I}_{N_r^{(lim)}} \otimes \mathbf{F}_{v_2^{(lim)'}, v_2^{(lim)''}} \tag{7.22}$$

Thus, using (7.12) we obtain

$$\begin{aligned} nnz(\mathbf{SDK}) &= nnz(\mathbf{F}_{N_R^{(lim)}} \otimes \mathbf{I}_{N_r^{(lim)}} \otimes \mathbf{F}_{v_2^{(lim)'}, v_2^{(lim)''}}) \\ &= nnz(\mathbf{F}_{N_R^{(lim)}}) \cdot nnz(\mathbf{I}_{N_r^{(lim)}}) \cdot nnz(\mathbf{F}_{v_2^{(lim)'}, v_2^{(lim)''}}) \\ &= (N_R^{(lim)})^2 \cdot N_r \cdot (v_2^{(lim)'} \cdot v_2^{(lim)''}) \end{aligned} \tag{7.23}$$

and application of (7.10) yields

$$\begin{aligned} sr(\mathbf{SDK}) &= nnz(\mathbf{SDK}) / ((N_R^{(lim)})^2 \cdot (N_r^{(lim)})^2 \cdot (v_2^{(lim)'} \cdot v_2^{(lim)''})) \\ &= 1 / N_r^{(lim)} \end{aligned} \tag{7.24}$$

Here the density of a DK-block solely depends on the parameter $N_r^{(lim)}$.

Owing to the associativity of a Kronecker product we can write (7.22) as

$$\mathbf{SDK} = \mathbf{F}_{N_R^{(lim)}} \otimes (\underbrace{\mathbf{I}_{N_r^{(lim)}} \otimes \mathbf{F}_{v_2^{(lim)'},v_2^{(lim)''}}}_{=:\mathbf{A}}) \tag{7.25}$$

and explain the structure of a DK-block in two steps:

1. $\mathbf{A}$ is block-diagonal (consisting of N_r blocks of dimension $(v_2^{(lim)'} \cdot v_2^{(lim)''})$

2. As $\mathbf{F}_{N_R^{(lim)}}$ is dense and has the dimension $N_R^{(lim)} \times N_R^{(lim)}$, the resulting $\mathbf{SDK}$ is a $(N_R^{(lim)} \times N_R^{(lim)})$ block matrix whose blocks have the structure of $\mathbf{A}$. Thus, there are $(N_R^{(lim)})^2 \cdot N_r$ non-zero v_2-blocks, which is obviously consistent with (7.23). The resulting sparsity pattern is illustrated in Figure 7.3.

Figure 7.3.: Sparsity and compact storage of a DK-block $(N_R^{(lim)} = 7, N_r^{(lim)} = 5)$

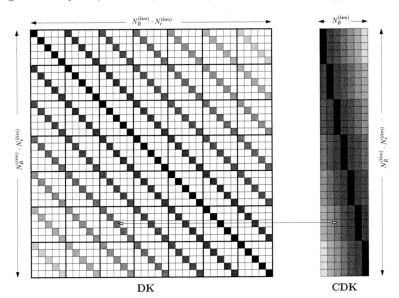

Although things are slightly more complicated for DK-blocks, we can employ the same ideas as for SO-blocks, as we can find a bijection σ between $struct(\mathbf{SDK})$ and the index set of **CDK** (see Figure 7.3):

$$\sigma \; : \; I' \times \{1, \ldots, N_R^{(lim)} \cdot v_2^{(lim)''}\} \; \longrightarrow \; struct(\mathbf{SDK}) \subset I' \times I''$$
$$(\tilde{i}, \tilde{j}) \; \longmapsto \; (\sigma_1(\tilde{i}, \tilde{j}) \, , \, \sigma_2(\tilde{i}, \tilde{j})) =: (i, j) \qquad (7.26)$$

$$i = \sigma_1(\tilde{i}, \tilde{j}) \;\; = \;\; \tilde{i} \qquad\qquad\qquad\qquad\qquad\qquad (7.27)$$

$$j = \sigma_2(\tilde{i}, \tilde{j}) \;\; = \;\; \left\lfloor \tilde{j}/v_2^{(lim)''} \right\rfloor \cdot (N_r^{(lim)} \cdot v_2^{(lim)''})$$
$$+ \;\; \left(\left\lfloor \tilde{i}/v_2^{(lim)'} \right\rfloor \bmod N_r^{(lim)} \right) \cdot v_2^{(lim)''} + \tilde{j} \bmod v_2^{(lim)''} \qquad (7.28)$$

The inverse map σ^{-1} is given by:

$$\sigma^{-1} \; : \; struct(\mathbf{SDK}) \subset I' \times I'' \; \longrightarrow \; I' \times \{1, \ldots, N_R^{(lim)} \cdot v_2^{(lim)''}\}$$
$$(i, j) \; \longmapsto \; (\, (\sigma^{-1})_1(i, j) \, , \, (\sigma^{-1})_2(i, j) \,) =: (\tilde{i}, \tilde{j}) \;\; (7.29)$$

$$\tilde{i} = (\sigma^{-1})_1(i, j) \;\; = \;\; i \qquad\qquad\qquad\qquad\qquad (7.30)$$

$$\tilde{j} = (\sigma^{-1})_2(i, j) \;\; = \;\; \left\lfloor j/(N_r^{(lim)} \cdot v_2^{(lim)''}) \right\rfloor \cdot v_2^{(lim)''} + j \bmod v_2^{(lim)''} \qquad (7.31)$$

7.2.2. Storage Scheme for the Hamiltonian Matrix Blocks

We are only interested in matrix blocks with non-zero entries, i.e. those blocks, whose addresses satisfy one of the rules in Table 6.13. Since accessing a K-block by its address $(N', K'), (N'', K'')$ is rather tedious and inefficient, we now derive a storage scheme which is better suited for the design and implementation of an efficient matrix-vector multiplication in a computer code. The natural and straightforward approach is to work with three one-dimensional arrays **JDIAG**, **JDK** and **JSO** that enable direct or indirect access (e.g. by means of pointers in a C-Code) to the DIAG-, DK- and SO-blocks. In the following, we will show how the address $(N', K'), (N'', K'')$ of a matrix block can be mapped to the corresponding index I in one of the three arrays (depending on the type of the block).

7.2.2.1. Addressing the DIAG-blocks

We set $K = K'$ and $N = N'$.
For every N-Block with $N \in \{N_{min}, \ldots, N_{max}\}$ there are $(N + 1)$ DIAG blocks, thus

$$\#\text{DIAG} \;\; = \;\; \sum_{\ell=N_{min}}^{N_{max}} (\ell + 1)$$

$$= \;\; \sum_{\ell=N_{min}+1}^{N_{max}} \ell + N_{max} + 1 \qquad\qquad (7.32)$$

DIAG-blocks altogether. For convenience we let the index of the JDIAG array run from zero, such that for any given pair (N, K) the index

$$I \in \left\{ 0, \ldots, \sum_{\ell=N_{min}+1}^{N_{max}} \ell + N_{max} \right\} \tag{7.33}$$

is obtained by the following relation:

$$I = \sum_{\ell=N_{min}+1}^{N} \ell + K \tag{7.34}$$

7.2.2.2. Addressing the DK-blocks

We set $K = \max\{K', K''\}$ and $N = N'$.
Differently from the DIAG-blocks there are N DK-blocks in an N-block associated with the quantum number $N \in \{N_{min}, \ldots, N_{max}\}$ resulting in

$$\#DK = \sum_{\ell=N_{min}}^{N_{max}} \ell \tag{7.35}$$

DK-blocks altogether. Here it is more convenient to let the index I run from 1, such that for a given pair (N, K) the corresponding index

$$I \in \left\{ 1, \ldots, \sum_{\ell=N_{min}}^{N_{max}} \ell \right\} \tag{7.36}$$

in **JDK** is given by

$$I = \sum_{\ell=N_{min}}^{N-1} \ell + K \tag{7.37}$$

7.2.2.3. Addressing the SO-blocks

We set $K = K'$ and $N = \max\{N', N''\}$.
There are $(N_{max} - 1)$ off-diagonal N-blocks for $N \in \{N_{min} + 1, \ldots, N_{max}\}$, each of them holding N SO-blocks. Hence, there are

$$\#SO = \sum_{\ell=N_{min}+1}^{N_{max}} \ell \tag{7.38}$$

SO-blocks altogether. Here again the index I starts at 0 and we can relate (N, K) to

$$I \in \left\{ 0, \dots, \sum_{\ell=N_{min}+1}^{N_{max}} \ell - 1 \right\} \tag{7.39}$$

by means of the following mapping:

$$I = \sum_{\ell=N_{min}+1}^{N-1} \ell + K \tag{7.40}$$

The Tables 7.5 and 7.6 for doublet $(S = 1/2)$ and quartet $(S = 3/2)$ spin multiplicity show how the DIAG-, DK-, and SO-blocks are adressed by means of (7.34), (7.37), (7.40).

Now combining compact storage and efficient addressing results in the storage scheme for a Hamiltonian J-block $\mathbf{H}^{(J,S,\Gamma_{rve})}$ depicted in Fig. 7.4:

Figure 7.4.: Storage scheme for the matrix blocks $(S = 1/2, J = 5/2)$

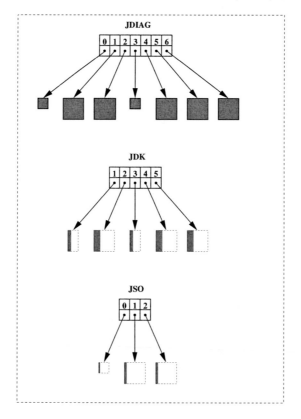

Table 7.5.: Addressing the Hamiltonian blocks ($S = 1/2$, $J = 5/2$)

		N	2			3			
		K	0	1	2	0	1	2	3
DIAG	I		0	1	2	3	4	5	6
DK	I			1	2		3	4	5
SO	I					0	1	2	

Table 7.6.: Addressing the Hamiltonian blocks ($S = 3/2$, $J = 7/2$)

	N	2			3				4					5					
	K	0	1	2	0	1	2	3	0	1	2	3	4	0	1	2	3	4	5
DIAG	I	0	1	2	3	4	5	6	7	8	9	10	11	12	13	14	15	16	17
DK	I		1	2		3	4	5		6	7	8	9		10	11	12	13	14
SO	I				0	1	2		3	4	5	6		7	8	9	10	11	

7.2.3. Matrix-Vector Multiplication Exploiting Compact Storage

We can now exploit our results from the previous subsections and derive an efficient algorithm for the matrix-vector multiplication of the Hamiltonian matrix. To this end we first formulate algorithms for the DK-block and SO-block multiplication that take advantage of the compact storage of blocks discussed in Sections 7.2.1.1 and 7.2.1.2.

The Algorithms **SO-mult** and **DK-mult** given below perform the multiplication of an SO-block or a DK-block, respectively. They require

- d', d'' (block dimension), $N_R^{(lim)}$, $N_r^{(lim)}$ (basis set information)

- TRANSP (multiplication by the transpose (**true/false**))

- **CSO** $\in \mathbb{R}^{d' \times v_2^{(lim)''}}$ resp. **CDK** $\in \mathbb{R}^{d' \times N_R^{(lim)} \cdot v_2^{(lim)''}}$, $\mathbf{x} \in \mathbb{R}^{d'}$ or $\mathbf{x} \in \mathbb{R}^{d''}$}

and compute

- $\mathbf{y} = $ **SO** $\cdot \mathbf{x}$ resp. $\mathbf{y} = $ **DK** $\cdot \mathbf{x}$, if TRANSP=**false** and $\mathbf{x} \in \mathbb{R}^{d''}$

- $\mathbf{y} = $ **SO**$^T \cdot \mathbf{x}$ resp. $\mathbf{y} = $ **SO**$^T \cdot \mathbf{x}$, if TRANSP=**true** and $\mathbf{x} \in \mathbb{R}^{d'}$

Algorithm 7.2: DK-block-vector multiplication exploiting compact storage

---

1   **function** $\mathbf{y} = \mathbf{DK\text{-}mult}(\,\mathbf{CDK}, \mathbf{x},\, k,\, l,\, N_R^{(lim)},\, N_r^{(lim)}, \text{TRANSP}\,)$

2   $v_2^{(lim)'} = k/(N_R^{(lim)} \cdot N_r^{(lim)})$

3   $v_2^{(lim)''} = l\,/(N_R^{(lim)} \cdot N_r^{(lim)})$

4   **if** *(TRANSP=**false**)* **then**

5   **for** $i = 1, \ldots, (N_R^{(lim)} \cdot N_r^{(lim)})$ **do**

6        $\hat{\mathbf{y}} := \mathbf{y}[\,(i-1) \cdot v_2^{(lim)'} + 1 : i \cdot v_2^{(lim)'}\,]$

7   **for** $j = 1, \ldots, N_R^{(lim)}$ **do**

8          $\widehat{\mathbf{CDK}} = \mathbf{CDK}[\,(i-1) \cdot v_2^{(lim)'} + 1 : i \cdot v_2^{(lim)'},\, (j-1) \cdot v_2^{(lim)''} + 1 : j \cdot v_2^{(lim)''}\,]$

9          $\hat{i} = (j-1) \cdot (N_r^{(lim)} \cdot v_2^{(lim)''}) + ((i-1) \cdot v_2^{(lim)''}) \bmod (N_r^{(lim)} \cdot v_2^{(lim)''})$

10         $\hat{\mathbf{x}} := \mathbf{x}[\,\hat{i} + 1 : \hat{i} + v_2^{(lim)''}\,]$

11         $\hat{\mathbf{y}} \mathrel{+}= \widehat{\mathbf{CDK}} \cdot \hat{\mathbf{x}}$

12   **end for**

13   **end for**

14   **else**

15   **for** $i = 1, \ldots, (N_R^{(lim)} \cdot N_r^{(lim)})$ **do**

16        $\hat{\mathbf{y}} := \mathbf{y}[\,(i-1) \cdot v_2^{(lim)''} + 1 : i \cdot v_2^{(lim)''}\,]$

17   **for** $j = 1, \ldots, N_R^{(lim)}$ **do**

18          $\hat{q} = (j-1) \cdot (N_r^{(lim)} \cdot v_2^{(lim)'}) + ((i-1) \cdot v_2^{(lim)'}) \bmod (N_r^{(lim)} \cdot v_2^{(lim)'})$

19          $\hat{r} = (\,(i-1)/N_r^{(lim)}\,) \cdot v_2^{(lim)''}$

20          $\widehat{\mathbf{CDK}} = \mathbf{CDK}[\,\hat{q} + 1 : \hat{q} + v_2^{(lim)'},\, \hat{r} + 1 : \hat{r} + v_2^{(lim)''}\,]$

21          $\hat{i} = (j-1) \cdot (N_r^{(lim)} \cdot v_2^{(lim)'}) + ((i-1) \cdot v_2^{(lim)'}) \bmod (N_r^{(lim)} \cdot v_2^{(lim)'})$

22         $\hat{\mathbf{x}} := \mathbf{x}[\,\hat{i} + 1 : \hat{i} + v_2^{(lim)'}\,]$

23         $\hat{\mathbf{y}} \mathrel{+}= \widehat{\mathbf{CDK}}^{T} \cdot \hat{\mathbf{x}}$

24   **end for**

25   **end for**

26   **end if**

27   **return** $\mathbf{y}$

To obtain the final matrix-vector multiplication (Algorithm 7.4) we simply have to replace the explicit DK-block and SO-block multiplications in Algorithm 7.1 by the corresponding block-multiplication Algorithms 7.2 and 7.3 and to use the storage scheme (Figure 7.4) derived in Section 7.2.2. Figure 7.5 impressively illustrate the advantage of exploiting sparsity for matrix-vector multiplication and storage (**MgCN** molecule, big basis set, see Table 6.5): Both memory costs and computing time grow linearly with the rotational quantum number J. By contrast, the quadratic growth of memory costs when storing the Hamiltonian matrix blocks $\mathbf{H}^{(J,S,\Gamma_{rve})}$ explicitly makes eigenvalue computations by means of direct solvers (see Section 3.2) intractable for values of $J > 9/2$ for the big basis (as defined in Table 6.5). Finally, the diagrams illustrate that it clearly pays to exploit the sparsity of the off-diagonal Hamiltonian K-blocks (and not only the block sparsity). A simple calculation shows that our scheme (Fig. 7.4) allows to store FBRs (for the **MgCN** molecule with respect to the big basis) up to $J = 31/2$ with an amount of 32 G memory available (cf. Fig. 7.5) on our SUN^{TM} Fire workstation.

Algorithm 7.3: SO-block-vector multiplication exploiting compact storage

1 **function** $\mathbf{y} = \text{SO-mult}(\ \mathbf{CSO}, \mathbf{x},\ d',\ d'',\ N_R^{(lim)},\ N_r^{(lim)}, \text{TRANSP}\ )$

2 $\quad v_2^{(lim)'} = d'/(N_R^{(lim)} \cdot N_r^{(lim)})$

3 $\quad v_2^{(lim)''} = d''/(N_R^{(lim)} \cdot N_r^{(lim)})$

4 **if** *(TRANSP=false)* **then**

5 $\quad$ | **for** $i = 1,\ldots,(N_R^{(lim)} \cdot N_r^{(lim)})$ **do**

6 $\quad$ | $\quad \hat{\mathbf{x}} := \mathbf{x}[\ (i-1) \cdot v_2^{(lim)''} + 1 : i \cdot v_2^{(lim)''}\ ]$

7 $\quad$ | $\quad \hat{\mathbf{y}} := \mathbf{y}[\ (i-1) \cdot v_2^{(lim)'} + 1 : i \cdot v_2^{(lim)'}\ ]$

8 $\quad$ | $\quad \widehat{\mathbf{CSO}} = \mathbf{CSO}[\ (i-1) \cdot v_2^{(lim)'} + 1 : i \cdot v_2^{(lim)'},\ 1 : v_2^{(lim)''}\ ]$

9 $\quad$ | $\quad \hat{\mathbf{y}} \mathrel{+}= \widehat{\mathbf{CSO}} \cdot \hat{\mathbf{x}}$

10 $\quad$ | **end for**

11 **else**

12 $\quad$ | **for** $i = 1,\ldots,(N_R^{(lim)} \cdot N_r^{(lim)})$ **do**

13 $\quad$ | $\quad \hat{\mathbf{x}} := \mathbf{x}[\ (i-1) \cdot v_2^{(lim)'} + 1 : i \cdot v_2^{(lim)'}\ ]$

14 $\quad$ | $\quad \hat{\mathbf{y}} := \mathbf{y}[\ (i-1) \cdot v_2^{(lim)''} + 1 : i \cdot v_2^{(lim)''}\ ]$

15 $\quad$ | $\quad \widehat{\mathbf{CSO}} = \mathbf{CSO}[\ (i-1) \cdot v_2^{(lim)'} + 1 : i \cdot v_2^{(lim)'},\ 1 : v_2^{(lim)''}\ ]$

16 $\quad$ | $\quad \hat{\mathbf{y}} \mathrel{+}= \widehat{\mathbf{CSO}}^{T} \cdot \hat{\mathbf{x}}$

17 $\quad$ | **end for**

18 **end if**

19 **return y**

Figure 7.5.: Comparison of the storage schemes (**MgCN** molecule, big basis)

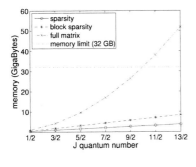

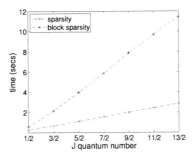

a) Storage costs   b) Matrix-vector multiplication timings

Algorithm 7.4: Matrix-vector multiplication exploiting sparsity

1 **function** $\mathbf{y} = \mathbf{mult}(\,\mathbf{H}, \mathbf{x}, \Gamma_{rve}, S, J, N_R^{(lim)}, N_r^{(lim)}, (v_2^a)^{(lim)}, (v_2^b)^{(lim)}, \text{ABC}\,)$

2 $\mathbf{y} = 0$

3 $N_{min} = |J - S|,\ N_{max} = J + S$

4 **for** $N' = N_{min}, \ldots, N_{max}$ **do**

5   **for** $K' = 0, \ldots, N'$ **do**

6    $i = \mathrm{pos_K}(\,\Gamma_{rve}, S, J, N', K', N_R^{(lim)}, N_r^{(lim)}, (v_2^a)^{(lim)}, (v_2^b)^{(lim)}, \text{ABC}\,)$

7    $k = \mathrm{dim_K}(\,\Gamma_{rve}, N', K', N_R^{(lim)}, N_r^{(lim)}, (v_2^a)^{(lim)}, (v_2^b)^{(lim)}, \text{ABC}\,)$

8   **for** $N'' = N_{min}, \ldots, N_{max}$ **do**

9   **for** $K'' = 0, \ldots, N''$ **do**

10     $j = \mathrm{pos_K}(\,\Gamma_{rve}, S, J, N'', K'', N_R^{(lim)}, N_r^{(lim)}, (v_2^a)^{(lim)}, (v_2^b)^{(lim)}, \text{ABC}\,)$

11     $l = \mathrm{dim_K}(\,\Gamma_{rve}, N'', K'', N_R^{(lim)}, N_r^{(lim)}, (v_2^a)^{(lim)}, (v_2^b)^{(lim)}, \text{ABC}\,)$

12     $\Delta N = |\,N' - N''\,|,\ \Delta K = |\,K' - K''\,|$

13   **if** $(\,\Delta N = 0 \ \text{\textit{and}} \ \Delta K = 0\,)$ **then** $\longleftarrow$————————————DIAG block ?

14      $I = \displaystyle\sum_{\ell=N_{min}+1}^{N'} \ell + K''$

15      $\mathbf{y}[i : i + k - 1] \mathrel{+}= \mathbf{JDIAG}[I] \cdot \mathbf{x}[j : j + l - 1]$

16   **end if**

17   **if** $(\,\Delta N = 0 \ \text{\textit{and}} \ \Delta K = 1\,)$ **then** $\longleftarrow$————————————DK block ?

18   **if** $(\,K' > K''\,)$ **then**

19       $I = \displaystyle\sum_{\ell=N_{min}}^{N'-1} \ell + K'$

20       $\mathbf{y}[i, \ldots, i + k - 1] \mathrel{+}= \text{DK-mult}(\,\mathbf{JDK}[I], \mathbf{x}[j, \ldots, j + l - 1], k, l,$

21               $N_R^{(lim)}, N_r^{(lim)}, \textbf{false}\,)$

22   **else**

23       $I = \displaystyle\sum_{\ell=N_{min}}^{N'-1} \ell + K''$

24       $\mathbf{y}[i : i + l - 1] \mathrel{+}= \text{DK-mult}(\,\mathbf{JDK}[I], \mathbf{x}[j : j + k - 1], k, l,$

25               $N_R^{(lim)}, N_r^{(lim)}, \textbf{true}\,)$

26   **end if**

27   **end if**

28   **if** $(\,\Delta N = 1 \ \text{\textit{and}} \ \Delta K = 0\,)$ **then** $\longleftarrow$————————————SO block ?

29   **if** $(\,N' > N''\,)$ **then**

30       $I = \displaystyle\sum_{\ell=N_{min}+1}^{N'-1} \ell + K'$

31       $\mathbf{y}[i : i + k - 1] \mathrel{+}= \text{SO-mult}(\,\mathbf{JSO}[I], \mathbf{x}[j : j + l - 1], k, l,$

32               $N_R^{(lim)}, N_r^{(lim)}, \textbf{false}\,)$

33   **else**

34       $I = \displaystyle\sum_{\ell=N_{min}+1}^{N''-1} \ell + K',$

35       $\mathbf{y}[i : i + l - 1] \mathrel{+}= \text{SO-mult}(\,\mathbf{JSO}[I], \mathbf{x}[j : j + k - 1], k, l,$

36               $N_R^{(lim)}, N_r^{(lim)}, \textbf{true}\,)$

37   **end if**

38   **end if**

39   **end for**

40   **end for**

41 **end for**

42 **end for**

43 **return** $\mathbf{y}$

7.3. Contraction Scheme and Contracted Basis

In this section we make more precise how the basis contraction according to the general description in Section 5.8.2 is realized for the concrete case of the *Double Renner Hamiltonian*. The contraction scheme is proposed in ODAKA'S PhD thesis [86] as a means to make the eigenvalue computations viable and it is employed in the related software DR. A detailed discussion along with numerical results may be found in [87]. The basic idea can be described as follows

- Compute eigensystems of the arising basic K-blocks in $\mathbf{H}_u$

$$(\mathbf{H}_u)_{(N,K)} = \mathbf{B}^{K,\Gamma_{vib},\eta} = \mathbf{V}^*_{N,K} \mathbf{\Lambda}_{N,K} \mathbf{V}_{N,K} \qquad (7.41)$$

  where we assume the eigenvalues to be ordered by ascending magnitude. Because of the multiple occurences of basic K-blocks (cf. Fact 6.10 and Corollary 6.11) this only needs to be done for one instance of $\mathbf{B}^{K,\Gamma_{vib},\eta}$ in each case.

- The contraction and its quality is controlled by an offset parameter E_{cont}, which we call *contraction limit*. More precisely, we only employ those columns of $\mathbf{V}_{N,K}$ whose related eigenvalues are less or equal than $\lambda_1 + E_{\text{cont}}$, i.e. $\mathbf{T}_{N,K} = \mathbf{V}_{N,K}[1:j]$, where j is the largest index such that $\lambda_j \leq \lambda_1 + E_{\text{cont}}$. The total projection is then obtained as the block-diagonal matrix $\mathbf{T}$ whose blocks are $\mathbf{V}_{N,K}[1:j]$. The larger one chooses E_{cont} the better the quality of the contracted Hamiltonian matrix will be.

- The contracted Hamiltonian matrix $\widetilde{\mathbf{H}}$ is now simply obtained by the *Rayleigh-Ritz projection*

$$\widetilde{\mathbf{H}}^{(J,S,\Gamma_{rve})} = \mathbf{T}^* \, \mathbf{H}^{(J,S,\Gamma_{rve})} \, \mathbf{T} \qquad (7.42)$$

These ideas are formalized by Algorithm 7.5 (Computation of the projection) and Algorithm 7.6 (Application of the projection), where the blocks $\mathbf{T}_{N,K}$ of the transformation matrix $\mathbf{T}$ are addressed using the scheme for the DIAG blocks from Section 7.2.2.1. Note that Algorithm 7.6 makes use of the block sparsity and returns the non-zero blocks of the contracted Hamiltonian matrix $\widetilde{\mathbf{H}}^{(J,S,\Gamma_{rve})}$. To this end it uses nearly the same storage scheme as for the uncontracted Hamiltonian matrix $\mathbf{H}^{(J,S,\Gamma_{rve})}$, the difference being that the contracted SO- and DK-blocks cannot be compressed any further since their sparsity structure is lost upon contraction.

Algorithm 7.5: Computing the projection for the contraction scheme

---

1 **function** $\mathbf{T} = \mathbf{projection}(\,\mathbf{H},\ \mathbf{H_u},\ E_{\text{cont}},\ S,\ J,\ \Gamma_{rve},$
$$N_R^{(lim)},\ N_r^{(lim)},\ (v_2^a)^{(lim)},\ (v_2^b)^{(lim)},\ \text{ABC}\,)$$

2 $N_{min} = |J - S|,\ N_{max} = J + S$

3 **for** $N = N_{min}, \ldots, N_{max}$ **do**

4   **for** $K = 0, \ldots, N$ **do**

5         $k = \dim_{\mathrm{K}}(\,\Gamma_{rve},\ N,\ K,\ N_R^{(lim)},\ N_r^{(lim)},\ (v_2^a)^{(lim)},\ (v_2^b)^{(lim)},\ \text{ABC}\,)$

6         $I = \sum\limits_{\ell = N_{min}+1}^{N} \ell + K$

7   **if** *(*CONTDIM$[I] = 0$*)* **then**

8   Determine Γ_{vib} and η as a function of N, K, Γ_{rve}
            from Tables 6.3 resp. 6.4

9             $\mathbf{B} := \mathbf{B}^{K, \Gamma_{vib}, \eta}$

10   Compute sorted eigendecomposition

$$\mathbf{B} = \mathbf{V}^* \cdot \mathbf{\Lambda} \cdot \mathbf{V}$$

$$\mathbf{V}^* \cdot \mathbf{V} = \mathbf{I_k},\ \mathbf{\Lambda} = \mathrm{diag}(\lambda_1, \lambda_2, \ldots, \lambda_k),\ \lambda_1 \leq \lambda_2 \leq \ldots \leq \lambda_k$$

11           $j = 0$

12   **while** $(\lambda_{j+1} \leq \lambda_1 + E_{\text{cont}})$ **and** $(j < k)$ **do**

13             $j := j + 1$

14   **end while**

15   **if** *(*$K > 0$*)* **then**

16             $N_{max} = N_{max}$

17   **else**

18             $N_{max} = N$

19   **end if**

20   **for** $N' = N, \ldots, N_{max}$ **do**

21             $I = \sum\limits_{\ell = N_{min}+1}^{N'} \ell + K$

22   CONTDIM$[I] = j$

23             $\mathbf{T}[I] = \mathbf{V}[:, 1 : j] \in \mathbb{R}^{k \times j}$

24   **end for**

25   **end if**

26   **end for**

27 **end for**

28 **return T**

Algorithm 7.6: Contracting the Hamiltonian matrix blocks

---

1   function $\widetilde{\mathbf{H}} = \text{contract}(\mathbf{H}, \mathbf{H}_u, E_{\text{cont}}, S, J, N_R^{(lim)}, N_r^{(lim)}, (v_2^a)^{(lim)}, (v_2^b)^{(lim)}, \text{ABC})$

2   $N_{min} = |J - S|, \ N_{max} = J + S$

3   $\mathbf{T} = \text{projection}(\mathbf{H}, \mathbf{H}_u, E_{\text{cont}}, \Gamma_{rve}, S, J, N_R^{(lim)}, N_r^{(lim)}, (v_2^a)^{(lim)}, (v_2^b)^{(lim)}, \text{ABC})$

4   for $N' = N_{min}, \ldots, N_{max}$ do

5   for $K' = 0, \ldots, N'$ do

6   for $N'' = N_{min}, \ldots, N_{max}$ do

7   for $K'' = 0, \ldots, N''$ do

8              $\Delta N = |N' - N''|, \ \Delta K = |K' - K''|$

9   if $(\Delta N = 0$ **and** $\Delta K = 0)$ then $\leftarrow$Contraction of the DIAG blocks

10                $I = \displaystyle\sum_{\ell=N_{min}+1}^{N'} \ell + K''$

11                $\mathbf{DIAG} := \mathbf{JDIAG}[I], \ U = \mathbf{T}[I]$

12                $\widetilde{\mathbf{DIAG}}[I] = \mathbf{U}^T \cdot \mathbf{DIAG} \cdot \mathbf{U}$

13   end if

14   if $(\Delta N = 0$ **and** $\Delta K = 1)$ then $\longleftarrow$Contraction of the DK blocks

15   if $(K' > K'')$ then

16                   $I' = \displaystyle\sum_{\ell=N_{min}}^{N'-1} \ell + K', \ \ I'' = \displaystyle\sum_{\ell=N_{min}}^{N'-1} \ell + K''$

17                   $\mathbf{DK} := \mathbf{JDK}[I'], \ U = \mathbf{T}[I'], \ V = \mathbf{T}[I'']$

18                   $\widetilde{\mathbf{DK}}[I] = \mathbf{U}^T \cdot \mathbf{DK} \cdot \mathbf{V}$

19   end if

20   end if

21   if $(\Delta N = 1$ **and** $\Delta K = 0)$ then $\longleftarrow$Contraction of the SO blocks

22   if $(N' > N'')$ then

23                   $I' = \displaystyle\sum_{\ell=N_{min}+1}^{N'-1} \ell + K', \ \ I'' = \displaystyle\sum_{\ell=N_{min}+1}^{N''-1} \ell + K'$

24                   $\mathbf{SO} := \mathbf{JSO}[I], \ U = \mathbf{T}[I'], \ V = \mathbf{T}[I'']$

25                   $\widetilde{\mathbf{SO}}[I] = \mathbf{U}^T \cdot \mathbf{SO} \cdot \mathbf{V}$

26   end if

27   end if

28   end for

29   end for

30   end for

31   end for

32   return $\widetilde{\mathbf{H}} = [\widetilde{\mathbf{DIAG}}, \widetilde{\mathbf{DK}}, \widetilde{\mathbf{SO}}]$

In analogy to Corollary 6.9 we can also explicitly specify the dimensions and positions of matrix blocks in the contracted Hamiltonian matrix $\widetilde{\mathbf{H}}_{(N',K'),(N'',K'')}$, where we exploit the information which is available in the auxiliary array CONTDIM determined in Alg. 7.5:

Corollary 7.7 (Dimensions and positions of contracted K-blocks)
Any K-block with the address $(N', K'), (N'', K'')$ within a contracted Hamiltonian J-block $\mathbf{H}^{(J,S,\Gamma_{rve})}$ is fully determined by

$$\widetilde{\mathbf{H}}^{(J,S,\Gamma_{rve})}_{(N',K'),(N'',K'')} = \widetilde{\mathbf{H}}^{(J,S,\Gamma_{rve})}[i : i + d' - 1, j : j + d'' - 1] \in \mathbb{R}^{d' \times d''} \qquad (7.43)$$

where

$$d' = CONTDIM[I'] \qquad\qquad i = \sum_{m=0}^{I'-1} CONTDIM[m] + 1 \qquad (7.44)$$

$$d'' = CONTDIM[I''] \qquad\qquad j = \sum_{m=0}^{I''-1} CONTDIM[m] + 1 \qquad (7.45)$$

and I' and I'' are the addresses of K-blocks as derived in Section 7.2.2.1, i.e.

$$I' = \sum_{\ell=N_{min}+1}^{N'} \ell + K' \qquad (7.46)$$

$$I'' = \sum_{\ell=N_{min}+1}^{N''} \ell + K'' \qquad (7.47)$$

Remark 7.8 (Properties of the contracted Hamiltonian J-block $\widetilde{\mathbf{H}}^{(J,S,\Gamma_{rve})}$)
- *By the Poincaré separation theorem (Corollary 2.19) it is known that the eigenvalues $\tilde{\lambda}_i$ of the contracted Hamiltonian matrix $\widetilde{\mathbf{H}}$ are upper bounds to the eigenvalues λ_i of the original Hamiltonian matrix $\mathbf{H}$, and thus, also upper bounds to the exact eigenvalues of the Double Renner Hamiltonian $\widehat{H}_{DR}$ defined by (6.1).*

- *A contracted Hamiltonian J-block $\widetilde{\mathbf{H}}^{(J,S,\Gamma_{rve})}$ exhibits the same block sparsity structure as its uncontracted counterpart $\mathbf{H}^{(J,S,\Gamma_{rve})}$ (cf. Table 6.13). The sparsity pattern of the off-diagonal SO- and DK-blocks as described in Sections 7.2.1.1 and 7.2.1.2, however, is destroyed in general.*

- *Note, that the "complete" transformation $\mathbf{T}$ (i.e. the contraction limit is chosen to be $E_{cont} = \infty$) does not turn the Hamiltonian DIAG-blocks into diagonal matrices, because they are obtained as perturbations of the basic diagonal K-blocks $\mathbf{B}^{K,\Gamma_{vib},\eta}$ (additional SO- and NK-terms, see Line 15 in Algorithm 6.7 which describes the construction of the Hamiltonian J-block $\mathbf{H}^{(J,S,\Gamma_{rve})}$). From (7.41) it follows that for a "completely" transformed diagonal K-block it holds*

$$
\begin{aligned}
\widetilde{\mathbf{H}}_{(N,K)} &= \mathbf{V}^*_{N,K} \mathbf{H}_{(N,K)} \mathbf{V}_{N,K} \\
&= \mathbf{V}^*_{N,K} \left(\mathbf{B}^{K,\Gamma_{vib},\eta} + (\mathbf{H}_{nk})_{(N,K)} + (\mathbf{H}_{so})_{(N,K)} \right) \mathbf{V}_{N,K} \\
&= \mathbf{V}^*_{N,K} \mathbf{B}^{K,\Gamma_{vib},\eta} \mathbf{V}_{N,K} + \mathbf{V}^*_{N,K} (\mathbf{H}_p)_{(N,K)} \mathbf{V}_{N,K} \\
&= \text{diag}(\lambda_1, \ldots, \lambda_k) + (\widetilde{\mathbf{H}}_p)_{(N,K)}
\end{aligned}
\qquad (7.48)
$$

The Frobenius norms of the perturbation blocks $(\mathbf{H}_p)_{(N,K)}$ are known to be small in comparison to the norm of the basic diagonal K-block (see also Table 7.1) and due to the invariance under orthogonal transformations this property carries over to the Frobenius norms of the transformed perturbation blocks $\| (\widetilde{\mathbf{H}}_p)_{(N,K)} \|_F$. Hence, the contracted diagonal K-blocks $\widetilde{\mathbf{H}}_{(N,K)}$ are "near-diagonal", i.e. the contraction scheme makes the contracted Hamiltonian J-block $\widetilde{\mathbf{H}}^{J,S,\Gamma_{rve}}$ even more diagonally dominant.

□

Table 7.7 shows the information determined by Algorithm 7.5 (i.e. the number of eigenvalues below the contraction limit E_{cont} per basic diagonal K-block $\mathbf{B}^{K,\Gamma_{vib},\eta}$) and Table 7.8 gives a survey of the sizes of the resulting contracted Hamiltonian J-blocks for different values of E_{cont}. Both choices, $E_{cont} = 5000 \text{ cm}^{-1}$ and $E_{cont} = 10000 \text{ cm}^{-1}$, reduce the arising J-block dimensions to a great deal and computing eigensystems by means of direct solvers (see Section 3.2) now becomes also feasible for rotational J-quantum numbers beyond $J = 9/2$ (see also Fig. 7.5).

Table 7.7.: Number of eigenvalues with $\lambda \leq \lambda_1 + E_{cont}$ in $\mathbf{B}^{K,\eta,\Gamma_{vib}}$ (**MgCN**, big basis)

Basic K-blocks			$\dim(\mathbf{B}^{K,\eta,\Gamma_{vib}})$	Number of $\lambda \leq \lambda_1 + E_{cont}$ for	
K	η	Γ_{vib}		$E_{cont} = 5000 \text{ cm}^{-1}$	$E_{cont} = 10000 \text{ cm}^{-1}$
0	a	A'	2976	164	817
0	b	A'	2976	147	801
1		A'	5952	274	1538
2		A'	5952	290	1593
3		A'	5952	308	1636
4		A'	5952	326	1679
5		A'	5952	343	1722
6		A'	5952	359	1763
7		A'	5952	374	1808

Table 7.8.: Sizes of contracted and uncontracted Hamiltonian J-blocks

$\mathbf{H}^{(J,S,\Gamma_{rve})}$		$\dim(\mathbf{H}^{(S,J,\Gamma_{rve})})$	$\dim(\widetilde{\mathbf{H}}^{(J,S,\Gamma_{rve})})$	
J	Γ_{rve}		$E_{cont} = 5000 \text{ cm}^{-1}$	$E_{cont} = 10000 \text{ cm}^{-1}$
1/2	A'	11904	585	3156
3/2	A'	23808	1149	6287
5/2	A'	35712	1747	9516
7/2	A'	47616	2381	12831
9/2	A'	59529	3050	16232
11/2	A'	71424	3752	19717
13/2	A'	83328	4485	23288

Figure 7.6.: Relative error of eigenvalues of $\widetilde{\mathbf{H}}^{(1/2,\,1/2,\,A')}$

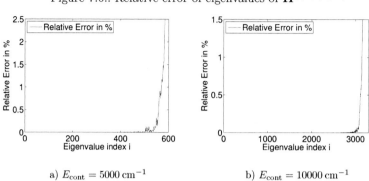

a) $E_{\mathrm{cont}} = 5000\ \mathrm{cm}^{-1}$ \qquad\qquad b) $E_{\mathrm{cont}} = 10000\ \mathrm{cm}^{-1}$

Figure 7.7.: Cosine of angles between exact eigenvectors and Ritz vectors

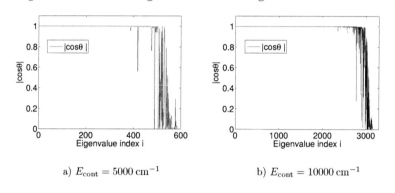

a) $E_{\mathrm{cont}} = 5000\ \mathrm{cm}^{-1}$ \qquad\qquad b) $E_{\mathrm{cont}} = 10000\ \mathrm{cm}^{-1}$

In Figures 7.6 and 7.7 we compare the quality of the exact eigenpairs of the uncontracted Hamiltonian matrices $\mathbf{H}^{(J,S,\Gamma_{rve})}$ with the *Ritz pairs* related to the projections leading to the contracted matrices $\widetilde{\mathbf{H}}^{(J,S,\Gamma_{rve})}$. Fig. 7.6 shows the relative errors of the *Ritz values* (as compared to the exact eigenvalues) and Fig. 7.7 shows the deviations of Ritz and exact eigenvectors in terms of $|\cos\phi|$, where ϕ is the "error angle". The corresponding values are obtained using the well-known relation between cosine and scalar product established in Lemma 2.28. These examples again reveal that assessing the quality of *Ritz vectors* is a delicate matter. The plots for both contraction limits show that the *Ritz vectors* sooner deviate from the related eigenvectors than the corresponding *Ritz values*. For $E_{\mathrm{cont}} = 5000\ \mathrm{cm}^{-1}$ the *Ritz values* around the index $i = 400$ have a very tiny relative error whereas there are already large "peaks" in the deviation of the *Ritz vectors*. On the other hand, one can recognize that below a certain index number ($i \approx 350$ for $E_{\mathrm{cont}} = 5000\ \mathrm{cm}^{-1}$ and $i \approx 2200$ for $E_{\mathrm{cont}} = 10000\ \mathrm{cm}^{-1}$) the *Ritz approximations* are of excellent quality which demonstrates the value of the contracted basis approach.

7.4. Direct Solvers

To assess the potential of the preconditioned JDQR methods in the context of our quantum chemical computations it is enlightening to see how optimized direct solvers perform when applied to the same problem class. We therefore first present the corresponding results for both, product basis and contracted basis calculations, which we obtain using the algorithmic combination of two-stage tridiagonalization and RRR tridiagonal eigensolver (see discussion in Chapter 3.2). Essentially, the timings only depend on the matrix size and for this reason, to demonstrate the limitations of direct approaches, we can restrict ourselves to the most time consuming case in our experiments, i.e. computations w.r.t. to the **MgCN** molecule and the big basis.

7.4.1. Product Basis Calculation

As we already know from the discussion in Section 3.2 the time complexity of eigenvalue computations based on the generic approach (Alg. 3.4) is governed by the tridiagonalization step for which the costs are known to be $\mathcal{O}(n^3)$. Thus, it is hardly suprising that the time required for the computation of a (partial) eigensystem of the Hamiltonian matrix $\mathbf{H}^{(J,S,\Gamma_{rve})}$ grows cubically with respect to the rotational quantum number J, upon which the problem size n depends linearly (cf. Fig. 6.4). Using the big basis (cf. Table 6.5) for the **MgCN**-molecule it is only possible to store Hamiltonian matrices for $J \leq 9/2$, because the SUN^{TM} Fire workstation employed in our experiments is equipped with 32 GB work space (cf. Fig. 7.5). Due to technical restrictions the limit for practical computations is even less, i.e. $J = 7/2$.

Notation 7.9 (Two-stage tridiagonalization + RRR = TST-RRR)
The combination of the two-stage tridiagonalization (cf. Section 3.2.1.2) and the RRR tridiagonal eigensolver (cf. Section 3.2.2.4) is our method of choice for the computation of (partial) eigensystems. For the sake of brevity we refer to it as TST-RRR, the corresponding driver routine in the SBR software by Bischof, Lang et al. [13] is called DSYEVT.   □

Table 7.9 shows the timings for computing 200 eigenpairs using the TST-RRR solver. The restriction on the number of eigenpairs seems somewhat arbitrary at first glance and is motivated by the fact that about 200 (exact number: 197) eigenvalues are less or equal than $E_{\max} = 5000$ cm^{-1} for $J = 1/2$. As this number increases for larger values of J (see Table 7.4) and in view of a better comparability with the JD computations to be discussed in the following sections we decided to leave the number of sought-after eigenvalues fixed. The results impressively illustrate the unfavorable time complexity $\mathcal{O}(n^3)$, for a matrix with the dimension $n = 47616$ more than one day is needed, although the problem size is rather moderate and the number of sought-after eigenpairs is small.

It is also interesting to have a closer look at the detailed timings for the involved intermediate steps which are listed in Table 7.10. The following itemization gives a survey of the related computational routines:

- xSYRDB (part of SBR, [13]) reduces the input matrix to banded form

- xSBRDX (part of SBR, [13]) tridiagonalizes the banded matrix obtained

- xSTEGR is the LAPACK [2] implementation [7] of the RRR tridiagonal eigensolver

- xSBACC (part of SBR, [13]) recovers the eigenvectors of the banded matrix

- xORMTR (BLAS [1] routine, included in LAPACK [2] [7]) recovers the eigenvectors of the original input matrix

Essentially, the overall timings in Table 7.9 result from summing up the corresponding particular timings in Table 7.10 (except for very tiny and negligible contributions due to the auxiliary routine xSY2BC which is only of technical importance and therefore not included in the results). The timings reveal that the main part of the computational effort is due to the band reduction step xSYRDB. By contrast, the costs for the tridiagonalization are tiny. Also the contribution of the algorithmic part that depends on the number of sought-after eigenpairs k_{max}, the back-transformation steps xSBACC and xORMTR, is rather small, at least for a small ratio $r = k_{max}/n$, say $r \leq 0.1$ (see discussion in Section 3.2.1.2 for more details).

Table 7.9.: Overall Timings for 200 eigenpairs computed by TST-RRR

J	Problem size n	Storage	Time required for 200 eigenpairs
1/2	11904	1.06 G	2470.68 secs = 0.69 h
3/2	23808	4.22 G	17782.44 secs = 3.39 h
5/2	35712	9.50 G	60612.72 secs = 16.84 h
7/2	47616	16.89 G	142644.83 secs = 39.62 h

Table 7.10.: Detailed timings for 200 eigenpairs computed by TST-RRR

J	Problem size n	Times in secs				
		dsyrdb	dsbrdx	dstegr	dsbacc	dormtr
1/2	11904	2104.07	277.68	4.72	31.13	53.02
3/2	23808	16197.21	1243.33	11.30	90.96	239.23
5/2	35712	57016.81	2844.03	18.03	176.70	556.58
7/2	47616	136116.67	5088.03	25.89	311.88	1101.64

7.4.2. Contracted Basis Calculation

In Section 7.3 we have described a contraction scheme for the Hamiltonian matrices arising in our context. A contracted basis calculation thus comprises the following three parts:

1. computing the projection (by means of Alg. 7.5)

2. contracting the original matrix by applying the projection (see Alg. 7.6)

3. computing the (full) eigensystem of the contracted matrix by means of the LA-PACK [2] RRR driver (cf. Section 3.2.2.4)

Furthermore, we have already analyzed the sizes of contracted matrices and the quality of approximate eigensystems in dependance of the parameter E_{cont}. Let us now turn our attention to the quantitative aspect in terms of computing times. Table 7.11 gives a survey for $E_{\text{cont}} = 5000\ \text{cm}^{-1}$ and a comparison with the corresponding product basis calculations (Table 7.9) reveals that the contracted counterparts are always cheaper, even for the problem with the smallest size. On the other hand, the choice $E_{\text{cont}} = 5000\ \text{cm}^{-1}$ may not always lead to the desired accuracy, especially if a larger number of approximate eigenpairs is sought-after (cf. Figs. 7.6 and 7.7). Table 7.12 shows what impact higher values for E_{cont} have on the computing time for contracted calculations w.r.t. $J = 1/2$. A possible danger when choosing E_{cont} too large is that the contracted calculation can be even more expensive in total than its uncontracted analogon as it is the case for the choices $E_{\text{cont}} = 15000\ \text{cm}^{-1}$ and $E_{\text{cont}} = 20000\ \text{cm}^{-1}$ in our example. Clearly, it is important to find a reasonable trade-off between accuracy and computing time.

Remark 7.10
Unfortunately, it is not possible to use one of the LAPACK [2] driver routines for partial eigensystems when computing the projections (see Alg. 7.5, Line 10), because the range of interest is determined by λ_1 and the offset parameter E_{cont}. For this reason we first always computed the complete eigensystems of the basic diagonal K-blocks $\mathbf{B}^{K,\Gamma_{\text{vib}},\eta}$ and determined the index j of the largest eigenvalue with $\lambda_j \leq E_{\text{cont}}$ by means of a simple while-loop (Alg. 7.5, Lines 12-14). For this reason the timings for the computation of the projection (third column in Table 7.12) are constant apart from minor errors in measurement. However, there should be no major difficulty in devising a driver routine which is specialized for our purpose and reduces the computational overhead. $\qquad\square$

Finally, one should be aware that the contribution of computing and applying the projection is often not negligible and may amount to a considerable percentage of the overall costs as the following example impressively illustrates:

Problem 7.11 (Contracted calculation for quantum numbers $J \geq 7/2$)
*We have seen that when using the big basis for the **MgCN** molecule it is no more feasible to carry out product basis calculations for $J \geq 7/2$ due to memory restrictions. Consequently, one is forced to resort to contracted (direct or iterative) calculations. We consider the case $J = 13/2$ for which the dimension of the Hamiltonian matrix is known to be $n = 83328$ (as per Alg. 6.3). Using the paramter $E_{\mathrm{cont}} = 10000 \mathrm{cm}^{-1}$ the contraction scheme leads to a smaller matrix with the size $n_{\mathrm{cont}} = 23288$. The pie chart depicted in Fig. 7.8 shows that 14% of the overall computing time is due to the computation and application of the projection. Because of the large problem dimension n_{cont} this example will be also of interest later on, when we discuss the use of preconditioned JDQR variants when applied to contracted problems.* □

Table 7.11.: Direct contracted calculation for $E_{\mathrm{cont}} = 5000 \ \mathrm{cm}^{-1}$

J	n	n_{cont}	Projection (Alg. 7.5)	Contraction (Alg. 7.6)	Eigsys. (RRR)
1/2	11904	585	988.28 secs	21.07 secs	1.26 secs
3/2	23808	1149	1769.68 secs	70.53 secs	6.26 secs
5/2	35712	1747	2576.47 secs	128.92 secs	21.99 secs
7/2	47616	2381	3362.88 secs	188.70 sces	52.92 secs

Table 7.12.: Effect of different contraction limits for $J = 1/2$

E_{cont}	n_{cont}	Projection (Alg. 7.5)	Contraction (Alg. 7.6)	Eigsys. (RRR)
5000	585	988.28 secs	21.07 secs	1.26 secs
10000	3156	987.21 secs	131.05 secs	121.64 secs
15000	6555	983.76 secs	327.68 secs	1044.62 secs
20000	9088	982.86 secs	504.53 secs	2917.35 secs

Figure 7.8.: Contracted calculation for $J = 13/2$ and $E_{\mathrm{cont}} = 10000 \ \mathrm{cm}^{-1}$

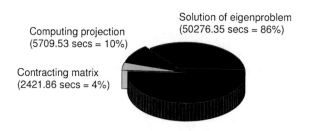

Computing projection
(5709.53 secs = 10%)

Solution of eigenproblem
(50276.35 secs = 86%)

Contracting matrix
(2421.86 secs = 4%)

7.5. JDQR Product Basis Calculation

In this section we come to the main results of our investigations. We will explain how to devise both memory and time efficient preconditioners for product basis Hamiltonian matrices $\mathbf{H}^{(J,S,\Gamma_{rve})}$, which are – apart from the matrix-vector multiplication – the second problem-dependent ingredient to be supplied by the user in order to use one of the preconditioned JDQR methods (Algorithms 4.7, 4.8 and 4.9). We will illustrate the use of our ideas by a couple of numerical results. To begin with, let us briefly sketch some general strategies to tackle the eigenvalue problem specified in Def. 7.3 by means of preconditioned JDQR methods since in contrast to direct eigensolvers (which can be used as "black boxes") the situation is less clear:

1. Compute as many eigenpairs as possible in "one batch":
   This is the most obvious approach, i.e. one constructs a fixed preconditioner $K \approx A - \tau I$ for a specific target value $\tau > 0$ and applies a preconditioned JDQR method. The discussion in Chapter 4.3.6, however, already anticipates that the quality of the preconditioner gradually deteriorates, and hence, the convergence speed decreases for eigenpairs to be computed in subsequent steps, especially if a relatively large number of eigenpairs has already been detected. A further drawback is the fact that the involved projections become increasingly expensive.

2. Use so-called "spectral windows", which is advocated in [47], for instance.
   Given a range $\mathcal{I} = [\alpha, \beta]$ which contains all eigenvalues of interest, the idea is to sub-divide $\mathcal{I}$ into intervals $\mathcal{I}_k = (\tau_k - \epsilon_k, \tau_k + \epsilon_k)$ ("spectral windows", radii $\epsilon_k > 0$) which need not necessarily be disjoint. One can now apply several instances of the JDQR method of choice, one for each target value τ_k, provided that efficient preconditioners K_k for $A_k = A - \tau_k I$ (which should be cheap to construct and to apply) are available. Unfortunately, this is often not true in practice, because the involved shifts $\tau_k > 0$ make the matrices A_k indefinite and as a general wisdom (see [11], [101], e.g.) it is difficult to find efficient preconditioners for such matrices. Besides, it is advisable (see the related discussion in Section 4.4.3) to use the refined or the harmonic preconditioned JDQR variant (Algorithms 4.8 and 4.9) for target values $\tau_k > 0$, because, as a consequence, interior eigenvalues are sought-after. A general technical problem arising in this context lies in the fact that certain eigenpairs may be missed or detected in an irregular order, which is an additional drawback. On the other hand, the inherent paralellism (e.g. use one processor for each window $\mathcal{I}_k$, cf. [47]) may be an argument in favor of the approach.

3. Improve the approach outlined in 1. by updating the preconditioner periodically, say after every 20 detected eigenpairs:
   In Section 4.3.5 we have seen that the auxiliary matrices $\widetilde{Y}$ (contains the preconditioned eigenvectors) and $\widetilde{H}$ (see Algorithms 4.7, 4.8, 4.9 and the preconditioned correction equation (4.103)) have to be updated whenever the preconditioner K

changes. Suppose that k (approximate) eigenpairs (λ_i, q_i) have already been detected. Then the following steps have to be carried out after the construction of an updated preconditioner K_{new}:

- compute $\widetilde{Y}_{new} = K_{new}^{-1} \widetilde{Q}$ ($k + 1$ preconditioner calls)
- compute $\widetilde{H}_{new} = \widetilde{Q}^* \widetilde{Y}_{new}$
- compute LU factorization of $\widetilde{H}_{new}$

Obviously, this approach is only viable if both, construction and application, of K_{new} are not too expensive and if the total number of sought-after eigenpairs is not too large. In [116] some techniques for updates of preconditioners based on algebraic techniques (ILU-type preconditioners) are outlined.

The discussion in the following sections will reveal that only the first strategy is a viable option for our purposes, because preconditioners for interior eigenvalues are not efficient enough.

7.5.1. Preconditioners for Exterior Eigenvalues

In Section 4.2.3.5 we have introduced the term "preconditioner" and pointed out its crucial importance as a convergence accelerator for Krylov methods. A state-of-the-art survey of existing types of preconditioning techniques may be found in [11]. Basically, one can distinguish between the following general approaches (cf. [44], [11]):

- preconditioners based on algebraic techniques:
  The construction is based on algebraic properties (i.e. sparsity pattern, matrix entries, etc.) of the matrix (for instance by means of incomplete factorizations like incomplete Cholesky, ILU, MILU, cf. [11])

- preconditioners based on hierarchy and multilevel techniqes:
  the preconditioner is constructed by transformation of the matrix to different scales (e.g. FFT, algebraic/geometric multigrid, wavelets)

- model and structure oriented preconditioners
  Unlike the above techniques, which are essentially "black boxes" one tries to exploit information from the underlying model

Note that these "classes" are not necessarily disjoint, since model oriented or multilevel based preconditioners in general also make use of algebraic techniques. In the following we will see that the choice of an appropriate approach primarily depends on what part of the spectrum of $B = A - \tau I$ one is interested in. If one is looking for exterior eigenvalues, then $\tau = 0$ and B is positive definite. We will identify near-optimal block preconditioners that exploit information of the model (quantum numbers, location of information), and hence, belong to the latter of the approaches listed above. They are conceptually simple, easy to construct and have moderate storage requirements. For interior eigenvalues the situation is much less favorable, the shift $\tau > 0$ makes B indefinite.We have already pointed out above that devising appropriate preconditioners for the indefinite case is rather difficult and often does not lead to satisfactory results. Unfortunately, this is also true for our concrete case and it will turn out that the techniques derived for exterior eigenvalues in general fail for the computation of interior eigenvalues. In our experiments it became clear that only sophisticated algebraic multilevel techniques (e.g. the ILUPACK software by Bollhöfer [15]) or extremely expensive block preconditioners are appropriate. Conversely, it is an often-made experience that preconditioners that do not lead to success for exterior eigenvalues will not work for interior eigenvalues either.

An important aspect when dealing with preconditioners is the analysis of costs and trade-offs, i.e. one has to answer the question whether it pays to incorporate preconditioning: Does a possibly faster convergence of the method in terms of iteration steps compensate for the costs of setting up and applying the preconditioner? This is typical of the assessment of Krylov methods designed for the solution of *one* linear system. In the context of *Jacobi-Davidson methods* working with a fixed preconditioner the situation is different, because *several* systems have to be solved (one in each step of the JD/JDQR iteration) so that the costs for constructing a preconditioner are often negligable, especially if a large number of eigenpairs is computed. Besides, we will see that preconditioning is mandatory for the Hamiltonian matrices we are dealing with, i.e. the unpreconditioned JDQR and JD methods (Algorithms 4.2 and 4.5) generally fail to converge. Consequently, the focus of our experiments will be on the comparison of preconditioners that make the method converge. In the following two sections we will comment more systematically on how to choose the preconditioner for exterior and interior eigenvalues.

7.5.1.1. Specification and Properties

We recall the following elementary and important types of preconditioners:

Definition 7.12 (Jacobi and Block Jacobi preconditioner)
Let $A, K \in \mathbb{C}^{n \times n}$ such that $K \approx A$. K is called

- Jacobi preconditioner, if $K = diag(a_{11}, a_{22}, \ldots, a_{nn})$ (i.e. K is a diagonal matrix with $k_{ii} = a_{ii}$)

- block Jacobi preconditioner, if $K = diag(A_{11}, A_{22}, \ldots, A_{nn})$ (i.e. K is a block-diagonal matrix with $K_{ii} = A_{ii}$ and $K_{ij} = 0$ for $i \neq j$)

When dealing with Hamiltonian matrices $\mathbf{H}^{(J,S,\Gamma_{rve})}$ it is advantageous to use the following terms:

Definition 7.13 (J-block, N-block and K-block preconditioners)
A block Jacobi preconditioner that approximates the Hamiltonian matrix $\mathbf{H}^{(J,S,\Gamma_{rve})}$ is called

- J-block preconditioner, *if it is the whole Hamiltonian matrix* $\mathbf{H}^{(J,S,\Gamma_{rve})}$

- N-block preconditioner, *if its diagonal blocks are the diagonal N-blocks of* $\mathbf{H}^{(J,S,\Gamma_{rve})}$

- K-block preconditioner, *if its diagonal blocks are the diagonal K-blocks of* $\mathbf{H}^{(J,S,\Gamma_{rve})}$

The Tables 7.1, 7.2 and 7.3 show that the main weight of the information is concentrated in the diagonal K-blocks (cf. the description of the matrix properties in Section 7.1). The off-diagonal DK- and SO-blocks can be regarded as a small perturbation E, and the Bauer-Fike theorem (Prop. 2.23) implies that the matrix made up of the diagonal K-blocks is a good approximation to the Hamiltonian matrix $\mathbf{H}^{(J,S,\Gamma_{rve})}$, in the sense that corresponding eigenvalues at most deviate by the norm of the perturbation E. This suggests to use K-block preconditioners, which are often also referred to as *Approximate Hamiltonian Preconditioners* (briefly: *AHP*) in the literature (cf. [92], e.g.). The obvious and severe disadvantage, however, are the costs for storing the LU factorizations of the diagonal K-blocks, which can be prohibitive, especially for product basis computations with respect to large rotational J quantum numbers and big vibrational basis sets. In case of the **MgCN**-molecule, for instance, the basis set defined in Table 6.5 leads to dense and huge (5952×5952)-blocks $(K \neq 0)$, which illustrates the problem. Besides, as a negative side-effect the preconditioning operations can be extremely time-consuming. We therefore have to look for cheaper and less memory consuming variants which are nevertheless not considerably worse than K-block preconditioners in terms of iteration steps. Towards this end we will pursue a combination of the following two strategies

1. exploit multiple occurences of information (see Fact 6.10 and Corollary 6.11)

2. exploit hidden sub-blocks in diagonal K-blocks.

Corollary 6.11 and the decomposition of the Hamiltonian in (6.24) suggest to use the basic diagonal K-blocks $\mathbf{B}^{K,\Gamma_{vib},\eta}$ introduced in Fact 6.10 as preconditioners. For large rotational J quantum numbers and spin multiplicity $S = 1/2$ this almost halves the storage requirements. However, it turns out, that this approximation is too crude and consequently the preconditioned standard JDQR method fails to converge. Fortunately, a slight modification of the idea leads to a preconditioner which is almost as efficient as the original K-block preconditioner and which has the same favorable storage properties as the original approach. This is made precise in the follwing

Definition 7.14 (Modified K-block preconditioner for $S = 1/2$)
For any given J quantum number we have $N_{min} = |J - S|$ and $N_{max} = J + S$ which implies $N_{max} = N_{min} + 1$ for $S = 1/2$. Let P be the K-block preconditioner as per Def. 7.13. A modified K-block preconditioner $\widetilde{P}$ is obtained as follows: Replace the blocks $P_{N_{max},K}$ by $P_{N_{min},K}$ for $0 < K \leq N_{min}$.

Remark 7.15
*The above definition seems somewhat restrictive at first glance, because we only consider the case $S = 1/2$. This is due to the fact that in our experiments only data for molecules with $S = 1/2$ (**MgCN** and **HOO**) was available. From our results to be presented in the following it may be expected that analogous ideas also work for molecules with spin multiplicities $S > 1/2$.* □

The following example clarifies the meaning of Def. 7.14

Example 7.16
Let $S = 1/2$ and $J = 5/2$. Then the blocks $P_{2,1}$ and $P_{2,2}$ (see Table 6.1 for an illustration) appear twice in the modified preconditioner $\widetilde{P}$, i.e.

- $\widetilde{P}_{2,1} = P_{2,1}$ *and* $\widetilde{P}_{3,1} = P_{2,1}$

- $\widetilde{P}_{2,2} = P_{2,2}$ *and* $\widetilde{P}_{3,2} = P_{2,2}$

The blocks $P_{2,0}$, $P_{3,0}$ and $P_{3,3}$ only appear once. □

As compared to ordinary K-Block preconditioners, storage and factorization costs reduce considerably. But we can even go a step further: Up to now, we have treated the diagonal K-Blocks $\mathbf{H}_{N,K}^{(J,S,\Gamma_{rve})}$ as dense matrices without any further structure. However, this is not entirely true as we shall see in the following: The hidden structure can be made visible by considering matrix entries whose moduli are greater than a certain threshold parameter γ, e.g. $\gamma = 0.01$. For the ease of presentation we employ the following notation:

Definition 7.17 (A greater than γ)
For an arbitrary matrix $A \in \mathbb{C}^{n \times m}$ we define

$$\mathbf{A}^{>\gamma} = \begin{cases} a_{ij} & , \quad \text{if } |a_{ij}| > \gamma \\ 0 & , \quad \text{otherwise} \end{cases} \tag{7.49}$$

Figures 7.9 and 7.10 show the sparsity plots of the matrix $\left(\mathbf{H}_{(1,1),(1,1)}^{(\frac{1}{2},\frac{1}{2},A')}\right)^{>0.01}$ for the big and the small basis as defined in Table 6.5. The interesting observation (which can also be made for the **HOO**-molecule) is the shimmering of the product basis structure. More precisely, one can recognize a surprisingly sharp partitioning into $(N_R \times N_R)$ sub-blocks which in turn motivates to use diagonal sub-blocks of a K-block, the number being a divisor of $N_R^{(lim)}$.

Figure 7.9.: Sparsity pattern of $\left(\mathbf{H}_{(1,1),(1,1)}^{(\frac{1}{2},\frac{1}{2},A')} \right)^{>0.01}$ (**MgCN** molecule, big basis)

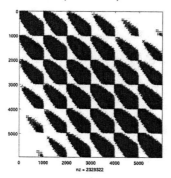

Figure 7.10.: Sparsity pattern of $\left(\mathbf{H}_{(1,1),(1,1)}^{(\frac{1}{2},\frac{1}{2},A')} \right)^{>0.01}$ (**MgCN** molecule, small basis)

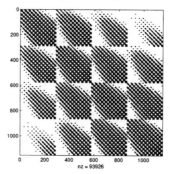

This leads to the following generalization of the block preconditioners specified in Def. 7.13 and Def. 7.14:

Definition 7.18 (K-block / modified K-block preconditioner, sub-blocks)
Let $\mathbf{H}^{(J,S,\Gamma_{rve})}$ be the FBR of $\widehat{H}_{DR}$ with respect to a given rotational quantum number J and a given vibrational basis with the sizes $N_R^{(lim)}$, $N_r^{(lim)}$, $(v_2^a)^{(lim)}$ and $(v_2^b)^{(lim)}$. Let be $n_b \in \mathbb{N}$ a divisor of $N_R^{(lim)}$, i.e. $n_b | N_R^{(lim)}$. Then we define

- a $K(n_b)$-block preconditioner as the matrix P_{n_b} obtained by cutting out n_b diagonal sub-blocks with equal size in every K-block $P_{N,K}$. Algorithm 7.7 formalizes the construction and Algorithm 7.8 specifies how a P_{n_b} is applied to a given vector x of appropriate dimension.

- a modified $K(n_b)$-Block preconditioner, as the matrix $\widetilde{P}_{n_b}$ obtained from P_{n_b} according to the principles from Definition 7.14. A formal description of how to set up and to apply $\widetilde{P}(n_b)$ is given in Algorithms 7.9 and 7.10.

Algorithm 7.7: Setting up K-block preconditioner using n_b sub-blocks

1   function
   PREC = setupprec(**H**, Γ_{rve}, S, J, $N_R^{(lim)}$, $N_r^{(lim)}$, $(v_2^a)^{(lim)}$, $(v_2^b)^{(lim)}$, ABC, n_b)
2   $\mathbf{y} = \mathbf{0}$
3   $N_{min} = |J - S|$, $N_{max} = J + S$
4   **for** $N = N_{min}, \ldots, N_{max}$ **do**
5   **for** $K = 0, \ldots, N$ **do**
6       $k = \dim_K(\Gamma_{rve}, N, K, N_R^{(lim)}, N_r^{(lim)}, (v_2^a)^{(lim)}, (v_2^b)^{(lim)}, \text{ABC})$
7       $k_b = k/n_b$
8       $I = \displaystyle\sum_{\ell=N_{min}+1}^{N} \ell + K$
9   **DIAG** := **JDIAG**$[I]$
10   **for** $j = 1, \ldots, n_b$ **do**
11        $\widehat{\textbf{DIAG}} = \textbf{DIAG}[(j-1) \cdot k_b + 1 : j \cdot k_b, (j-1) \cdot k_b + 1 : j \cdot k_b]$
12        $\widehat{\textbf{DIAG}} = \mathbf{P} \cdot \mathbf{L} \cdot \mathbf{U}$
13        $\textbf{PREC}[I][j] = [\, \mathbf{P}, \mathbf{L}, \mathbf{U} \,]$
14   **end for**
15   **end for**
16   **end for**
17   **return PREC**

Algorithm 7.8: K-block preconditioner using n_b sub-blocks

1   function $\mathbf{y} = $ prec(**PREC**, $\mathbf{x}$, Γ_{rve}, S, J, $N_R^{(lim)}$, $N_r^{(lim)}$, $(v_2^a)^{(lim)}$, $(v_2^b)^{(lim)}$, ABC, n_b)
2   $\mathbf{y} = \mathbf{0}$
3   $N_{min} = |J - S|$, $N_{max} = J + S$
4   **for** $N = N_{min}, \ldots, N_{max}$ **do**
5   **for** $K = 0, \ldots, N$ **do**
6       $i = \text{pos}_K(\Gamma_{rve}, S, J, N, K, N_R^{(lim)}, N_r^{(lim)}, (v_2^a)^{(lim)}, (v_2^b)^{(lim)}, \text{ABC})$
7       $k = \dim_K(\Gamma_{rve}, N, K, N_R^{(lim)}, N_r^{(lim)}, (v_2^a)^{(lim)}, (v_2^b)^{(lim)}, \text{ABC})$
8       $k_b = k/n_b$
9       $I = \displaystyle\sum_{\ell=N_{min}+1}^{N} \ell + K$
10   **for** $j = 1, \ldots, n_b$ **do**
11        $[\mathbf{P}, \mathbf{L}, \mathbf{U}] := \textbf{PREC}[I][j]$
12        $\mathbf{y}[i + (j-1) \cdot k_b : i + j \cdot k_b - 1] = \mathbf{U}^{-1}\mathbf{L}^{-1}\mathbf{P}^T \cdot \mathbf{x}[i + (j-1) \cdot k_b : i + j \cdot k_b - 1]$
13   **end for**
14   **end for**
15   **end for**
16   **return y**

Example 7.19 (Preconditioners with respect to different basis sets)

1. For the big basis (cf. Table 6.5) we have $N_R^{(lim)} = 6$ and according to Def. 7.18 one can use $n_b \in \{1, 2, 3, 6\}$ as a sub-block parameter such that there are 8 possible $K(n_b)$-block and modified $K(n_b)$-block preconditioners.

2. In case of the small basis (cf. Table 6.5) it holds $N_R^{(lim)} = 4$. Because of $n_b | N_R^{(lim)}$ one can use $n_b \in \{1, 2, 4\}$, and hence, there are 6 possible combinations, a survey and illustration of which is given in Table 7.13 .

$\square$

Remark 7.20

- Obviously, the choice $n_b = 1$ leads to the preconditioners introduced in Def. 7.13 and Def. 7.18.

- At first glance the restriction on the number of sublocks n_b as a divisor of $N_R^{(lim)}$ in Def. 7.18 seems somewhat artificial because any divisor s of the K-block dimension d as per Alg. 6.1 is possible, too. However, our results show that these choices are in general less efficient or may even lead to failure of convergence.

- The Algorithms 7.7 and 7.8 for ordinary $K(n_b)$-block preconditioners make use of the ideas presented in Section 7.2 in order to store and address the corresponding LU factorizations (see Fig. 7.11). The storage scheme for modified $K(n_b)$-block preconditioners employed by the Algorithms 7.9 and 7.10 is a straight forward adaption of these ideas and is illustrated in Fig. 7.12

$\square$

Figure 7.11.: Storage scheme for the $K(n_b)$-block preconditioner ($S = 1/2$, $J = 5/2$, $n_b = 4$)

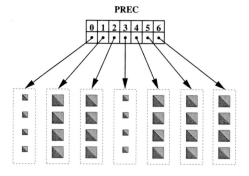

Figure 7.12.: Storage scheme for the modified K(n_b)-block preconditioner ($S = 1/2$, $J = 5/2$, $n_b = 4$)

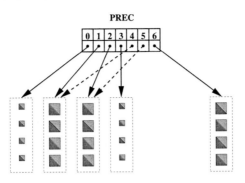

Finally, it is clear that the times for both setting up (LU factorizations of the involved blocks) and applying a K(n_b)-block preconditioner depend linearly upon the rotational quantum number J (see Fig. 7.13). However, savings for modified K(n_b)-block preconditioners can only be realized as far as their construction is concerned, whereas the timings for the application of modified and standard K(n_b)-block preconditioners are identical. This is illustrated in Fig. 7.13, Part a) and b).

Figure 7.13.: Timings for $K(n_b)$-block preconditioners (big basis, **MgCN**)

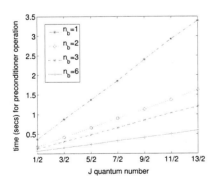

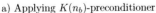

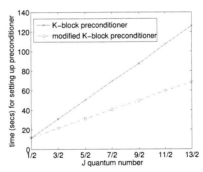

a) Applying $K(n_b)$-preconditioner   b) Setting up $K(3)$-preconditioner

---

Algorithm 7.9: Setting up mod. K-block precond. using n_b sub-blocks, $S = 1/2$

---

1 function

   $\textbf{PREC} = \textbf{setupprec}(\, \textbf{H},\ \Gamma_{rve},\ S,\ J,\ N_R^{(lim)},\ N_r^{(lim)},\ (v_2^a)^{(lim)},\ (v_2^b)^{(lim)},\ \text{ABC},\ n_b \,)$

2 $y = 0$

3 $S = 1/2$

4 $N_{min} = |J - S|,\ N_{max} = J + S$

   /* LU-Factorize representatives of "double" occurences   */

5 for $K = 1, \ldots, N_{min}$ do

6 $\quad k = \dim_K(\, \Gamma_{rve},\ N_{min},\ K,\ N_R^{(lim)},\ N_r^{(lim)},\ (v_2^a)^{(lim)},\ (v_2^b)^{(lim)},\ \text{ABC} \,)$

7 $\quad k_b = k/n_b$

8 $\quad I_1 = \displaystyle\sum_{\ell=N_{min}+1}^{N_{min}} \ell + K$

9 $\quad I_2 = \displaystyle\sum_{\ell=N_{max}+1}^{N_{max}} \ell + K$

10 $\quad \textbf{DIAG} := \textbf{JDIAG}[I]$

11 $\quad$ for $j = 1, \ldots, n_b$ do

12 $\quad\quad \widehat{\textbf{DIAG}} = \textbf{DIAG}[(j-1)\cdot k_b + 1 : j\cdot k_b, (j-1)\cdot k_b + 1 : j\cdot k_b]$

13 $\quad\quad \widehat{\textbf{DIAG}} = \textbf{P}\cdot\textbf{L}\cdot\textbf{U}$

14 $\quad\quad \textbf{PREC}[I_1][j] = [\,\textbf{P},\textbf{L},\textbf{U}\,]$

15 $\quad\quad \textbf{PREC}[I_2][j] = [\,\textbf{P},\textbf{L},\textbf{U}\,]$

16 $\quad$ end for

17 end for

   /* LU-Factorize "single" occurences   */

18 for $(N, K) \in \{\, (N_{min}, 0),\ (N_{max}, 0),\ (N_{max}, N_{max}) \,\}$ do

19 $\quad k = \dim_K(\, \Gamma_{rve},\ N,\ K,\ N_R^{(lim)},\ N_r^{(lim)},\ (v_2^a)^{(lim)},\ (v_2^b)^{(lim)},\ \text{ABC} \,)$

20 $\quad k_b = k/n_b$

21 $\quad I = \displaystyle\sum_{\ell=N_{min}+1}^{N} \ell + K$

22 $\quad$ for $j = 1, \ldots, n_b$ do

23 $\quad\quad \widehat{\textbf{DIAG}} = \textbf{DIAG}[(j-1)\cdot k_b + 1 : j\cdot k_b, (j-1)\cdot k_b + 1 : j\cdot k_b]$

24 $\quad\quad \widehat{\textbf{DIAG}} = \textbf{P}\cdot\textbf{L}\cdot\textbf{U}$

25 $\quad\quad \textbf{PREC}[I][j] = [\,\textbf{P},\textbf{L},\textbf{U}\,]$

26 $\quad$ end for

27 end for

28 return $\textbf{PREC}$

Algorithm 7.10: Modified K-block preconditioner using n_b sub-blocks for $S = 1/2$

---

1 function $\mathbf{y} = \mathbf{prec}(\mathbf{PREC}, \mathbf{x}, \Gamma_{rve}, S, J, N_R^{(lim)}, N_r^{(lim)}, (v_2^a)^{(lim)}, (v_2^b)^{(lim)}, \text{ABC}, \text{n}_b)$

2 $\mathbf{y} = 0$

3 $= 1/2$

4 $N_{min} = |J - S|$, $N_{max} = J + S$

    /* Solve for "double" occurences   */

5 for $K = 1, \ldots, N_{min}$ do

6      $i_1 = \text{pos}_K(\Gamma_{rve}, S, J, N_{min}, K, N_R^{(lim)}, N_r^{(lim)}, (v_2^a)^{(lim)}, (v_2^b)^{(lim)}, \text{ABC})$

7      $i_2 = \text{pos}_K(\Gamma_{rve}, S, J, N_{max}, K, N_R^{(lim)}, N_r^{(lim)}, (v_2^a)^{(lim)}, (v_2^b)^{(lim)}, \text{ABC})$

8      $k = \dim_K(\Gamma_{rve}, N, K, N_R^{(lim)}, N_r^{(lim)}, (v_2^a)^{(lim)}, (v_2^b)^{(lim)}, \text{ABC})$

9      $k_b = k/n_b$

10      $I = \sum\limits_{\ell=N_{min}+1}^{N_{min}} \ell + K$

11   for $j = 1, \ldots, n_b$ do

12          $\mathbf{x_1} = \mathbf{x}[(i_1 + (j-1) \cdot k_b : i_1 + j \cdot k_b - 1]$

13          $\mathbf{x_2} = \mathbf{x}[(i_2 + (j-1) \cdot k_b : i_2 + j \cdot k_b - 1]$

14          $\mathbf{X} = [\mathbf{x_1}, \mathbf{x_2}]$

15          $[\mathbf{P}, \mathbf{L}, \mathbf{U}] := \mathbf{PREC}[I][j]$

16          $\mathbf{Y} = [\mathbf{y_1}, \mathbf{y_2}] = \mathbf{U}^{-1}\mathbf{L}^{-1}\mathbf{P}^T \cdot \mathbf{X}$

17          $\mathbf{y}[(i_1 + (j-1) \cdot k_b : i_1 + j \cdot k_b - 1] = \mathbf{y_1}$

18          $\mathbf{y}[(i_2 + (j-1) \cdot k_b : i_2 + j \cdot k_b - 1] = \mathbf{y_2}$

19   end for

20 end for

    /* Solve for "single" occurences   */

21 for $(N, K) \in \{ (N_{min}, 0), (N_{max}, 0), (N_{max}, N_{max}) \}$ do

22      $i = \text{pos}_K(\Gamma_{rve}, S, J, N, K, N_R^{(lim)}, N_r^{(lim)}, (v_2^a)^{(lim)}, (v_2^b)^{(lim)}, \text{ABC})$

23      $k = \dim_K(\Gamma_{rve}, N, K, N_R^{(lim)}, N_r^{(lim)}, (v_2^a)^{(lim)}, (v_2^b)^{(lim)}, \text{ABC})$

24      $k_b = k/n_b$

25      $I = \sum\limits_{\ell=N_{min}+1}^{N} \ell + K$

26   for $j = 1, \ldots, n_b$ do

27          $[\mathbf{P}, \mathbf{L}, \mathbf{U}] := \mathbf{PREC}[I][j]$

28          $\mathbf{y}[(i + (j-1) \cdot k_b : i + j \cdot k_b - 1] = \mathbf{U}^{-1}\mathbf{L}^{-1}\mathbf{P}^T \cdot \mathbf{x}[(i + (j-1) \cdot k_b : i + j \cdot k_b - 1]$

29   end for

30 end for

31 return $\mathbf{y}$

Table 7.13.: Exploited information and storage ($J = 5/2$, $S = 1/2$, $N_R^{(lim)} = 4$)

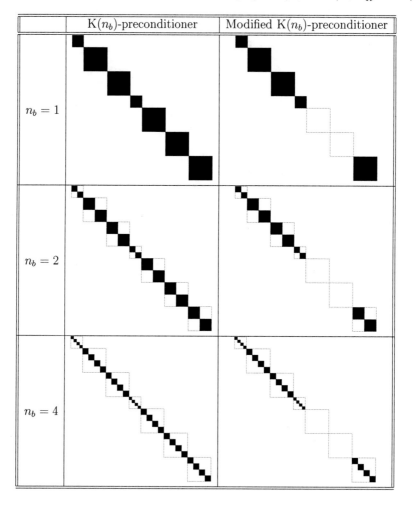

7.5.1.2. Numerical Results

First of all, the plots in Figure 7.14 show that the unpreconditioned standard JDQR method (Alg. 4.5) fails to converge (when computing exterior eigenvalues) and that employing a Jacobi preconditioner is not sufficient either (in spite of the diagonal dominance of the Hamiltonian matrices). This again shows that appropriate preconditioners are neccessary for the success of JDQR methods.

Figure 7.14.: Failure of the JDQR method (**MgCN**, $J = 1/2$, big basis)

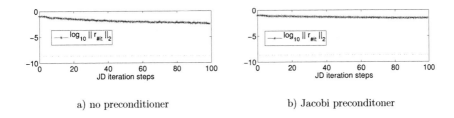

a) no preconditioner   b) Jacobi preconditoner

Furthermore, we have repeatedly pointed out that the preconditioned JDQR variants are capable of computing a relatively large number of eigenpairs although the preconditioner is held fixed. This phenomenon is impressively confirmed by our experiments in which we applied the preconditioned JDQR method (using a K(1)-preconditioner) in order to determine all eigenpairs below $E_{max} = 5000 \text{ cm}^{-1}$. The results in Table 7.14 show that the JDQR methods are capable of computing several hundreds of eigenpairs.

Table 7.14.: JDQR product basis calculation, **MgCN**, $E_{max} = 5000 \text{ cm}^{-1}$

J	n	Eigenpairs	It.steps	Mat-Vec mults	Computing time
1/2	11904	192	2823	13076	2.81 h
3/2	23808	374	6638	32804	16.94 h
5/2	35712	531	10685	54484	47.34 h
7/2	47616	668	14716	76529	170.96 h
9/2	59520	786	18598	98017	349.12 h

On the other hand, the computational effort to be invested is rather high. This has several reasons

- to play it safe we employed the K(1)-preconditioner as it is the best available and reasonable block approximation to the Hamiltonian matrix. However, it will turn out that it is also the least efficient choice.

- the computation becomes increasingly expensive because all detected eigenpairs have to be projected out ("deflation", see Section 4.3.1).

- the efficiency of the fixed preconditioner gradually deteriorates (cf. discussion in Section 4.3.6) in terms of iteration steps (outer loop) and matrix-vector multiplications required for the computation of one approximate eigenpair. This is nicely illustrated in Fig. 7.15. The oscillatory behavior can be explained as follows: If a large number of steps has to be invested to determine an eigenpair with the desired accuracy the subspace will also be rich in information on eigenpairs to be detected in subsequent steps. This in turn has the consequence that the effort will be considerably smaller.

- we did not make use of any parallelization, which we will briefly discuss in Section 7.7 and which can reduce the computing times to a great deal.

Figure 7.15.: Deterioration of the $K(1)$-preconditioner ($J = 9/2$, big basis, **MgCN**)

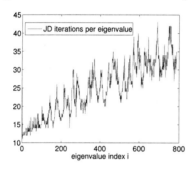 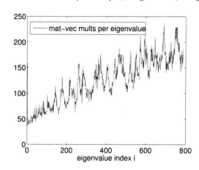

a) JD steps per eigenvalue   b) matrix-vector multiplications per eigenvalue

Let us now come to the answer of the crucial question of which of the block preconditioners one should actually make use of. To this end, we computed 200 eigenpairs for $J = 9/2$ (**MgCN**-molecule, big basis). This problem is reasonably large ($n = 59520$) on the one hand and allows for analyzing the differences between standard and modified block preconditioners on the other hand (for smaller values of J possible effects may not be distinct enough). The following observations can be made from the results in Table 7.15:

- Rather surprisingly, modified block preconditioners are always about as efficient as (often even slightly better than) their standard counterparts.

- Clearly, the larger one chooses the blocking parameter n_b the less information is included in the corresponding $K(n_b)$-block preconditioner which consequently leads to a larger number of iteration steps and matrix-vector multiplications in the computational process.

- On the other hand, the results impressively show that the modified $K(6)$ preconditioner is the most efficient choice (only about 75 % of the corresponding time for the standard $K(1)$-preconditioner is needed).

- The explanation for these contradictory ovservations is given in the last column of Table 7.15 where the ratio between the time required for a matrix-vector multiplication and a preconditioner operation is listed. For $n_b = 1$ a preconditioner call is two times as expensive as a matrix-vector multiplication whereas only a relatively small fraction of 0.2 is required for $n_b = 6$. Hence, the loss of information is obviously compensated for by the cheaper preconditioner call.

Table 7.15.: Efficiency of Preconditioners for $J = 9/2$, 200 EV, big basis, **MgCN**

Preconditioner	#it	#mv	Computing time	ratio prec/mv
std. $K(1)$	2330	12224	17.45 h	1.9980
mod. $K(1)$	2345	12358	17.43 h	1.9980
std. $K(2)$	2860	14668	14.41 h	0.5694
mod. $K(2)$	2681	14619	15.08 h	0.5694
std. $K(3)$	2736	15115	13.66 h	0.4286
mod. $K(3)$	2733	15035	13.42 h	0.4286
std. $K(6)$	3053	17248	13.13 h	0.2012
mod. $K(6)$	3032	17140	12.82 h	0.2012

The above observations are made independent of the J quantum number, the basis set and the type of molecule. Hence, one can clearly recommend to employ the modified $K(N_R^{(lim)})$ preconditioner, and consequently, we will restrict our analysis to this choice in what follows. The Tables 7.16 and 7.17 contain the detailed results for the computation of 200 eigenpairs for both the **MgCN** and the **HOO** molecule. Additionally, the timings in the last column are visualized in Fig. 7.16. We can make the following observations:

- The number of iteration steps and matrix-vector multiplications is almost constant, independent of the J quantum number. This seems to be inherent to the structure of the problem.

- Consequently, the computing time depends linearly upon J as illustrated in Fig. 7.16 (in contrast to the cubic behavior for direct solvers).

- A comparison with the corresponding direct calculation (cf. Table 7.9) shows that the *point of break even* is reached for $J = 5/2$, i.e. starting from this value direct computations are either far more expensive or no more feasible.

Table 7.16.: JDQR calculation, 200 EV, **MgCN**, big basis, modified $K(6)$ preconditioner

J	n	It.steps	Mat-Vec mults	Computing time
1/2	11904	3079	17340	1.90 h
3/2	23808	3078	17365	4.71 h
5/2	35712	3050	17212	7.32 h
7/2	47616	3035	17194	10.16 h
9/2	59520	3032	17140	12.82 h
11/2	71424	3098	17672	16.41 h
13/2	83328	3113	17800	19.23 h

Table 7.17.: JDQR calculation, 200 EV, **HOO**, big basis, modified $K(8)$ preconditioner

J	n	It.steps	Mat-Vec mults	Computing time
1/2	5120	2589	13450	0.33 h
3/2	10240	2557	13255	0.78 h
5/2	15360	2538	13122	1.42 h
7/2	20480	2498	12905	1.92 h
9/2	25600	2492	12795	2.47 h
11/2	30720	2510	13024	3.12 h

Figure 7.16.: Linear time behavior of preconditioned JDQR for 200 eigenpairs

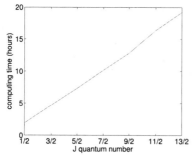
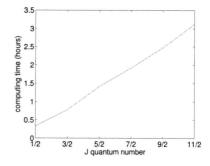

a) **MgCN**, modified $K(6)$-preconditioner

b) **HOO**, modified $K(8)$-preconditioner

7.5.2. Preconditioners for Interior Eigenvalues

As we have repeatedly pointed out computing *interior eigenvalues* by means of iterative projection methods rather often becomes a complicated matter. One reason is that the standard *Rayleigh-Ritz procedure* (Alg. 3.12) may yield bad or even unusable approximations (see Section 3.3.1.1 for examples and an analysis of this problem). To improve the situation we introduced the refined and the harmonic Rayleigh-Ritz procedure (Algs. 3.13 and 3.14) which we applied to a small model problem (Problem 4.7, $J = 1/2$ and big basis) in Section 4.4.3 (embedded in the corresponding preconditioned JDQR variants, Algs. 4.8 and 4.9). The results obtained (Result 4.10) demonstrate the use of the alternative projection methods when looking for interior eigenvalues because the computation becomes more effective in terms of matrix-vector opterations and computing time. However, these results must not mislead to the conclusion that the proper choice of the extraction method may serve as a general remedy to arising convergence difficulties. A closer look at the following analogous problem will reveal that this is not sufficient in general.

Problem 7.21 (Interior eigenvalues for $J = 5/2$)
*We consider an **MgNC** molecule and want to compute its rovibronic engery levels with respect to the rotational quantum number $J = 5/2$, the parameters $S = 1/2$, $\Gamma_{rve} = A'$ and a large vibrational basis set ("big basis", cf. Table 6.5) whose size is determined by the parameters*

$$N_r^{(lim)} = 16, N_R^{(lim)} = 6, (v_2^a)^{(lim)} = 31, (v_2^b)^{(lim)} = 31. \qquad (7.50)$$

From Section 6.3 (cf. Alg. 6.3) we know that the resulting FBR $\mathbf{H}^{(\mathbf{J},\mathbf{S},\mathbf{\Gamma}_{rve})}$ has the dimension $n = 35712$. We are interested in the 10 eigenpairs that are closest to the target value $\tau = 0.022782$. The preconditioner must approximate the shifted Hamiltonian matrix, i.e. we require $K \approx \mathbf{H}^{(\mathbf{J},\mathbf{S},\mathbf{\Gamma}_{rve})} - \tau I$. □

Applying the three preconditioned JDQR variants to the problem and making use of the $K(1)$-block preconditioner (as per Def. 7.18) we obtain the following results:

Result 7.22 (Interior eigenvalues, failure of the $K(1)$-block preconditioner)
As pointed out in Section 7.5.1.2 the $K(1)$-block preconditioner does not the contain the most information of all $K(n_b)$-block preconditioners. However, regardless of the extraction method all JDQR variants fail to compute the 10 sought-after eigenpairs within 200 JD iteration steps (outer loop) as the convergence plots (Figures 7.17, 7.18 and 7.19) clearly show.

Figure 7.17.: Failure of standard extraction, $K(1)$-block preconditioner

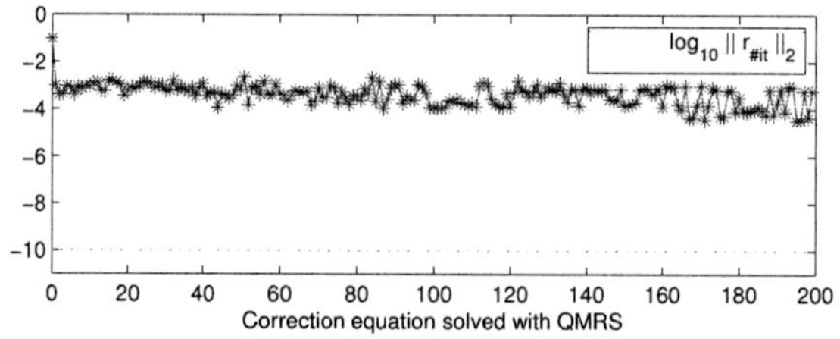

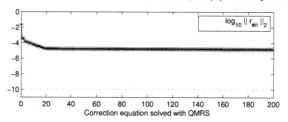

Figure 7.18.: Failure of refined extraction, $K(1)$-block preconditioner

Figure 7.19.: Failure of harmonic extraction, $K(1)$-block preconditioner

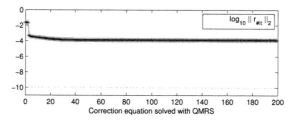

□

Obviously, it is not the extraction method we have to blame for the failure of our methods. One reason lies in the choice of the preconditioner: The $K(1)$-block preconditioner works perfectly well for *all* eigenproblems where exterior eigenvalues are sought-after, but it seems to be problematic for interior eigenvalues near an arbitrary target value τ. A possible way to cure the lack of convergence might be to increase the maximum number of "inner iterations" (i.e. the maximum number of iterations steps carried out by the Krylov solver for one "outer iteration" (i.e. one pass of the JD-loop). Of course, we cannot expect the remaining standard and modified $K(n_b)$ variants ($n_b > 1$) to work better as they contain even less information.

As far as block preconditioners are concerned, it remains to analyze the N-block and

J-block preconditioners proposed in Def. 7.13, and towards this end, we employ the
N-block preconditioner to tackle Problem 7.21 (we omit the *J*-block preconditioner as
it leads to analogous results):

Result 7.23 (Interior eigenvalues, *N*-block preconditioner)
*For the sake of simplicity we restrict ourselves to the refined and harmonic JDQR vari-
ants. As opposed to the situation for the K-block preconditioner both of them converge
rapidly (less than 35 matrix-vector multiplications per eigenvalue) to the desired eigenvalues without any remarkable difficulty.*

Figure 7.20.: Interior eigenvalues obtained by refined extraction

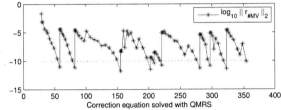

Figure 7.21.: Interior eigenvalues obtained by harmonic extraction

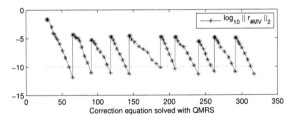

□

The *N*-block preconditioner makes the JDQR methods converge, which in combination
with Result 7.22 shows that, obviously, the contribution of the DK blocks can no more be
neglected for interior eigenvalues. However, this is unsatisfactory for practical purposes,
because the LU factorizations of 2 huge matrix blocks are required and the ratio prec/mv
of the time required for a preconditioner solve and a matrix-vector multiplication (see
discussion for exterior eigenvalues in Section 7.5.1) is extremely unfavorable as compared
to the cheap $K(6)$-block preconditioner. Unfortunately, as long as we restrict ourselves to
block preconditioners we cannot take advantage of the sparsity structure (block sparsity
and sparsity of the DK-blocks), which would reduce the storage costs considerably.

What alternative options do we have at hand? A possible means to obtain precondition-ers that exploit sparsity and include information of interest are algebraic techniques that rely on so-called *incomplete LU factorizations* (briefly: ILU). The topic has emerged a vital field of research and is beyond the scope of this thesis. Therefore, we can only motivate the general principle and refer to [101], [11] and [9] for a detailed description of the construction and the theory behind it. Basically, the idea is to compute a sparse lower triangular matrix L and a sparse upper triangular matrix U so that the residual matrix $R = LU - A$ satisfies certain constraints. A rather simple way to do so is to "drop" certain elements that are below a predefined threshold (*drop tolerance*) and to compute the LU factorization of the resulting matrix. There is a multitude of other possible approaches, all of them having in common that they attempt to make the resid-ual matrix R small in some sense. For our experiments we make use of a sophisticated multilevel approach which was developped by BOLLHÖFER and SAAD (see the refer-ences in [15] for more information) and for which a state-of-the-art software package, the ILUPACK library [15], is available. Essentially, it can be used as a "black box" which makes it interesting for our purposes. The user only needs to supply the matrix in CSR (compressed sparse row) storage format (see [101] for a specification) and has to specify a threshold parameter for the construction of the preconditioner. The computed factorization is stored using sparse techniques. For the sake of simplicity we always chose the complete J-block as input matrix for the ILUPACK software, so that the SO-blocks are taken into account as well.

The following example illustrates the general principle of the ILUPACK preconditioner for a smaller problem for which enlightening sparsity plots are available. The experiment was carried out in the MATLAB® environment using SLEIJPEN's JDQR implementation [112] in combination with BOLLHÖFERS's ILUPACK toolbox [15] for MATLAB®.

Result 7.24 (ILUPACK-preconditioner for a small problem)
We consider the Hamiltonian matrix $\mathbf{H}^{(1/2,\,3/2,\,A'')}$ for the **MgCN** *molecule with respect to $J = 3/2$ and the small basis (cf. Table 6.5). From Alg. 6.3 we know that the problem dimension is $n = 4608$. Besides, it is easy to check that the number of non zero elements amounts to $nnz(\mathbf{H}^{(1/2,\,3/2,\,A'')}) = 2665728$ which corresponds to a sparsity ratio (7.9) of $sr(\mathbf{H}^{(1/2,\,3/2,\,A'')}) \approx 0.2472$. The software takes advantage of the symmetry of the input matrix such that only the lower triangular part (i.e. $nnz = 2665728$ elements) has to be stored. We are interested in the 10 eigenvalues closest to the target value $\tau = 0.01961$. For the construction of the preconditioner we employed the parameters $\sigma = 0.01$ (drop tolerance) and $\ell = 10$ (ellbow factor) (see [15] for details on the meaning of the parameters). In Figure 7.24 one can recognize that the refined JDQR variant discovers the ten eigenvalues rather quickly (less than 25 matrix-vector multiplications in average are required per eigenvalue). The Figures 7.22 and 7.23 show the sparsity plots for the (shifted) Hamiltonian matrix $\mathbf{H}_\tau := \mathbf{H}^{(1/2,\,3/2,\,A'')} - \tau \cdot \mathbf{I}$ and the resulting 2-level preconditioner determined by the ILUPACK software which is symmetric as well. Due to fill-ins the number of non-zero elements in the lower triangular part increases to $nnz = 2994838$.*

Figure 7.22.: Sparsity plot of $\mathbf{H}_\tau := \mathbf{H}^{(1/2,\,3/2,\,A'')} - \tau \cdot \mathbf{I}$, small basis

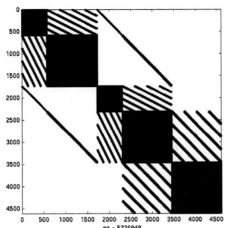

Figure 7.23.: Sparsity plot of the ILUPACK preconditioner for $\mathbf{H}_\tau$

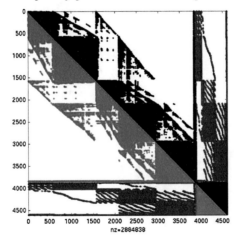

The preconditioner results from a recursive application of the PILUC strategy (inverse-based ILU that controls $\|L_k\|^{-1}$ and $\|U_k\|^{-1}$, where k denotes the index of the level and U_k resp. L_k denote the upper and lower unit triangular factors of the ILU of the leading block B_k) along with a suitable preordering (permutation) of the original input matrix A. The first level of the scheme is obtained as

$$\hat{P}^* A \hat{Q} = \begin{pmatrix} B & F \\ E & C \end{pmatrix} \approx \begin{pmatrix} L_B & 0 \\ L_E & I \end{pmatrix} \begin{pmatrix} D_B & 0 \\ 0 & S_C \end{pmatrix} \begin{pmatrix} U_B & U_F \\ 0 & I \end{pmatrix} \tag{7.51}$$

where

$$S_C \approx C - L_E D_B U_F \qquad (7.52)$$

is an approximation to the exakt Schur complement of C in $\hat{P}^* A \hat{Q}$. A second level of the scheme now may be obtained by applying the PILUC strategy to S_C (the blocks L_E and U_F which are also colored red in Fig. 7.23 are discarded in the course of the recursion). The factors L_k and U_k are colored green resp. blue in Fig. 7.23.

For a full account on the theory and the algorithms involved see [15] and the references therein.

Figure 7.24.: Interior eigenvalues obtained by refined extraction

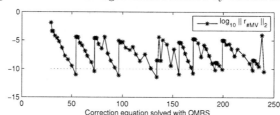

Correction equation solved with QMRS

□

The ILUPACK approach may be an interesting alternative to the J- and N-block preconditioners, because it works reliably and because the exploitation of sparsity allows to tackle relatively sparse problems as the following example demonstrates. In contrast to the previous experiment this time the computation again was carried out using a compiled stand-alone code which combines our JACDAV software (see Appendix A.4 for more information) with the ILUPACK library [15] (C/FORTRAN 77 code).

Result 7.25 (ILUPACK-preconditioner for large problems)

*We consider the Hamiltonian matrix for the **MgCN** molecule with respect to the big basis (see Table 6.5) and $J = 9/2$. As we know from Alg. 6.3 the problem dimension is $n = 59529$, which implies that 26.4 G memory is required for storing $\mathbf{H}^{(1/2,\,9/2,\,A')}$. Again 10 eigenvalues closest to the target value $\tau = 0.022782$ are sought-after. The convergence history (Fig. 7.25) shows that the refined JDQR variant (Alg. 4.8) detects the sought-after eigenpairs. However, it also becomes clear that the computational effort to be invested in terms of matrix-vector operations is rather high (c.a. 100 matrix-vector multiplications in avarage per eigenvalue).*

Figure 7.25.: Interior eigenvalues obtained by refined extraction

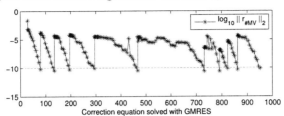

Correction equation solved with GMRES

□

Remark 7.26
As already pointed out in Section 4.4.4, one can use the JADAMILU software by BOLL-
HÖFER *and* NOTAY *[16],[17] which automatically generates suitable ILUPACK precon-
ditioners for the use in the Jacobi-Davidson method.*   □

We now collect the observations and conclusions from the Results 7.22, 7.23, 7.24 and
7.25 in the following summary:

Summary 7.27
*The problems arising in the computation of interior eigenvalues are due to the superpo-
sition of two general numerical difficulties:*

1. *the difficulty to extract sensible approximations to interior eigenpairs from a subspace*

2. *the difficulty to devise efficient preconditioners for indefinite matrices*

*In our experiments it became evident that the second of these difficulties is the deci-
sive one. It is no more possible to neglect the contribution of the DK-blocks, i.e. we
either have to employ expensive block preconditioners (N-block, J-block) or we have to
resort to algebraic approaches such as the ILUPACK software. Both preconditioning approaches along with the proper choice of the extraction method (refined/harmonic) and
the Krylov solver (GMRES) lead to convergence when looking for interior eigenvalues
near arbitrary target values τ. However, the price one has to pay is high:*

- *storage and set up costs are very high*

- *the ratio of time required for one preconditioner operation and one matrix-vector
  multiplication is extremely unfavorable.*

- *especially for higher values of J and target values τ that make the shifted matrix
  $\mathbf{H}_\tau$ highly indefinite much more effort in terms of matrix-vector multiplications
  per eigenvalue has to be invested (e.g. about 3 times as many in Result 7.25 as in
  Result 7.24)*

- *the preconditioners exhibit only little (N-block preconditioners) or no inherent parallelism at all (ILUPACK and J-block preconditioners) which is an additional severe drawback.*

Altogether, we can draw the conclusion that the computation is far more complicated than for the exterior part of the spectrum and only advisable for a very small number of eigenpairs. Nevertheless, it is worth while to point out the available options, because the knowledge of eigenvalues at arbitrary interior points of the spectrum has important applications. E.g. it enables one to check the quality of contraction schemes by computing some reference eigenpairs of the corresponding product basis problem.   □

7.5.3. Comparison with Other Methods

In what follows, we give a brief survey of how the JDQR variants compete with other iterative methods that make use of restart techniques:

- *Davidson's method* (see Section 3.3.2.3)

- *Olsen's method* (see Section 4.2.2.3)

- the *IRL method* (see Section 3.3.2.2)

To be fair, it should be pointed out that the algorithmic framework of Davidson's and Olsen's method is almost identical to the one of the JDQR method (apart from the subspace expansion) whereas the restart technique used by IRLM is rather different (see Section 3.3.2.2). This is also reflected in the numerical results. We shall see that it is sufficient to consider the computation of 100 eigenpairs for a medium-sized problem ($J = 1/2$, **MgCN** molecule, big basis) in order to outline the main differences.

First of all, it is no big surprise that *Davidson's method* fails to converge (Fig. 7.26) as the Jacobi preconditioner does not succeed for the preconditioned JDQR variants either (see Fig. 7.14).

Figure 7.26.: Failure of Davidson's method

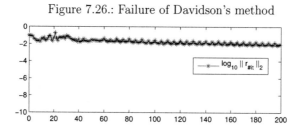

Let us now turn our attention to a comparison of Olsen's and the preconditioned JDQR method. For our experiments we employed the $K(6)$-block preconditioner. According to the results in Table 7.18 Olsen's method is about as efficient as the JDQR method for the computation of $k_{max} = 100$ eigenpairs but no more competetive if a larger number, e.g. $k_{max} = 200$ is sought-after. A closer analysis of the required preconditioner solves and matrix-vector multiplications for both methods (see Figure 7.27) reveals that Olsen's method is more effective in terms of matrix-vector multiplication and preconditioner solves per eigenvalue in the very beginning of the iteration process. Whereas the number of matrix-vector multiplications per eigenvalue required by Olsen's method is smaller or equal as compared to JDQR for a rather long time the corresponding number of preconditioner solves very soon develops to the disadvantage of Olsen's method. The reason why Olsen's method is still competetive for 100 eigenpairs in our example lies in the fact that the $K(6)$-block preconditioner is extremely cheap and has a very favorable time ratio prec/mv for a preconditioner operation and a matrix-vector multiplication (cf. Table 7.15). Summing up, it can be said that Olsen's method may be an interesting alternative if a relatively small number of eigenpairs (about 20 in our case) is sought-after.

Figure 7.27.: Olsen vs. JDQR: Number of operations required per eigenvalue

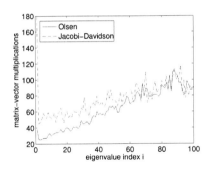
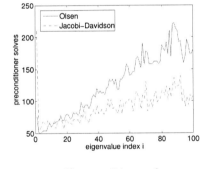

a) matrix-vector multiplications   b) preconditioner solves

Table 7.18.: Timings for JDQR/Olsen

	k_{max}	Computing time (secs)	#MV	#PREC
Olsen	100	2817.76	6299	12639
JDQR	100	2690.50	7572	8995
Olsen	200	9520.46	19369	38879
JDQR	200	6501.36	17340	20655

For our experiments with the IRL method we made use of the ARPACK library [75] which comprises robust and state-of-the-art implementations of both, the IRL and the IRA method. The advantage with the IRL approach is that the user only has to supply a routine for matrix-vector multiplications and need not worry about any preconditioning (provided that only exterior eigenvalues are of interest). However, the restart concept requires that the maximal size of the subspace $ncv = k + p$ (see the description of the method in Section 3.3.2.2) be at least equal to the number of sought-after eigenvalues, because the complete Lanczos factorization must "fit". This is a fundamental difference with the approaches of the Jacobi-Davidson family (JDQR, Olsen and Davidson) where the maximal size of the search space $\mathcal{K}$ can always be chosen to be a small constant (e.g. $m_{\max} = 20$), regardless of the problem under consideration. By contrast, the proper choice of the maximum size of the search space ncv turns out to be of crucial importance for the success of the ARPACK software, and unfortunately, there is no way to predict an optimal value ncv in relation to the number of sought-after eigenvalues k_{max}. Furthermore, in our experiments it turned out that obtaining a relatively large number of $k_{max} = 100$ converged eigenvalues by ARPACK is extremely time consuming and far more expensive than the corresponding JDQR computation. Therefore, we had to parallelize the code for the matrix-vector multiplication (by means of OpenMP^{TM}, see discussion in Section 7.7) and had to work with 8 processors in order to obtain converged results within a reasonable period of time. In Figure 7.28 we see what happens when we employ a fixed value of $ncv = 1000$ for different numbers of sought-after eigenpairs ($k_{max} = 20, 40, 60, 80, 100$) on the one hand, and the effect of different choices of ncv for the computation of 100 eigenpairs on the other hand. It is important to point out that one has to be careful with the results obtained by ARPACK as one can see in Fig. 7.29: The software only yields reasonable results (i.e. 100 fully converged eigenpairs with respect to a residual tolerance of $tol = 10^{-8}$) for choices of $ncv \geq 700$.

Figure 7.28.: ARPACK: Influence of the parameter ncv on the convergence behavior

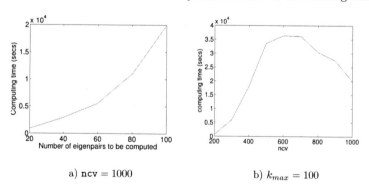

a) $ncv = 1000$ b) $k_{max} = 100$

Figure 7.29.: Accuracy of eigenpairs determined by ARPACK depending on ncv

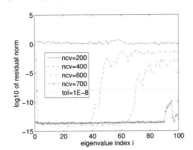

Altogether, it can be seen that ARPACK is rather difficult to use and not competetive in our context. However, it may be still an alternative for a very small number of $k_{max} \ll 100$ exterior eigenpairs.

7.6. JDQR Contracted Basis Calculation

The description in the following can be kept rather brief as we can take over many of the ideas for the product basis calculation without any significant modification. The main differences are collected in the following itemization:

- the sparsity of the off-diagonal DK- and SO-blocks (see Sections 7.2.1.1 and 7.2.1.2) is in general destroyed in the resulting contracted matrix blocks (cf. Alg. 7.6). Consequently, only the block sparsity of the product basis problem carries over to the contracted Hamiltonian matrix $\widetilde{\mathbf{H}}^{(J,S,\Gamma_{rve})}$ and can be exploited in a suitable algorithm for matrix-vector multiplication (Alg. 7.12). To do so, one can adopt the algorithm (Alg. 7.1) for product basis problems and take over the storage scheme for addressing the Hamiltonian blocks (see Section 7.2.2 and Fig. 7.4). Instead of the Algorithms 6.1 and 6.4 for product basis problems we make use of the information stored in the auxiliary array CONTDIM[] (set up in Alg. 7.5) in order to determine the dimensions and positions of the contracted blocks according to Corollary 7.7.

- analogously, the "hidden" structure of the diagonal K-blocks (see Section 7.5.1.1, Figs. 7.9 and 7.10) is destroyed upon contraction, such that it is no more sensible to employ $K(n_b)$-block preconditioners with $n_b > 1$ sub-blocks. Using the information in CONTDIM[] we can derive corresponding algorithms for set-up and application of (modified) K-block preconditioners for the contracted case (see Algs. 7.11 and 7.11 for standard K-block preconditioners, the modified versions are obtained analogously).

Result 7.28 (JDQR contracted basis calculation)
The results in terms of computing time, number of preconditioner solves and matrix-vector multiplications along with some supplementary information are collected in Table 7.19.

The most important and interesting observations and conclusions are listed in the following itemization:

- *As opposed to product basis problems Jacobi preconditioning leads to success when applying the preconditioned JDQR variants to contracted basis problems (a contraction limit of $E_{cont} = 5000$ cm^{-1} seems to be sufficient). This is hardly surprising as the diagonal K-blocks become near-diagonal upon contraction (see Remark 7.8).*

- *Jacobi preconditioning is still not sufficient for the computation of interior eigenvalues, but K-block preconditioning makes the refined/harmonic method converge.*

- *The results for the computation of 200 exterior eigenvalues for $J = 9/2$, (big basis) and $E_{cont} = 10000$ cm^{-1} show that it is also worthwhile to apply JDQR type methods to contracted basis problems as a comparison with the results for the corresponding direct-product and direct-contracted problems (see Table 7.9 and Figure 7.8) shows. In other words, the point of break even is reached, which is due to the fact that the size $n_{cont} = 23808$ of the contracted basis problem is rather big and the preconditioner extremely cheap.*

□

Algorithm 7.11: Setting up K-block preconditioner (contracted calculation)

---

1 **function** PREC $=$ setupprec($\widetilde{\mathbf{H}}$, S, J, CONTDIM[])
2 $y = 0$
3 $N_{min} = |J - S|$, $N_{max} = J + S$
4 **for** $N = N_{min}, \ldots, N_{max}$ **do**
5   **for** $K = 0, \ldots, N$ **do**
6           $I = \sum\limits_{\ell=N_{min}+1}^{N} \ell + K$
7   DIAG $:=$ $\widetilde{\mathbf{JDIAG}}[I]$
8           $\widehat{\mathbf{DIAG}} = \mathbf{P \cdot L \cdot U}$
9   PREC$[I] = [\,\mathbf{P}, \mathbf{L}, \mathbf{U}\,]$
10   **end for**
11 **end for**
12 **return** PREC

Algorithm 7.12: Matrix-vector multiplication for the contracted Hamiltonian matrix

1   **function** $\mathbf{y} = \text{mult}(\, \widetilde{\mathbf{H}}, \mathbf{x},\, S,\, J,\, \text{CONTDIM}[])$

2    $\mathbf{y} = \mathbf{0}$

3    $N_{min} = |J - S|,\ N_{max} = J + S$

4   **for** $N' = N_{min}, \ldots, N_{max}$ **do**

5   **for** $K' = 0, \ldots, N'$ **do**

6        $I' = \displaystyle\sum_{\ell=N_{min}+1}^{N'} \ell + K',\ i = \displaystyle\sum_{m=0}^{I'-1} \text{CONTDIM}[m] + 1,\ k = \text{CONTDIM}[I']$

7   **for** $N'' = N_{min}, \ldots, N_{max}$ **do**

8   **for** $K'' = 0, \ldots, N''$ **do**

9            $I'' = \displaystyle\sum_{\ell=N_{min}+1}^{N''} \ell + K'',\ j = \displaystyle\sum_{m=0}^{I''-1} \text{CONTDIM}[m] + 1,\ l = \text{CONTDIM}[I'']$

10            $\Delta N = |\, N' - N''\,|,\ \Delta K = |\, K' - K''\,|$

11   **if** ($\Delta N = 0$ **and** $\Delta K = 0$) **then** $\longleftarrow$————DIAG block ?

12              $\mathbf{y}[i : i + k - 1] += \widetilde{\mathbf{JDIAG}}[I'] \cdot \mathbf{x}[j : j + l - 1]$

13   **end if**

14   **if** ($\Delta N = 0$ **and** $\Delta K = 1$) **then** $\longleftarrow$————DK block ?

15   **if** ($K' > K''$) **then**

16                  $I = \displaystyle\sum_{\ell=N_{min}}^{N'-1} \ell + K'$

17                  $\mathbf{y}[i, \ldots, i + k - 1] += \widetilde{\mathbf{JDK}}[I] \cdot \mathbf{x}[j, \ldots, j + l - 1]$

18   **else**

19                  $I = \displaystyle\sum_{\ell=N_{min}}^{N'-1} \ell + K''$

20                  $\mathbf{y}[i : i + l - 1] += \widetilde{\mathbf{JDK}}[I]^T \cdot \mathbf{x}[j : j + k - 1]$

21   **end if**

22   **end if**

23   **if** ($\Delta N = 1$ **and** $\Delta K = 0$) **then** $\longleftarrow$————SO block ?

24   **if** ($N' > N''$) **then**

25                  $I = \displaystyle\sum_{\ell=N_{min}+1}^{N'-1} \ell + K'$

26                  $\mathbf{y}[i : i + k - 1] += \widetilde{\mathbf{JSO}}[I] \cdot \mathbf{x}[j : j + l - 1]$

27   **else**

28                  $I = \displaystyle\sum_{\ell=N_{min}+1}^{N''-1} \ell + K',$

29                  $\mathbf{y}[i : i + l - 1] += \widetilde{\mathbf{JSO}}[I]^T \cdot \mathbf{x}[j : j + k - 1]$

30   **end if**

31   **end if**

32   **end for**

33   **end for**

34   **end for**

35   **end for**

36   **return** $\mathbf{y}$

Algorithm 7.13: K-block preconditioner (contracted calculation)

---

1 **function** $\mathbf{y} = \text{prec}(\ \mathbf{PREC}, \mathbf{x},\ S,\ J,\ \text{CONTDIM}[]\ )$

2 $\mathbf{y} = \mathbf{0}$

3 $N_{min} = |J - S|,\ N_{max} = J + S$

4 **for** $N = N_{min}, \ldots, N_{max}$ **do**

5 $\quad$ **for** $K = 0, \ldots, N$ **do**

6 $\qquad I = \sum\limits_{\ell = N_{min}+1}^{N} \ell + K$

7 $\qquad i = \sum\limits_{m=0}^{I-1} \text{CONTDIM}[m] + 1$

8 $\qquad k = \text{CONTDIM}[I]$

9 $\qquad [\mathbf{P}, \mathbf{L}, \mathbf{U}] := \mathbf{PREC}[I]$

10 $\qquad \mathbf{y}[(i : i + k - 1] = \mathbf{U}^{-1}\mathbf{L}^{-1}\mathbf{P}^{T} \cdot \mathbf{x}[(i : i + k - 1]$

11 $\quad$ **end for**

12 **end for**

13 **return y**

7.7. Parallelization

In the previous sections we have seen that JDQR calculations are still very time-consuming for higher J quantum numbers – in spite of the existing advantages over other iterative methods and direct approaches. Therefore, it is important and interesting to investigate the parallelization of JDQR product basis computations (the ideas also apply for contracted calculations). Essentially, one can distinguish between two general parallelization paradigms (for more details see [49], e.g.):

- parallelization on shared memory architectures
  all processors p_i operate and have access on the same memory unit M, in other words they "share" it. An example of such an architecture is provided by the SUN^{TM} Fire workstation on which we carried out our calculations and where the workspace of 32 GB is shared by 8 processors. A shared memory system is relatively simple to program since all processors share a single view of data. On the other hand, the CPU-to-memory connection very often becomes a bottelneck because many modern CPUs need fast access to memory which is in general obviated by the restricted memory bandwidth.

- parallelization on distributed memory architectures
  each processor p_i operates on a memory block M_i of its own. In contrast to shared memory architectures the main issue is the communication of information and data between the single processors. In general the parallelization is sophisticated and requires much more technical effort (suitable partitioning of the data into memory units and assignment to the processors).

Table 7.19.: JDQR contracted calculation (**MgCN**, big basis)

J	n_{cont}	k_{max}	E_{cont}	#MV	#PREC	time (secs)	Target value τ	Prec.
1/2	585	200	5000 cm^{-1}	7030	9842	13.01	0	Jacobi
1/2	585	200	5000 cm^{-1}	6998	9050	17.80	0.02781669	Block
1/2	585	200	5000 cm^{-1}	2538	13122	no conv.	0.02781669	Jacobi
13/2	23288	200	10000 cm^{-1}	10309	12651	9502.44	0	Jacobi

In case of our JDQR type methods the success of these paradigms, of course, necessarily depends on the inherent parallelism of the user-supplied procedures for matrix-vector multiplications and preconditioner solves. The latter rather often obviates a successful parallelization, especially when interior eigenvalues are sought-after as the available preconditioners (N-block, J-block and ILUPACK preconditioners) exhibit little or no parallel structure.The parallelization of JDQR type methods on distributed memory architectures is examined in the PhD thesis by GOEKE [47]. Unfortunately, only little attention is paid to the construction and parallelization of suitable preconditioners, i.e. only examples with simple Jacobi preconditioning or no preconditioning at all are considered. However, GOEKE'S ideas might be of interest for large contracted basis calculations because Jacobi preconditioning is sufficient for exterior eigenpairs as we have seen in the previous section.

In the following, we will focus on the shared memory parallelization of the preconditioned standard JDQR algorithm for the computation of exterior eigenpairs. Therefore, we only need to consider the block preconditioners specified in Def. 7.18. The parallelization approach for the (modified) $K(n_b)$-block preconditioners and the matrix-vector multiplication is rather obvious and straight-forward. The basic idea is to assign one processor to each K-block row, provided that enough processors are available. In case of the **MgCN** molecule, for instance, this is only possible for J quantum numbers $J \leq 5/2$ on our SUN^{TM} Fire workstation with 8 processors, because there are at most 7 K-block rows altogether (see discussion in Section 7.2.2, Formula (7.32) and Table 7.5). Of course, it is also possible to treat problems for $J > 5/2$. However, now more than one K-block row has to be assigned to one processor. In our experiments we made use of the following parallelization techniques:

- parallelization using OpenMP^{TM} (see [3],[22])
  For the sake of simplicity we employed the automatic for-loop parallelization of OpenMP^{TM} by marking the corresponding `for`-loop sections in Alg. 7.4 (matrix-vector multiplication) and Algs. 7.9 and 7.10 (set-up and application of $K(n_b)$-block preconditioner) with `#pragma omp parallel for` (see [3],[22] for details). The advantage with this approach is that the user need not further specify the assignment of processors and can leave the organizational details to the OpenMP^{TM} software. On the other hand the results obtained may not always be fully satisfactory.

- parallelization using the parallel SUN^{TM} BLAS [1] routines (contained in the SUN^{TM} Performance Library).
  Essentially, this is a black-box technique which exploits the parallelism of the BLAS [1] routines but does not take advantage of any additional available information on the structure of the problem. It is switched on by means of the option `-XPARALLEL` when linking the code. Note that this technique is specific to SUN^{TM} architectures and only applicable when a parallelized version of the SUN^{TM} performance library is available.

In our analysis we proceed in two steps:

1. We first examine the speed-up

$$s_p = \frac{\text{time required for 1 processor}}{\text{time required for p processors}} \qquad (7.53)$$

of the user-supplied matrix-vector routines depending on the number of processors p for the **MgCN** molecule (big basis) with respect to different J quantum numbers which is visualized by the plots in Figures 7.30-7.36. Table 7.20 gives a survey of the related speedups and timings for both the OpenMP^{TM} and the parallel SUN^{TM} BLAS [1] parallelization.

The following observations can be made:

- the OpenMP^{TM} technique is clearly superior in all cases which is hardly surprising because existing information on inherent parallelism is exploited

- due to the obvious lack of parallelism in the implementation of the SUN^{TM} BLAS [1] routines for solving triangular systems no speed-up is achieved for the preconditioner solves. Note, however, that this is no general restriction because there are parallel implementations of triangular solvers available.

- the plots show the typical speed-up behavior for shared memory parallelization which is due to the limited memory band width, i.e. the maximal attainable speed-up is about $s_p = 4.5$ in fortitious cases (e.g. for $J = 13/2$, see Fig. 7.36).

- there are characteristic "bends" in the speed-up plots for the OpenMP^{TM} approach. These arise when the number of K-block rows is not "compatible" to the number of available processors, i.e. when OpenMP^{TM} does not succeed in finding an appropriate assignment between available processors and K-block rows to be distributed.

2. Finally, we are now in a position to assess the parallelization of the preconditioned JDQR method by means of OpenMP^{TM} .

To this end we consider the computation of 200 eigenpairs for the **MgCN** molecule with respect to the big basis and $J = 9/2$. The plot in Figure 7.37 and the timings in Table 7.21 show that the speed-up behavior of the preconditioned JDQR method is almost identical to the corresponding single matrix-vector operations (see Fig. 7.34). The maximum speed-up of $s_p = 3.5$ is achieved for $p = 6$ processors. Obviously, the impact of the serial component in the preconditioned JDQR method (i.e. the part of the algorithm which cannot be parallelized) on the speed up behavior is almost negligable. Of course, analogous observations can also be made for all other values of J.

Figure 7.30.: Speedup for matrix-vector operations ($J = 1/2$)

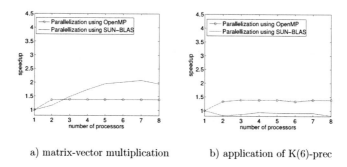

       a) matrix-vector multiplication   b) application of K(6)-prec

Figure 7.31.: Speedup for matrix-vector operations ($J = 3/2$)

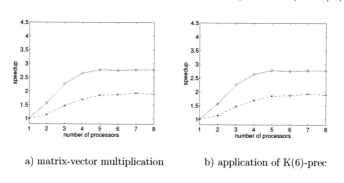

       a) matrix-vector multiplication   b) application of K(6)-prec

Figure 7.32.: Speedup for matrix-vector operations ($J = 5/2$)

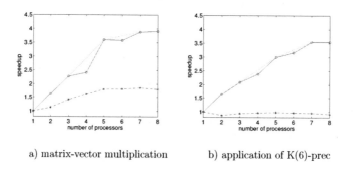

       a) matrix-vector multiplication   b) application of K(6)-prec

Figure 7.33.: Speedup for matrix-vector operations ($J = 7/2$)

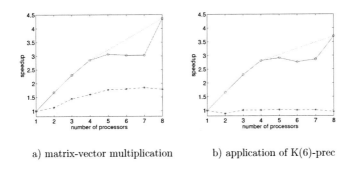

a) matrix-vector multiplication   b) application of K(6)-prec

Figure 7.34.: Speedup for matrix-vector operations ($J = 9/2$)

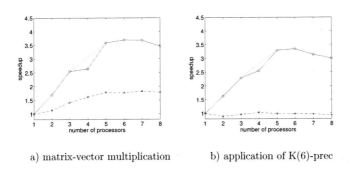

a) matrix-vector multiplication   b) application of K(6)-prec

Figure 7.35.: Speedup for matrix-vector operations ($J = 11/2$)

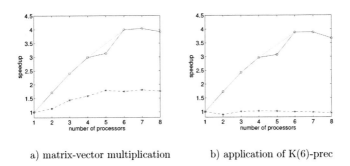

a) matrix-vector multiplication   b) application of K(6)-prec

Figure 7.36.: Speedup for matrix-vector operations ($J = 13/2$)

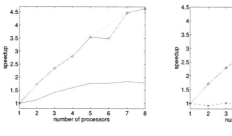 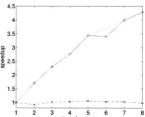

    a) matrix-vector multiplication   b) application of K(6)-prec

Table 7.21.: Computing times for 200 eigenpairs by JDQR ($J = 9/2$)

Number of processors	Computing time (h)	speedup
1	12.82	1
2	7.69	1.67
3	5.2	2.47
4	4.85	2.64
5	3.78	3.39
6	3.54	3.62
7	3.58	3.58
8	3.77	3.40

Figure 7.37.: Speedup for computation of 200 eigenpairs by JDQR ($J = 9/2$)

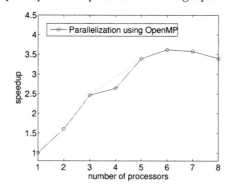

Table 7.20.: Computing times and speedups for matrix-vector routines

J	Number of processors	OpenMP[TM] [3]				parallel SUN[TM] BLAS [1]			
		preconditioning		MV multiplication		preconditioning		MV multiplication	
		t_{prec} (secs)	s_{prec}	t_{mv} (secs)	s_{mv}	t_{prec} (secs)	s_{prec}	t_{mv} (secs)	s_{mv}
1/2	1	0.0640	1.0000	0.2580	1.0000	0.0640	1.0000	0.2600	1.0000
	2	0.0480	1.3333	0.2220	1.3684	0.0780	0.8205	0.1900	1.3684
	3	0.0460	1.3913	0.1740	1.3830	0.0740	0.8649	0.1880	1.3830
	4	0.0460	1.3913	0.1480	1.3830	0.0680	0.9412	0.1880	1.3830
	5	0.0460	1.3913	0.1320	1.3830	0.0700	0.9143	0.1880	1.3830
	6	0.0480	1.3913	0.1280	1.3830	0.0700	0.9143	0.1880	1.3830
	7	0.0460	1.3913	0.1240	1.3830	0.0700	0.9143	0.1880	1.3830
	8	0.0460	1.3913	0.1320	1.3830	0.0780	0.8205	0.1880	1.3830
3/2	1	0.1480	1.0000	0.6720	1.0000	0.1460	1.0000	0.6640	1.0000
	2	0.0940	1.5745	0.4300	1.5628	0.1700	0.8588	0.5840	1.1370
	3	0.0660	2.2424	0.2980	2.2550	0.1520	0.9605	0.4520	1.4690
	4	0.0620	2.3871	0.2540	2.6457	0.1440	1.0139	0.3920	1.6939
	5	0.0540	2.7407	0.2420	2.7769	0.1500	0.9733	0.3580	1.8547
	6	0.0520	2.8462	0.2440	2.7541	0.1540	0.9481	0.3540	1.8757
	7	0.0520	2.8462	0.2420	2.7769	0.1520	0.9605	0.3440	1.9302
	8	0.0520	2.8462	0.2420	2.7769	0.1660	0.8795	0.3500	1.8971
5/2	1	0.2340	1.0000	1.1100	1.0000	0.2320	1.0000	1.1040	1.0000
	2	0.1420	1.6479	0.6760	1.6420	0.2660	0.8722	0.9760	1.1311
	3	0.1120	2.0893	0.4880	2.2746	0.2460	0.9431	0.7820	1.4118
	4	0.0980	2.3878	0.4600	2.4130	0.2400	0.9667	0.6780	1.6283
	5	0.0780	3.0000	0.3080	3.6039	0.2360	0.9831	0.6080	1.8158
	6	0.0740	3.1622	0.3100	3.5806	0.2400	0.9667	0.6060	1.8218
	7	0.0660	3.5455	0.2860	3.8811	0.2420	0.9587	0.5920	1.8649
	8	0.0660	3.5455	0.2840	3.9085	0.2520	0.9206	0.6080	1.8158
7/2	1	0.3260	1.0000	1.5520	1.0000	0.3280	1.0000	1.5460	1.0000
	2	0.1960	1.6633	0.9280	1.6724	0.3680	0.8913	1.3700	1.1285
	3	0.1420	2.2958	0.6740	2.3027	0.3240	1.0123	1.0700	1.4449
	4	0.1160	2.8103	0.5440	2.8529	0.3240	1.0123	0.9660	1.6004
	5	0.1120	2.9107	0.5060	3.0672	0.3200	1.0250	0.8740	1.7689
	6	0.1180	2.7627	0.5120	3.0312	0.3240	1.0123	0.8620	1.7935
	7	0.1140	2.8596	0.5120	3.0312	0.3240	1.0123	0.8380	1.8449
	8	0.0880	3.7045	0.3560	4.3596	0.3440	0.9535	0.8680	1.7811
9/2	1	0.4080	1.0000	1.9920	1.0000	0.4100	1.0000	1.9880	1.0000
	2	0.2500	1.6320	1.1740	1.6968	0.4600	0.8913	1.7560	1.1321
	3	0.1780	2.2921	0.7820	2.5473	0.4280	0.9579	1.4020	1.4180
	4	0.1600	2.5500	0.7540	2.6419	0.3980	1.0302	1.2280	1.6189
	5	0.1240	3.2903	0.5560	3.5827	0.4160	0.9856	1.1140	1.7846
	6	0.1220	3.3443	0.5400	3.6889	0.4180	0.9809	1.1220	1.7718
	7	0.1300	3.1385	0.5420	3.6753	0.4240	0.9670	1.0880	1.8272
	8	0.1360	3.0000	0.5760	3.4583	0.4400	0.9318	1.1120	1.7878
11/2	1	0.5280	1.0000	2.4440	1.0000	0.5000	1.0000	2.4280	1.0000
	2	0.3060	1.7255	1.4300	1.7091	0.5580	0.8961	2.1520	1.1283
	3	0.2180	2.4220	1.0180	2.4008	0.4960	1.0081	1.6900	1.4367
	4	0.1780	2.9663	0.8180	2.9878	0.4880	1.0246	1.5320	1.5849
	5	0.1720	3.0698	0.7800	3.1333	0.4920	1.0163	1.3600	1.7853
	6	0.1360	3.8824	0.6120	3.9935	0.5040	0.9921	1.3900	1.7468
	7	0.1360	3.8824	0.6060	4.0330	0.5120	0.9766	1.3480	1.8012
	8	0.1440	3.6667	0.6240	3.9167	0.5280	0.9470	1.3840	1.7543
13/2	1	0.5920	1.0000	2.8620	1.0000	0.5980	1.0000	2.8600	1.0000
	2	0.3460	1.7110	1.6540	1.7304	0.6500	0.9200	2.5340	1.1287
	3	0.2580	2.2946	1.2280	2.3306	0.5880	1.0170	2.0200	1.4158
	4	0.2140	2.7664	1.0200	2.8059	0.5820	1.0275	1.7900	1.5978
	5	0.1720	3.4419	0.8080	3.5421	0.5660	1.0565	1.6160	1.7698
	6	0.1740	3.4023	0.8220	3.4818	0.5800	1.0310	1.6120	1.7742
	7	0.1480	4.0000	0.6400	4.4719	0.5820	1.0275	1.5560	1.8380
	8	0.1380	4.2899	0.6180	4.6311	0.6080	0.9836	1.6020	1.7853

8. Summary and Outlook

In this concluding chapter we now summarize the essential results of the investigations in this thesis. We have shown how to apply the preconditioned *Jacobi-Davidson type methods* (preconditioned JDQR methods) to both, product basis and contracted basis problems in order to compute energy levels of triatomic molecules that exhibit the *Double Renner effect*. To this end we devised efficient matrix-vector multipliation algorithms and storage schemes for the arising matrices. In case of product basis (FBR) problems it is possible to avoid conventional storage techniques (CSR, CSC etc.) and to exploit the regular sparsity pattern which is due to the fact that many of the matrix blocks may be written as sums of Kronecker products. Of course, similar techniques carry over to Kronecker products with more than 3 factors as it is the case for molecules with $N > 3$ nuclei. Unfortunately, one cannot directly take advantage of the Kronecker product structure, because the potential matrix cannot be expressed a sum of Kronecker products in our case, which would reduce the computational complexity of a matrix-vector multiplication to a great deal (see discussion in Section 5.8.1, Scenario 1).

In general, the user has to provide suitable preconditioners in order to make *Jacobi-Davidson type methods* work. This is also true of our situation, i.e. the method fails to converge without any preconditioning. The appropriate choice of a preconditioner depends on

- the type of the problem (product basis or contracted basis)

- the part of the spectrum (interior eigenvalues or exterior eigenvalues)

For product basis problems we can make the following recommendations:

- use the modified/standard block preconditioners $K(n_b)$ introduced in Section 7.5.1. They save memory by taking advantage of returning information (multiple occurences of basic K-blocks, see Corollary 6.11 in Section 6.4.3) and using sub-blocks the number being a divisor of the cut-off number for the vibrational basis $N_R^{(lim)}$

- use ILU-type preconditioners (e.g. the ILUPACK software by BOLLHÖFER) for arbitrary interior eigenvalues

For contracted basis problems things are getting considerably simpler (at least if the contraction limit E_{cont} is sufficiently large).

- it is now sufficient to use Jacobi preconditioning to obtain exterior eigenvalues

- K-block and modified K-block preconditioners now also work for interior eigenvalues

What are the advantages of using *Jacobi-Davidson type methods* for the computation of energy levels?

- it is possible to tackle much larger problems, because the Hamiltonian matrices need not be stored explicitly and because the sparsity structure can be exploited very efficiently (as explained above).

- the *point of break even* (i.e. the problem size when *Jacobi-Davidson type methods* become more efficient than direct methods in terms of computing time, see Result 7.28 and Section 7.5.1.2) is perceivable and already reached for medium problem sizes provided that the number of sought-after eigenpairs is not too large.

- *Jacobi-Davidson type methods* are in general superior to iterative projection methods that rely on the same framework (Davidson, Olsen). ARPACK (the state-of-the-art implementation of the IRL method) is clearly inferior for more than $k_{max} = 50$ sought-after eigenpairs.

- *Jacobi-Davidson type methods* may be employed as a means of verification, i.e. to assess the quality of contraction schemes when direct methods are no more feasible. To this end one can compute a couple of selected eigenvalues of the original product basis problem near the contraction limit.

The following drawbacks are typical of iterative projection methods:

- *Jacobi-Davidson type methods* are only suited for the computation of small and medium-sized partial eigensystems (up to about 200-500 eigenpairs). This is due to the fact that the projections involved become increasingly expensive and that the preconditioner gradually deteriorates

- the computation of interior eigenpairs is feasible, but rather complicated and expensive. Refined and harmonic extraction methods improve the convergence behavior but do not cure the lack of efficient and cheap preconditioners.

Summing up it can be said that the *Jacobi-Davidson type methods* presented and developped in this thesis offer a valuable alternative to direct (product and contracted) approaches and should be more popular in the context of eigenvalue computations arising in Theoretical Spectroscopy.

To conclude with, let us now give an outlook along with some suggestions for future research:

- As a next step, it would be interesting to see how our methods perform for eigenvalue problems coming from molecules with $N > 3$ nuclei

- In this context it might be worthwhile to analyze to what extent the potential matrix can be approximated by sums of Kronecker products (see [127] for hints on a possible realization). This would allow for an efficient matrix-vector multiplication as outlined in Scenario 1 in Section 5.8.1. Maybe it could also serve as a starting point for devising even more efficient preconditioners.

- In our investigations it turned out that efficient and paralellizable preconditioners for highly indefinite matrices are of crucial importance as they arise when looking for interior eigenvalues. This issue is of general interest and not only specific to our situation.

- Last but not least, it might be also rewarding to analyze whether or not hierarchical structures can be exploited in devising preconditioners. First experiments in which we tried to employ contracted Hamiltonian matrices for the construction of preconditioners to the original FBR matrix were not successful.

Part IV.

Appendices and Surveys

A. Appendix

A.1. Conventions for the Usage of Fonts

For the sake of lucidity and a better readability we decided to use

- capital letters in normal fonts for software libraries, e.g. LAPACK [2], BLAS [1], SBR [12], MATLAB® [5] and OpenMP™ [3]

- capital letters in normal fonts for programming languages, i.e. FORTRAN 77 [85] and the C programming language [67]

- typewriter font for names of particular routines in software libraries, e.g. xSYEV (LAPACK [2] routine for computing complete eigensystems), xGEMV (BLAS routine for computing general matrix-vector products), xSYBTRD (SBR routine for the two-stage tridiagonalization) where $x \in \{S, D, Z, C\}$. For our purposes we almost always have $x = D$ because we are dealing with data made up of double precision entries. See [7] for further information on LAPACK's [2] naming conventions.

- bold letters for matrices in Chapters 5,6 and 7 in order to better distinguish matrices and operators, e.g. $\mathbf{H}^{(J,S,\Gamma_{rve})}$ for the Hamiltonian matrix with respect to the triple (J, S, Γ_{rve}).

- bold letters when referring to molecules, e.g. **MgNC** molecule, **ABC**-type molecule

- mathtype font in connection with a hat for operators, e.g. $\hat{H}$, $\hat{P}_R$, etc.

A.2. Romanization of Russian Names

The transcription of slavonic names in cyrillic letters into latin alphabet is often a source of confusion as it depends on the pronunciation rules of the target language. Besides, the ignorance of Russian pronunciation rules also often leads to inappropriate transliteration results. The following two examples illustrate possible ambiguities and inconsistencies:

- Пафнутий Львович Чебышёв
  The transliteration of the surname leads to a great variety of possible results, e.g. *Chebyshev* (in English), *Tchebychev* (in French) and *Tschebyscheff, Tschebyschew or Tschebyschow* (in German). Only the latter of these reproduces the Russian original adequately.

- Борис Григорьевич Галёркин

  This surname of this Russian mathematician is most commonly romanized as *Galerkin*, although only the transliterations *Galjorkin* (in German) resp. *Galyorkin* (in English) are appropriate and correspond to the actual Russian pronunciation.

To overcome these problems slavic philologists commonly make use of *scientific transliteration* (also called the *International Scholarly System*) which allows for a non-ambiguous reconstruction of the original word. We decided to employ the ISO 9 standard [4] which is widely identical to the scientific translation system but has the advantage that it provides a one-to-one correspondence between Cyrillic letters and latin letters (with diacritic marks). Table A.1 summarizes the transliteration (in compliance with the ISO 9 norm) of the Russian names in this thesis.

A.3. Mathematical Notation

Apart from very rare exceptions we make use of the commonly accepted and widespread notation standards. In some situations, for the sake of convenience, we make use of MATLAB®style notation (see List of Symbols) when referring to sub-matrices or partial columns resp. rows of matrices.

In computer science it is very common to use the so-called Big Oh notation in order to give a rough measure for time complexity of algorithms and their memory consumption. Since we also make use of it at several occasions it is appropriate to give a precise definition:

Definition A.1 (Big Oh notation, Landau notation)
Let $f(x)$ and $g(x)$ be two functions defined on some subset of $\mathbb{R}$. We say

$$f(x) = \mathcal{O}(\,g(x)\,) \qquad as \qquad x \to \infty \tag{A.1}$$

if and only if

$$\limsup_{x \to \infty} \left| \frac{f(x)}{g(x)} \right| < \infty \tag{A.2}$$

We are almost always concerned with the time complexity of matrix algorithms (especially eigensolvers and methods for the tridiagonalization of Hermitian matrices). In this context it is usual to employ the number of rows (resp. the number of columns) $n \in \mathbb{N}$ of the matrix as the problem size. The statement "Algorithm X has the time complexity $\mathcal{O}(g(n))$" then actually means that for the (unknown) function $f(n)$ that exactly measures the time complexity of X it holds $f(x) \in \mathcal{O}(g(n))$.

Table A.1.: Transliteration of Russian Names according to ISO 9 [4]

Original Russian Name	English Transcription	ISO 9 Romanization
Пафнутий Львович Чебышёв	Pafnuty Lvovich Chebyshev	*Pafnutij L'vovič Čebyšëv*
Семён Аранович Гершгорин	Semyon Aranovich Gershgorin	*Semën Aranovič Geršgorin*
Вера Николаевна Кублановская	Vera Nikolaevna Kublanovskaya	*Vera Nikolaevna Kublanovskaâ*
Алексей Николаевич Крылов	Alexei Nikolaevich Krylov	*Aleksej Nikolaevič Krylov*
Борис Григорьевич Галёркин	Boris Grigorevich Galerkin	*Boris Grigor'evič Galërkin*
Георгий Иванович Петров	Georgi Ivanovich Petrov	*Georgij Ivanovič Petrov*
Сергей Львович Соболев	Sergei Lvovich Sobolev	*Sergej L'vovič Sobolev*

Table A.2.: Technical specification of the V1280 SUN™ Fire compute server

Number of processors	8
Processor architecture	UltraSPARC IIICu, Superscalar SPARC V9, ECC Protected
Cache per processor	Level 1: Parity protecte 32 KB instruction and 64 KB data on chip (single-bit errors are corrected) Level 2: 8 MB external cache
Main memory	32 GB
Platform name	SUNW, Netra-T12

A.4. Technical Details and Implementation Issues

For our numerical experiments we developped and employed a software JACDAV written in the C programming language [67]. Essentially, it is geared to the techniques and ideas of the MATLAB® software by SLEIJPEN [112] and the C code JDBSYM by GEUS [46]. As repeatedly already pointed out in this thesis, it incorporates several important additional features and switches for our experiments:

- Davidson and Olsen type subspace expansion (see discussion in Sections 4.2.2.2 and 4.2.2.3)

- harmonic and refined extraction as described in the Algs. 3.13 and 3.14

- options for mixing JD type and Olsen type subspace expansion

To make the JD methods work we had to provide C code for the obligatory plug-ins

- matrix-vector multiplication

- preconditioning

Furthermore we developped C code for

- the contraction scheme described in Section 7.3

- the test of direct-product and direct-contracted calculations in Sections 7.4.1 - 7.4.2 (using the SBR driver xSYEVT for two-stage tridiagonalization and the the LAPACK [2] RRR routine xSTEGR for the application of the RRR method)

We carried out our numerical experiments on a SUN™ Fire workstation (see Table A.2 for a concise technical specification) with 32 G main memory and 8 processors.

Let us now briefly comment on some technicalities with respect to the implementation in the C programming language. As the BLAS and LAPACK routines (see [7], [1] and [2] for a full account and an extensive documentation) are written in FORTRAN 77 (see [85] for a documentation of the programming language) one has to provide an adapter in order to be able to call the routines from a C code. This is necessary because the FORTRAN 77 [85] compiler almost always (depending on the flag NOF77UNDERSCORE) appends an an underscore _ to the routine's name and using the original name would lead to problems with the linker. The adapter is realised by the header fortran.h in Listing A.1. The macro F77 appends – if required – an underscore, e.g. F77(dsymv) corresponds to dsymv_. For the sake of simplicity it is convenient to pre-define some often-required int-valued and double-valued constants at the beginning of the program, such as 1, -1, 0.0, 1.0 and -1.0 (see Listing A.4). Furthermore, one has to define prototypes for all BLAS and LAPACK [2] routines that are called from the C code, which is done in the header files `blas.h` (Listing A.2) and `lapack.h` (Listing A.3). Finally, Listing A.5 demonstrates some of the required memory allocations in terms of the problemsize n, the maximum number of sought-after eigenvalues k_{max} and the maximum dimension m_{max} of the search space $\mathcal{K}$ which corresponds to `jmax` in the C code (see also the related survey in Section 4.4). The Listings A.6 and A.7 show how the skew projektion

$$\widetilde{P} = I - \widetilde{Y}\widetilde{H}^{-1}\widetilde{Q}^*$$

and the projected preconditioned operator

$$\widetilde{A} = (I - \widetilde{Y}\widetilde{H}^{-1}\widetilde{Q}^*)K^{-1}(A - \theta I)$$

which are crucial parts of our JDQR type algorithms (see general discussion in Section 4.3.6.1, Equations (4.101), (4.102), (4.103) and Algs. 4.7, 4.8 and 4.9) are realized in a C code by means of BLAS and LAPACK [2] routines. We tested our implementation JACDAV of the JD-type methods on both Linux and SUN^{TM} Solaris platforms. Listings A.8 and A.9 give a brief summary of the essential compiler and linker calls to be carried out in order to produce executable binaries. In case of Linux platforms it is important not to forget to link the library `libg2c` by means of `-lg2c` to make sure that the FORTRAN 77 [85] routines are called properly. Our numerical experiments regarding the computation of rovibronic energy levels, of course, were carried out on the SUN^{TM} Fire workstation because of 32 G heap space and 8 processors being available which also leads to the possibility of paralellizing the calculation by means of OpenMP^{TM} (see [22] and [3] for more details).

For hints on how to connect ILUPACK routines to our code see the project's homepage [15] and the information given therein. For the use of the SBR FORTRAN 77 [85] routines in the context of TST-RRR calculations (cf. Section 7.4.1) one can follow the same principles as explained above for LAPACK [2] and BLAS. The FORTRAN 77 [85] code including all required interface information may be obtained from web site of the corresponding paper [13].

Listing A.1: FORTRAN [85] adapter (`fortran.h`)

```
1   #ifndef FORTRAN_H
2   #define FORTRAN_H
3
4   #ifdef NOF77UNDERSCORE
5   #define F77(s)  s
6   #define F77_(s)  s
7   #else
8   #define F77(s)  s ## _
9   #define F77_(s)  s ## _
10  #endif
11
12  #endif
```

Listing A.2: C interface for employed BLAS [1] routines (`blas.h`)

```
1  #ifndef BLAS_H
2  #define BLAS_H
3
4  #include "fortran.h"
5
6  /* BLAS-1 functions */
7  extern double F77(dasum) (int* n, double x[], int* incx);
8  extern double F77(ddot) (int* n, double x[], int* incx, double y[],
9          int* incy);
10 extern double F77(dnrm2) (int* n, double x[], int* incx);
11 extern int F77(idamax) (int* n, double x[], int* incx);
12
13 /* BLAS-1 subroutines */
14 extern void F77(daxpy) (int* n, double* a, double x[], int* incx,
15          double y[], int* incy);
16 extern void F77(dcopy) (int* n, double x[], int* incx, double y[],
17          int* incy);
18 extern void F77(drot) (int* n, double* x, int* incx, double* y, int* incy,
19                        double* c, double* s);
20 extern void F77(dscal) (int* n, double* a, double x[], int* incx);
21
22 /* BLAS-2 subroutines */
23 extern void F77(dgemv) (char* trans, int* m, int* n, double* alpha,
24          double a[], int* lda, double x[], int* incx, double* beta,
25          double y[], int* incy, int len_trans);
26
27
28 extern void F77(dsymv) (char* uplo, int* n, double* alpha,
29                        double a[], int* lda, double x[],
30                        int* incx, double* beta, double y[],
31                        int* incy);
32 extern void F77(dtrsv) (char* uplo, char* trans, char* diag, int* n,
33                        double a[], int* lda, double x[], int* incx);
34 /* BLAS-3 subroutines */
35 extern void F77(dgemm) (char* transa, char* transb, int* m, int* n, int* k,
36          double* alpha, double a[], int* lda, double b[], int* ldb,
37          double* beta, double c[], int* ldc, int len_transa,
38          int len_transb);
39
40 #endif
```

Listing A.3: C interface for employed LAPACK [2] routines (`lapack.h`)

```
1  #ifndef LAPACK_H
2  #define LAPACK_H
3
4  #include "fortran.h"
5
6  extern void F77(dsyev) (char* jobz, char* uplo, int* n, double a[],
7          int* lda, double w[], double work[], int* lwork, int* info,
8          int len_jobz, int len_uplo);
9
```

```
10  extern void F77(dgetrs) (char* trans, int* n, int* nrhs, double a[],
11          int* lda, int ipiv[], double b[], int* ldb, int* info,
12          int len_trans);
13
14  extern void F77(dgetrf) (int* m, int* n, double a[], int* lda, int ipiv[],
15          int* info);
16
17
18  extern void F77(dsytrf) (char *UPLO, int *N, double *A, int *LDA,
19                          int IPIVOT[], double WORK[], int* LDWORK,
20                          int* INFO);
21
22  extern void F77(dsptrf) (char *UPLO, int *N, double *A,
23                          int IPIVOT[], int* INFO);
24
25  extern void F77(dsytrs) (char *UPLO, int *N, int *NRHS, double *A,
26                          int *LDA, int IPIVOT[], double *B, int* LDB,
27                          int* INFO);
28
29  extern void F77(dsptrs) (char *UPLO, int *N, int *NRHS, double *A,
30                          int IPIVOT[], double *B, int* LDB,
31                          int* INFO);
32
33  extern void F77(dlarnv) (int *IDIST, int *ISEED, int *N, double *X);
34
35  extern void F77(dsyevx) (char* jobz, char* range, char* uplo, int* n,
36          double a[], int* lda, double* vl, double* vu, int* il, int* iu,
37          double* abstol, int* m, double w[], double z[], int* ldz,
38          double work[], int* lwork, int iwork[], int ifail[], int* info,
39          int len_jobz, int len_range, int len_uplo);
40
41  extern void F77(dgesvd) (char* jobu, char* jobvt, int* m, int* n,
42                          double a[], int* lda, double* sing,
43                          double u[], int* ldu,
44                          double vt[], int* ldvt,
45                          double work[], int* ldwork,
46                          int* info);
47
48  extern void F77(dlacpy)(char *UPLO, int *M, int *N, double *A, int *LDA,
49                          double *B, int *LDB, int len_uplo);
50
51  extern void F77(dlaset)(char *UPLO, int *M, int *N, double *ALPHA,
52                          double *BETA, double *A, int *LDA, int len_uplo );
53
54  extern double F77(dlamch_) (char* name, int len_name);
55
56  extern int F77(ilaenv)(int *ISPEC, char *NAME, char *OPTS, int *N1,
57                          int *N2, int *N3, int *N4, int len_name,
58                          int len_opts);
59
60  #endif
```

Listing A.4: Constants for LAPACK [2] /BLAS [1]

```
1  static double DMONE = −1.0, DZER = 0.0, DONE = 1.0;
2  static int MONE = −1, ONE = 1;
```

Listing A.5: Memory allocation for matrices and vectors

```
1   double* Q=NULL;          // matrix of converged eigenvectors
2   double* Y=NULL;          // matrix of preconditioned eigenvectors Y=K^{−1}Q
3
4   double* H=NULL;          // H=Q^{*}K^{−1}Q = Q^{*}*Y
5   double* Hlu=NULL;        // matrix of the LU factorization of H
6   int*    Hpiv=NULL;       // information on pivoting in LU factorization of H
7
8   double *temp1=NULL,
9          *temp2=NULL,
10         *temp3=NULL;      // auxiliary vectors
11
12  Q = (double *)malloc(n * kmax * sizeof(double));
13  Y = (double *)malloc(n * kmax * sizeof(double));
14
15  assert( Q && Y);
16
17  H    = (double *)malloc(kmax * kmax * sizeof(double));
18  Hlu  = (double *)malloc(kmax * kmax * sizeof(double));
19  Hpiv = (int    *)malloc(kmax * sizeof(int));
20
21  assert( H && Hlu && Hpiv);
22
23  temp1 = (double*)malloc( n * sizeof(double));
24  temp2 = (double*)malloc( n * sizeof(double));
25  temp3 = (double*)malloc( n * sizeof(double));
26
27  assert( temp1 && temp2 && temp2 );
28
29  F77(dlaset)("a", &kmax, &kmax, &DZER, &DZER , H,    &jmax, 1);
30  F77(dlaset)("a", &kmax, &kmax, &DZER, &DZER , Hlu, &jmax, 1);
```

Listing A.6: Application of the skew projection $\widetilde{P} = I − \widetilde{Y}\,\widetilde{H}^{-1}\widetilde{Q}^{*}$

```
1  void SkewProj(double* Q, double* Y, double* Hlu, int Hpiv[] ,
2                double* r, double* rtemp1, double* rtemp2, int n,
3                int k, int kmax)
4  {
5    int info;
6    if (k>0) // Q is a non−empty matrix
7      {
8          // rtemp=Q'*r
9          F77(dgemv)("t", &n, &k,  &DONE, Q, &n, r, &ONE, &DZER,
10                     rtemp1, &ONE, 1);
11         // rtemp=H^{−1} * Q' * r
12         F77(dgetrs)("n", &k, &ONE, Hlu, &kmax, Hpiv, rtemp1, &kmax, &info, 1);
13         if (info!=0)
```

```
14      {
15          printf(" Error ␣ solving ␣the ␣LES ␣ in ␣SkewProj ␣ . . . . ␣\n" );
16          exit (0);
17      }
18      // rtemp=Y * H^{-1} * Q' * r
19      F77(dgemv)("n" , &n, &k, &DONE, Y, &n, rtemp1 , &ONE, &DZER,
20              rtemp2 , &ONE, 1);
21      // r = r - Y * H^{-1} * Q' * r
22      F77(daxpy)(&n, &DMONE, rtemp2 , &ONE, r , &ONE );
23      }
24  }
```

Listing A.7: Application of the operator $\widetilde{A} = (I - \widetilde{Y}\widetilde{H}^{-1}\widetilde{Q}^*)K^{-1}(A - \theta I)$

```
1  void mvp(double theta , double* Q, double* Y, double *Hlu, int Hpiv [] ,
2          double* v, double* u, double* vtemp1 ,
3          double* vtemp2 ,
4          void (*domatvec )(double*, double*),
5          void (*doprecon )(double*, double*), int n, int k, int kmax)
6  {
7      double SCAL;
8      F77(dlacpy )("a" , &n, &ONE, v, &n, u, &n, 1);    // v ——> u
9      // v = A*u - theta * u
10     domatvec(u,v); MV++;
11     SCAL=-theta ;
12     F77(daxpy)(&n, &SCAL, u, &ONE, v, &ONE );
13     doprecon (v,vtemp1 ); PS++;
14     F77(dlacpy )("a" , &n, &ONE, vtemp1, &n, v, &n, 1); // vtemp1 ——> v
15     F77(dlaset )("a" , &n, &ONE, &DZER, &DZER , vtemp1 , &n, 1);
16     F77(dlaset )("a" , &n, &ONE, &DZER, &DZER , vtemp2 , &n, 1);
17     SkewProj (Q, Y, Hlu, Hpiv, v, vtemp1 , vtemp2 , n, k, kmax);
18 }
```

Listing A.8: Compiling and Linking of `jacdav.c` and `DR_test.c` under Linux

```
1  gcc −Iinclude −c jacdav.c
2  gcc −Iinclude −c DR_test.c
3  gcc −o DR_TEST jacdav.o DR_test.o −lm −lblas −llapack −lg2c
```

Listing A.9: Compiling and Linking of `jacdav.c` and `DR_test.c` under SUN[TM] Solaris

```
1  cc −Iinclude −c jacdav.c −xarch=v9b −xautopar −xopenmp −xloopinfo
2  cc −Iinclude −c DR_test.c −xarch=v9b −xautopar −xopenmp −xloopinfo
3  cc −o DR_TEST DR_test.o jacdav.o
4      −xarch=v9b −xlic_lib=sunperf −lm −xparallel
```

A.5. Input Files for DR

Finally, for the sake of completeness, we also present the input files which we employed for the program runs of ODAKA'S software DR in order to produce the data of the matrix blocks $\mathbf{H}^{(J,S,\Gamma_{rve})}$ for both the **MgCN** and the **HOO** molecule. These provide several switches to control important parameters, such as

- the sizes of the vibrational basis which are specified by the cut-off numbers $N_R^{(max)}$, $N_r^{(max)}$, $(v_2^a)^{(max)}$ and $(v_2^b)^{(max)}$ as per Def. 6.3 (cf. Lines 28, 29, 31 and 33 in the input file A.10 for the **MgCN** molecule)

- the maximum J quantum number for which matrix blocks have to be computed (see Line 27). It holds $\texttt{MAXJ} = 2 \cdot J_{\max}$

- the masses of the involved nuclei (see Line 6) and information on the molecular equilibrum geometry (see Lines 21 and 24)

- several thresholds (see Lines 41 - 48) and further parameters, such as the maximum number of iteration steps (see Lines 34 - 40) and the number of nodes (see Lines 30 and 32) to control the arising numerical integration schemes

- the contraction limit CONTMAX in Line 49 to control the contraction scheme

Listing A.10: DR input file for the **MgCN** molecule w.r.t. big basis

```
 1  TITLE
 2  SIGMA=1   is lower potential
 3  SIGMA=2   is the upper potential
 4  TITLE
 5  'MASSES'
 6  12.0D+00,14.00307401D+00,23.98504187D+00
 7  LAMBDA        1
 8  MULTI         1
 9  XSO          39.D00
10  'znorenner'
11  F
12  'zstartfrom_J'
13  F
14  'zopposite'
15  F
16  'zabbf,zabb'
17  F
18  F
19  NSP           6
20  Ntau          8
21  RE1(bohr)  2.20779841252920D00
22  DISS1 (h )2.9D+01
23  WE1   (h )0.105D-01
24  RE2(bohr)  4.56695008781160D00
25  DISS2 (h )0.5D+00
```

26	WE2	(h)	0.25D−02
27	MAXJ		7
28	MAXV2A		30
29	MAXV2B		30
30	NPNT1		15
31	NMAX1		5
32	NPNT2		30
33	NMAX2		15
34	NSTINT		9999
35	NSERIN		230
36	NSERP		5
37	NSERQ		20
38	NSTNIN		1000
39	NPNTB		140
40	NSPB		50
41	THRSH1		0.50000000D+04
42	THRSH2		1.00000000D−06
43	THRSH3		1.00000000D+06
44	THRSH4		1.00000000D+03
45	THRSH5		1.00000000D−14
46	THRSH6		1.00000000D+10
47	THRSH7		1.00000000D+10
48	THRSH8		1.00000000D+10
49	CONTMAX		5.0D3
50	'Vmin '		
51	1.0D+02		
52	'Vmax '		
53	1.0D+04		
54	'adif '		
55	1.0D+02		
56	'bdif '		
57	1.5D+00		
58	'cdif '		
59	3.0D+02		
60	'znumpot '		
61	F		
62	'zanaepot '		
63	T		
64	'zpoteq '		
65	F		
66	'zsingle '		
67	F		
68	'RHOMAX,PNM1 '		
69	2.00D+00,1.0D−2		
70	'zhamilvv '		
71	T		
72	'zmgCNharmf '		
73	F		
74	'zHOO_f '		
75	F		
76	'zHOOBOWMAN_f '		
77	F		

```
 78  'zmgCNf'
 79  F
 80  'zmgNCf'
 81  F
 82  'zMORBID'
 83  F
 84  'zmgNCCNf'
 85  T
 86  'zmgNCCNMIDDLE'
 87  F
 88  'zHCNf'
 89  F
 90  'zHCNbowmanf'
 91  F
 92  'zMinimuEP'
 93  F
 94  'zvmin'
 95  F
 96  'zprintb'
 97  F
 98  'zcheck'
 99  F
100  'ZPOTCHECK'
101  F
102  'znohamilv'
103  F
104  'zjdqz'
105  F
106  'zmat_'
107  T
108  'zbass'
109  F
110  'ziham'
111  F 10
112  'zihamp1'
113  F 11
114  'zjout'
115  F 500
```

Listing A.11: DR input file for the **HOO** molecule w.r.t. big basis

```
 1  TITLE
 2  SIGMA=1  is  lower  potential
 3  SIGMA=2  is  the  upper  potential
 4  TITLE
 5  'MASSES'
 6  15.99491463,15.99491463,1.00782505
 7  LAMBDA        1
 8  MULTI         1
 9  XSO          −160.1D00
10  'znorenner'
11  F
12  'zstartfrom_J'
```

```
13  F
14  'zopposite'
15  F
16  'zabbf,zabb'
17  T
18  T
19  NSP          6
20  Ntau         8
21  RE1(bohr)  2.5390409D00
22  DISS1 (h )1.75D-01
23  WE1     (h )4.000D-3
24  RE2(bohr)  2.4840281D00
25  DISS2 (h )1.5D+01
26  WE2     (h )1.25D-02
27  MAXJ        45
28  MAXV2A      35
29  MAXV2B      35
30  NPNT1       30
31  NMAX1       25
32  NPNT2       35
33  NMAX2       15
34  NSTINT    8000
35  NSERIN     200
36  NSERP        5
37  NSERQ       20
38  NSTNIN    2000
39  NPNTB      140
40  NSPB        50
41  THRSH1      0.50000000D+04
42  THRSH2      1.00000000D-05
43  THRSH3      1.00000000D+05
44  THRSH4      1.00000000D+03
45  THRSH5      1.00000000D-15
46  THRSH6      1.00000000D+10
47  THRSH7      1.00000000D+10
48  THRSH8      1.00000000D+10
49  CONTMAX     2.5D4
50  'Vmin'
51  1.0D+03
52  'Vmax'
53  4D+05
54  'adif'
55  2.0D+02
56  'bdif'
57  1.1D+00
58  'cdif'
59  1.0D+03
60  'znumpot'
61  F
62  'zanaepot'
63  T
64  'zpoteq'
```

```
65    F
66    'zsingle'
67    F
68    'RHOMAX,PNM1'
69    2.00D+00,1.0D-2
70    'zhamilvv'
71    T
72    'zmgCNharmf'
73    F
74    'zHOO_f'
75    T
76    'zHOOBOWMAN_f'
77    F
78    'zmgCNf'
79    F
80    'zmgNCf'
81    F
82    'zMORBID'
83    F
84    'zmgNCCNf'
85    F
86    'zmgNCCNMIDDLE'
87    F
88    'zHCNf'
89    F
90    'zHCNbowmanf'
91    F
92    'zMinimuEP'
93    F
94    'zvmin'
95    F
96    'zprintb'
97    F
98    'zcheck'
99    F
100   'ZPOTCHECK'
101   F
102   'znohamilv'
103   T
104   'zjdqz'
105   F
106   'zmat_'
107   F
108   'zbass'
109   F
110   'ziham'
111   F 10
112   'zihamp1'
113   F 11
114   'zjout'
115   F 500
```

Bibliography

[1] BLAS (Basic Linear Algebra Subroutines) [online]. Available from: `http://www.netlib.org/blas/` [cited 09 October 2008]. 30, 43, 45, 210, 244, 245, 249, 257, 260, 262, 264, 291

[2] LAPACK (Linear Algebra Package) [online]. Available from: `http://www.netlib.org/lapack/` [cited 04 October 2008]. 3, 5, 9, 26, 40, 42, 43, 44, 45, 47, 48, 51, 55, 56, 57, 70, 210, 211, 257, 260, 261, 262, 264, 291

[3] Official website for OpenMP™ [online]. Available from: `http://www.openmp.org` [cited 09 October 2008]. 185, 244, 249, 257, 261

[4] *ISO 9:1995. Information and documentation - Transliteration of Cyrillic characters into Latin characters - Slavic and non-Slavic languages*, pages 230–245. Bibliotheks- und Dokumentationswesen. Beuth, Berlin, 2002, ISBN 3-410-15311-X. 258, 259, 286

[5] Official website for MATLAB® [online]. 2009. Available from: `http://www.mathworks.com/products/matlab/` [cited 14 March 2009]. 8, 115, 116, 257

[6] AHLFORS, L. V. *Complex Analysis*. McGraw Hill Higher Education, 3rd edition, 1978, ISBN 978-0070006577. 14

[7] ANDERSON, E., BAI, Z., BISCHOF, C., BLACKFORD, S., DEMMEL, J., DONGARRA, J., CROZ, J. D., GREENBAUM, A., HAMMARLING, S., MCKENNEY, A., OSTROUCHOV, S., AND SORENSEN, D. *LAPACK Users' Guide*. SIAM, Philadelphia, Third edition, 1999, ISBN 0-89871-294-7. 3, 40, 42, 43, 48, 51, 55, 56, 57, 70, 210, 257, 260

[8] BAI, Z., DEMMEL, J., DONGARRA, J., RUHE, A., AND VAN DER VORST, H., Editors. *Templates for the Solution of Algebraic Eigenvalue Problems: A Practical Guide*. SIAM, Philadelphia, 2000, ISBN 0-89871-471-0. Available from: `http://www.cs.ucdavis.edu/~bai/ET/contents.html` [cited 12 October 2008]. 4, 58, 71, 74, 84, 92

[9] BARRETT, R., BERRY, M., CHAN, T. F., DEMMEL, J., DONATO, J., DONGARRA, J., EIJKHOUT, V., POZO, R., ROMINE, C., AND VAN DER VORST, H. A. *Templates for the Solution of Linear Systems: Building Blocks for Iterative Methods*. SIAM, Philadelphia, 2nd edition, 1994, ISBN 978-0898713282. Available from: `http://www.netlib.org/linalg/html_templates/Templates.html`. 85, 87, 232

[10] BEHNKE, H., AND GOERISCH, F. Inclusions for Eigenvalues of Selfadjoint Problems. In HERZBERGER, J., Editor, *Topics in Validated Computations*, volume 5 of *Studies in Computational Mathematics*, pages 277–323. Elsevier, 1994, ISBN 978-0444816856. Proceedings of the IMACS-GAMM International Workshop on Validated Computation, Oldenburg, Germany, 30 August - 3 September 1993. 151

[11] BENZI, M. Preconditioning Techniques for Large Linear Systems: A Survey. *Journal of Computational Physics*, 182:418–477, 2002. doi:10.1006/jcph.2002. 7176. 89, 213, 214, 232

[12] BISCHOF, C. H., LANG, B., AND SUN, X. A Framework for Symmetric Band Reduction. *ACM Transactions on Mathematical Software*, 26(4):581–601, December 2000. doi:10.1145/365723.365735. 5, 44, 45, 257

[13] _____. Algorithm 807: The SBR Toolbox – Software for Successive Band Reduction. *ACM Transactions on Mathematical Software*, 26(4):602–616, December 2000. doi:10.1145/365723.365736. 5, 44, 55, 56, 209, 210, 261

[14] BJÖRCK, Å. Numerics of Gram-Schmidt Orthogonalization. *Linear Algebra and Its Applications*, 197/198:297–316, 1994. doi:10.1016/0024-3795(94)90493-6. 30

[15] BOLLHÖFER, M. ILUPACK V2.1, software package providing multilevel ILU preconditioners [online]. 2006. Available from: http://www-public.tu-bs.de/~bolle/ilupack/ [cited 09 October 2008]. 8, 116, 215, 232, 234, 261

[16] BOLLHÖFER, M., AND NOTAY, Y. JADAMILU, software package for the Jacobi-Davidson method incorporating multilevel ILU preconditioning [online]. 2006. Available from: http://homepages.ulb.ac.be/~jadamilu/ [cited 09 October 2008]. 115, 235

[17] _____. JADAMILU: a software code for computing selected eigenvalues of large sparse symmetric matrices. *Computer Physics Communications*, 177(12):951–964, 2007. Available from: http://mntek3.ulb.ac.be/pub/docs/reports/pdf/2007_CPC.pdf [cited 19 October 2008], doi:10.1016/j.cpc.2007.08.004. 115, 235

[18] BRAMLEY, M. J., AND CARRINGTON, JR., T. A general discrete variable method to calculate vibrational energy levels of three- and four-atom molecules. *Journal of Chemical Physics*, 99(11):8519–8541, 1993. doi:10.1063/1.465576. 6, 151, 155

[19] _____. Calculation of triatomic vibrational eigenstates: Product or contracted basis sets, Lanczos or conventional eigensolvers? What is the most efficient combination? *Journal of Chemical Physics*, 101(10):8494–8507, November 1994. doi:10.1063/1.468110. 4, 6, 151, 152, 156, 157

[20] BUNKER, P. R., AND JENSEN, P. *Computational Molecular Spectroscopy*. John Wiley & Sons, LTD, 1st edition, 2001, ISBN 978-0471489986. 151, 152

[21] ――――. *Molecular Symmetry and Spectroscopy*. NRC Research Press (Canada), 2nd edition, 2008, ISBN 978-0660175195. 123, 140, 142, 145, 152, 161, 163, 167

[22] CHANDRA, R., MENON, R., DAGUM, L., KOHR, D., MAYDAN, D., AND MC-DONALD, J. *Parallel Programming in OpenMP^{TM}*. Morgan Kaufmann, 2000, ISBN 1558606718. 185, 244, 261

[23] CHATELIN, F. *Spectral approximation of linear operators*. Computer Science and Applied Mathematics. Academic Press, 1983, ISBN 978-0121706203. 150

[24] CONWAY, J. B. *A Course in Functional Analysis*. Springer, Berlin, 4th edition, 1997, ISBN 978-0387972459. 124, 125, 127

[25] CROUZEIX, M., PHILIPPE, B., AND SADKANE, M. The Davidson method. *SIAM Journal on Scientific Computing*, 15(1):62–76, 1994. doi:10.1137/0915004. 76

[26] CULLUM, J. K., AND WILLOUGHBY, R. A. *Programs*, volume 2 of *Lanczos Algorithms for Large Symmetric Eigenvalue Computations*. Birkhäuser, Boston, 1985, ISBN 978-0817630584. 71

[27] ――――. *Theory*, volume 1 of *Lanczos Algorithms for Large Symmetric Eigenvalue Computations*. Birkhäuser, Boston, 1985, ISBN 978-0817630584. 71

[28] CUPPEN, J. J. M. A divide and conquer method for the symmetric eigenproblem. *Numerische Mathematik*, 36:177–195, 1981. doi:10.1007/BF01396757. 49

[29] DAVIDSON, E. R. The iterative calculation of a few of the lowest eigenvalues and corresponding eigenvectors of large real symmetric matrices. *Journal of Computational Physics*, 17(1):87–94, 1975. doi:10.1016/0021-9991(75)90065-0. 6, 70, 75

[30] ――――. Monster matrices: Their eigenvalues and eigenvectors. *Computers in Physics*, 7(5):519–522, Sep/Oct 1993. 6, 76

[31] DEMMEL, J., AND VESELIĆ, K. Jacobi's method is more accurate than QR. *SIAM Journal on Matrix Analysis and Applications*, 13(4):1204–1245, 1992. doi:10.1137/0613074. 54, 56

[32] DHILLON, I. S. *A new $\mathcal{O}(n^2)$ Algorithm for the Symmetric Tridiagonal Eigenvalue / Eigenvector Problem*. PhD thesis, University of California, Berkeley, 1997 [cited 12 October 2008]. Available from: http://www.cs.utexas.edu/users/inderjit/public_papers/thesis.pdf [cited 12 October 2008]. 5, 6, 51, 56

[33] DHILLON, I. S., AND PARLETT, B. N. Multiple representations to compute orthogonal eigenvectors of symmetric tridiagonal matrices. *Linear Algebra and its Applications*, 387:1–28, 2004. Available from: http://www.cs.utexas.edu/users/inderjit/public_papers/reptree.pdf, doi:10.1016/j.laa.2003.12.028. 51

[34] ──────. Orthogonal eigenvectors and relative gaps. *SIAM Journal on Matrix Analysis and Applications*, 25(3):858–899, 2004. Available from: http://www.cs.utexas.edu/users/inderjit/public_papers/relgaps.pdf, doi:10.1137/S0895479800370111. 51

[35] DONGARRA, J. J., HAMMARLING, S., AND SORENSEN, D. C. Block reduction of matrices to condensed forms for eigenvalue computations. *Journal of Computational and Applied Mathematics*, 27(1–2):215–227, 1989. (LAPACK Working Note #2). doi:10.1016/0377-0427(89)90367-1. 44

[36] FENG, S., AND JIA, Z. A Refined Jacobi-Davidson Method and Its Correction Equation. *Computers and Mathematics with Applications*, 49:417–427, 2005. doi:10.1016/j.camwa.2003.01.018. 70, 81, 104

[37] FOKKEMA, D. R., SLEIJPEN, G. L. G., AND VAN DER VORST, H. A. Jacobi-Davidson style QR and QZ algorithms for the partial reduction of matrix pencils. *SIAM journal on Scientific Computing*, 20:94–125, 1998. Available from: http://www.math.uu.nl/people/sleijpen/Reprints/SISC2098.ps.gz [cited 19 October 2008], doi:10.1137/S1064827596300073. 90, 94, 95, 97, 98, 102

[38] FRANCIS, J. G. F. The QR transformation, A Unitary Analogue to the LR Transformation – Part 1. *The Computer Journal*, 4(3):265–271, 1961. doi:10.1093/comjnl/4.3.265. 45

[39] ──────. The QR transformation – Part 2. *The Computer Journal*, 4(4):332–345, 1962. doi:10.1093/comjnl/4.4.332. 45

[40] FREUND, R. W., AND NACHTIGAL, N. M. QMR: A quasi-minimal residual method for non-hermitian linear systems. *Numerical Mathematics*, (60):315–339, 1991. doi:10.1007/BF01385726. 88

[41] ──────. Software for simplified Lanczos and QMR algorithms. *Applied Numerical Mathematics*, 3(19):319–341, 1994. Special issue on iterative methods for linear equations (Atlanta, GA, 1994). doi:10.1016/0168-9274(95)00089-5. 88

[42] FROMMER, A. Iterationsverfahren. Lecture notes, Bergische Universität Wuppertal, Fachgruppe Mathematik, 2003 [cited 14 October 2008]. Available from: http://www.math.uni-wuppertal.de/~frommer/manuscripts/iterationen.ps.gz [cited 14 October 2008]. 87

[43] _____. Algorithmen auf Graphen und Dünn besetzte Matrizen. Lecture notes, Bergische Universität Wuppertal, Fachgruppe Mathematik, 2004 [cited 14 October 2008]. Available from: http://www.math.uni-wuppertal.de/~frommer/manuscripts/DBM.ps.gz [cited 14 October 2008]. 116, 191

[44] GENSEBERGER, M. *Domain decomposition in the Jacobi-Davidson method for eigenproblems.* PhD thesis, Universiteit Utrecht, 2001, ISBN 90-6196-507-1. Available from: http://homepages.cwi.nl/~genseber/Thesis_M_Genseberger.pdf.gz. 6, 82, 111, 116, 214

[45] GEUS, R. *The Jacobi-Davidson algorithm for solving large sparse symmetric eigenvalue problems with application to the design of accelerator cavities.* PhD thesis, Eidgenössische Technische Hochschule Zürich, 2002 [cited 09 October 2008]. Available from: http://e-collection.ethbib.ethz.ch/view/eth:26147 [cited 09 October 2008]. 6, 116

[46] _____. JDBSYM package for the Jacobi-Davidson method, library in the C programming language, this reference is part of [54] [online]. 2002. Available from: http://www.win.tue.nl/casa/research/topics/jd/software.html [cited 09 October 2008]. 8, 116, 260

[47] GOEKE, A. *Parallele Teilraumverfahren zur Bestimmung von Eigenwerten im Inneren des Spektrums.* PhD thesis, Zentralinstitut für Angewandte Mathematik, Forschungszentrum Jülich, 2000 [cited 13. October 2008]. Available from: http://www.fz-juelich.de/jsc/docs/autoren2000/goeke [cited 13. October 2008]. 62, 63, 70, 116, 213, 244

[48] GOLUB, G. H., AND VAN DER VORST, H. A. Eigenvalue computation in the 20th century. *Journal of Computational Applied Mathematics*, 123(1–2):35–65, 2000. doi:10.1016/S0377-0427(00)00413-1. 4

[49] GOLUB, G. H., AND VAN LOAN, C. F. *Matrix Computations.* Johns Hopkins Studies in the Mathematical Sciences. The Johns Hopkins University Press, Baltimore, 1996, ISBN 978-0801854149. 3, 26, 27, 30, 34, 36, 37, 38, 42, 46, 50, 53, 54, 71, 72, 242

[50] GREENBAUM, A. *Iterative Methods for Solving Linear Systems*, volume 17 of *Frontiers in Applied Mathematics*. SIAM, Philadelphia, 1997, ISBN 978-0898713961. 85, 87, 89, 116

[51] HEHRE, W. J., RADOM, L., VON SCHLEYER, P. R., AND POPLE, J. A. *Ab Initio Molecular Orbital Theory.* John Wiley & Sons, New York, 2nd edition, 1986, ISBN 978-0471812418. 140

[52] HERNÁNDEZ, V., ROMÁN, J. E., TOMÁS, A., AND VIDAL, V. A Survey of Software for Sparse Eigenvalue Problems [online]. December 2005. Available from: http://www.grycap.upv.es/slepc [cited 09 October 2008]. 58, 76

[53] HISLOP, P. D., AND SIGAL, I. M. *Introduction to Spectral Theory. With Applications to Schrödinger Operators*, volume 113 of *Applied Mathematical Sciences*. Springer, Berlin, 1st edition, 1995, ISBN 978-0387945019. 124, 127, 129, 131, 132, 138

[54] HOCHSTENBACH, M. Jacobi-Davidson Gateway [online]. Available from: http://www.win.tue.nl/casa/research/topics/jd/ [cited 19 October 2008]. 7, 115

[55] HOCHSTENBACH, M. E., AND NOTAY, Y. The Jacobi–Davidson method. *GAMM Mitteilungen*, 29(2):368–382, 2006. Available from: http://www.win.tue.nl/~hochsten/pdf/jdgamm.pdf [cited 13. October 2008]. 84

[56] HORN, R. A., AND JOHNSON, C. R. *Matrix analysis*. Cambridge University Press, Reprint edition, 1990, ISBN 978-0521386326. 15, 21

[57] HYLLERAAS, E. A., AND UNDHEIM, B. Numerische Berechnung der $2S$-Terme von Ortho- und Par-Helium. *Zeitschrift für Physik*, 65(11–12):759–772, 1930. doi:10.1007/BF01397263. 149

[58] JACOBI, C. G. J. Ueber ein leichtes Verfahren, die in der Theorie der Säcularstörungen vorkommenden Gleichungen numerisch aufzulösen. *Journal für die reine und angewandte Mathematik*, 30:51–94, 1846. Available from: http://www.digizeitschriften.de/resolveppn/GDZPPN002144522 [cited 13. October 2008]. 52, 77

[59] JENSEN, P. The Nonrigid Bender Hamiltonian for Calculating the Rotation-Vibration Energy Levels of a Triatomic Molecule. *Computer Physics Reports*, 1(1):1–55, 1983. doi:10.1016/0167-7977(83)90003-5. 152

[60] ———. A new morse oscillator-rigid bender internal dynamics (MORBID) Hamiltonian for triatomic molecules. *Journal of Molecular Spectroscopy*, 128(2):478–501, 1988. doi:10.1016/0022-2852(88)90164-6. 152

[61] ———. Theoretische Spektroskopie. Lecture notes, Bergische Universität Wuppertal, Fachgruppe Chemie, 1995 [cited 09 October 2008]. Available from: http://elpub.bib.uni-wuppertal.de/edocs/dokumente/fb09/vorlesung/jensen/v090003.pdf [cited 09 October 2008]. 121, 123, 142, 145, 152, 161

[62] ———. Einführung in die Methoden der Quantenchemie. Lecture notes, Bergische Universität Wuppertal, Fachgruppe Chemie, 1997 [cited 09 October 2008]. Available from: http://elpub.bib.uni-wuppertal.de/edocs/dokumente/fb09/vorlesung/jensen/v090001.pdf [cited 09 October 2008]. 140

[63] JIA, Z. Refined iterative algorithms based on Arnoldi's process for large unsymmetric eigenproblems. *Linear Algebra and Its Applications*, 259(1):1–23, 1997. doi:10.1016/S0024-3795(96)00238-8. 70

[64] ———. The Convergence of Harmonic Ritz Values, Harmonic Ritz Vectors and Refined Harmonic Ritz Vectors. *Mathematics of Computation*, 74(251):1441–1456, 2004. `doi:10.1090/S0025-5718-04-01684-9`. 59, 68

[65] JIA, Z., AND STEWART, G. W. An Analysis of the Rayleigh-Ritz Method for Approximating Eigenspaces. *Mathematics of Computation*, 70(234):637–647, 2001. `doi:10.1090/S0025-5718-00-01208-4`. 64

[66] JORDAN, T. F. *Linear Operators for Quantum Mechanics*. John Wiley & Sons Inc., 1969, `ISBN 978-0471450412`. 124

[67] KERNIGHAN, B., AND RITCHIE, D. *The C Programming Language*. Prentice Hall, 2nd edition, 1988, `ISBN 0-13-110362-8`. 8, 9, 116, 257, 260

[68] KLAHN, B., AND BINGEL, W. A. The Convergence of the Rayleigh-Ritz Method in Quantum Chemistry. *Theoretica Chimica Acta*, 44(1):27–43, 1977. ISSN 1432-881X (Print) 1432-2234 (Online). `doi:10.1007/BF00548027`. 150

[69] KUBLANOVSKAYA, V. N. On some algorithms for the solution of the complete eigenvalue problem. *Zhurnal Vychislitelnoi Matematiki i Matematicheskoi Fiziki*, 1:637–657, 1962. In Russian. Translation in *USSR Computational Mathematics and Mathematical Physics*. 45

[70] LANCZOS, C. An iteration method for the solution of the eigenvalue problem of linear differential and integral operators. *Journal of Research of the National Bureau of Standards*, 45(4):255–282, 1950. Available from: `http://nvl.nist.gov/pub/nistpubs/jres/045/4/V45.N04.A01.pdf` [cited 13. October 2008]. 72

[71] LANG, S. *Linear Algebra*. Undergraduate Texts in Mathematics. Springer, Berlin, 3rd edition, 2004, `ISBN 978-0387964126`. 13, 14, 16, 17

[72] ———. *Algebra*. Graduate Texts in Mathematics. Springer, Berlin, 3rd revised edition, 2005, `ISBN 978-0387953854`. 41

[73] LEHOUCQ, R. B. *Analysis and Implementation of an Implicitly Restarted Iteration*. PhD thesis, Rice University, Houston, Texas, 1995. Also available as Technical report TR95-13, Department of Computational and Applied Mathematics. 72

[74] ———. The Computation of Elementary Unitary Matrices. *ACM Transactions on Mathematical Software*, 22(4):393–400, 1996. `doi:10.1145/235815.235817`. 26

[75] LEHOUCQ, R. B., MASCHHOFF, K., SORENSEN, D. C., AND YANG, C. ARPACK Software [online]. 1996. Available from: `http://www.caam.rice.edu/software/ARPACK/` [cited 09 October 2008]. 8, 74, 238

[76] LEHOUCQ, R. B., SORENSEN, D. C., AND YANG, C. *ARPACK users' guide: Solution of Large-Scale Eigenvalue Problems with Implicitly Restarted Arnoldi Methods (Software, Environment, Tools)*. SIAM, Philadelphia, 1998, ISBN 978-0898714074. 74, 96

[77] LORD RAYLEIGH (J. W. STRUTT). On the calculation of the frequency of vibration of a system in its gravest mode, with an example from hydrodynamics. *The Philosophical Magazine*, 47:556–572, 1899. 62

[78] MACDONALD, J. K. L. Successive Approximations by the Rayleigh-Ritz Variation Method. *Physical Review*, 43(10):830–833, March 1933. doi:10.1103/PhysRev.43.830. 149

[79] MARCUS, M. *Finite Dimensional Multilinear Algebra, Part I*. Pure and Applied Mathematics. Marcel Dekker, Inc. New York, 1973. 35

[80] MEISE, R., AND VOGT, D. *Einführung in die Funktionalanalysis*. Vieweg Studium. Aufbaukurs Mathematik. Vieweg, Wiesbaden, 1992, ISBN 978-3528072629. 124, 125, 126

[81] MORGAN, R. B. Computing Interior Eigenvalues of Large Matrices. *Linear Algebra and its Applications*, 154/156:289–309, 1991. doi:10.1016/0024-3795(91)90381-6. 68

[82] NOTAY, Y. Combination of Jacobi-Davidson and conjugate gradients for the partial symmetric eigenproblem. *Numerical Linear Algebra and its Applications*, 9:21–44, 2002. Available from: http://mntek3.ulb.ac.be/pub/docs/reports/pdf/2002_NLAA_1.pdf [cited 19 October 2008], doi:10.1002/nla.246. 116

[83] ———. MATLAB® code for the JDCG algorithm [online]. 2002. Available from: http://mntek3.ulb.ac.be/pub/docs/jdcg/ [cited 09 October 2008]. 116

[84] ———. Is Jacobi–Davidson Faster than Davidson? *SIAM Journal on Matrix Analysis and Applications*, 26(2):522–543, 2005. Available from: http://mntek3.ulb.ac.be/pub/docs/reports/pdf/2005_SIMAX.pdf [cited 19 October 2008], doi:10.1137/S0895479803430941. 83

[85] NYHOFF, L., AND LEETSMA, S. *FORTRAN 77 for Engineers and Scientists with an Introduction to Fortran 90*. Prentice Hall, 4th edition, 1995, ISBN 0-13-363003-X. 5, 8, 115, 159, 257, 260, 261, 291

[86] ODAKA, T. E. *The Double Renner Effect*. PhD thesis, Bergische Universität Wuppertal, 2003 [cited 09 October 2008]. Available from: http://elpub.bib.uni-wuppertal.de/edocs/dokumente/fb09/diss2003/odaka/index.html [cited 09 October 2008]. 1, 5, 159, 160, 162, 167, 174, 175, 176, 177, 179, 181, 185, 186, 203

[87] ODAKA, T. E., MELNIKOV, V. V., JENSEN, P., HIRANO, T., LANG, B., AND LANGER, P. A Theoretical Study of the Double Renner Effect for $\tilde{A}^2\Pi$ MgNC/MgCN: Higher Excited Rotational States. *Journal of Chemical Physics*, 126(9):094301 (9 pages), 2007. `doi:10.1063/1.2464094`. 5, 7, 9, 45, 56, 155, 164, 203

[88] OLSEN, J., JØRGENSEN, P., AND SIMONS, J. Passing the one-billion limit in full configuration-interaction (FCI) calculations. *Chemical Physics Letters*, 169(6):463–472, 1990. `doi:10.1016/0009-2614(90)85633-N`. 83

[89] PAIGE, C. C., PARLETT, B. N., AND VAN DER VORST, H. A. Approximate Solutions and Eigenvalue Bounds from Krylov Subspaces. *Numerical Linear Algebra with Applications*, 2(2):115–133, 1995. `doi:10.1002/nla.1680020205`. 66, 68

[90] PAIGE, C. C., AND SAUNDERS, M. A. Solution of Sparse Indefinite Systems of Linear Equations. *SIAM Journal on Numerical Analysis*, 12(4):617–629, 1975. `doi:10.1137/0712047`. 87

[91] PARLETT, B. *The Symmetric Eigenvalue Problem*. Classics in Applied Mathematics. Cambridge University Press, reprint edition, 1998, ISBN 978-0898714029. 3, 4, 14, 23, 25, 34, 37, 40, 42, 46, 62, 71, 72, 81

[92] POIRIER, B., AND CARRINGTON, JR., T. Accelerating the calculation of energy levels and wave functions using an efficient preconditioner with the inexact spectral transform method. *Journal of Chemical Physics*, 114(21):9254–9264, 2001. `doi:10.1063/1.1367396`. 6, 216

[93] PUTTIN, R. Modulare und parallele Implementierung des Jacobi-Davidson-Verfahrens. Master's thesis, Forschungszentrum Jülich, Zentralinstitut für Angewandte Mathematik, December 2005. Available from: `http://www.fz-juelich.de/zam/docs/autoren2005/puttin` [cited 09 October 2008]. 112

[94] QUATERONI, A., AND SALERI, F. *Scientific Computing with MATLAB® and Octave*. Springer, 2006, ISBN 978-3-540-32612-0. 8

[95] QUINTUS HORATIUS FLACCUS. Epistulae. In SHACKLETON BAILEY, D. R., Editor, *Quinti Horatii Flacci Opera*, Bibliotheca Scriptorum Graecorum Et Romanorum Teubneriana, Stutgardiae, 2008. de Gruyter, ISBN 978-3110202922. 1

[96] REED, M., AND SIMON, B. *Functional Analysis*, volume I of *Methods of Modern Mathematical Physics*. Academic Press, 1972, ISBN 978-0125850506. 124, 127, 130, 135

[97] ———. *Fourier Analysis, Self-Adjointness*, volume II of *Methods of Modern Mathematical Physics*. Academic Press, 1975, ISBN 978-0125850025. 124, 128, 129, 131, 138

[98] RENNER, R. Zur Theorie der Wechselwirkung zwischen Elektronen- und Kern-bewegung bei dreiatomigen, stabförmigen Molekülen. *Zeitschrift für Physik – A Hadrons and Nuclei*, 92(3):172–193, 1934. doi:10.1007/BF01350054. 159

[99] RITZ, W. Über eine neue Methode zur Lösung gewisser Variationsprobleme der mathematischen Physik. *Journal für die reine und angewandte Mathematik*, 135:1–61, 1909. Available from: http://www.digizeitschriften.de/resolveppn/GDZPPN002166739 [cited 13. October 2008]. 62

[100] SAAD, Y. *Numerical methods for large eigenvalue problems.* Algorithms and Architectures for Advanced Scientific Computing. Manchester University Press, 1991, ISBN 978-0719033865. 4, 15, 19, 23, 37, 58, 72

[101] _____. *Iterative Methods for Sparse Linear Systems.* Cambridge University Press, 2003, ISBN 978-0898715347. 85, 89, 116, 191, 213, 232

[102] SAAD, Y., AND SCHULTZ, M. H. GMRES: A generalized minimal residual algorithm for solving nonsymmetric linear systems. *SIAM Journal on Scientific and Statistical Computing*, 7(3):856–869, 1986. doi:10.1137/0907058. 86

[103] SADKANE, M. *Analyse Numérique de la Méthode de Davidson.* PhD thesis, UER mathématiques et Informatique, Rennes, France, 1989. 76

[104] SARKAR, P., POULIN, N., AND TUCKER CARRINGTON, J. Calculating rovibrational energy levels of a triatomic molecule with a simple lanczos method. *Journal of Chemical Physics*, 110(21):10269–10274, 1999. doi:10.1063/1.478960. 6, 151

[105] SCHRÖDINGER, E. Quantisierung als Eigenwertproblem I. *Annalen der Physik*, 384(4):361 – 376, 1926. doi:10.1002/andp.19263840404. 2

[106] _____. Quantisierung als Eigenwertproblem II. *Annalen der Physik*, 384(6):489 – 527, 1926. doi:10.1002/andp.19263840602. 2

[107] _____. Quantisierung als Eigenwertproblem III. *Annalen der Physik*, 385(13):437 – 490, 1926. doi:10.1002/andp.19263851302. 2

[108] _____. Quantisierung als Eigenwertproblem IV. *Annalen der Physik*, 386(18):109 – 139, 1926. doi:10.1002/andp.19263861802. 2

[109] SCHWABL, F. *Quantum Mechanics.* Advanced Texts in Physics. Springer, 3rd edition, 2002, ISBN 978-3540431091. 1, 122, 123, 137, 138, 140

[110] SCOTT, D. S. The Advantages of Inverted Operators in Rayleigh-Ritz Approximations. *SIAM Journal on Scientific and Statistical Computing*, 3(1):68–75, 1982. doi:10.1137/0903006. 59, 63

[111] SHARMA, C. S., AND SRIRANKANATHAN, S. A coordinate-free treatment of the minimax and maximini theorems for eigenvalues of self-adjoint operators on a Hilbert space. *Journal of Physics A: Mathematical and General*, 8:1853–1862, 1975. doi:10.1088/0305-4470/8/12/002. 150

[112] SLEIJPEN, G. L. G. MATLAB® code for the JDQR algorithm [online]. 2000. Available from: http://www.math.uu.nl/people/sleijpen/JD_software/JDQR.html [cited 09 October 2008]. 8, 115, 116, 232, 260

[113] SLEIJPEN, G. L. G., AND VAN DEN ESHOF, J. On the use of harmonic Ritz pairs in approximating internal eigenpairs. *Linear algebra and its Applications*, 358(1-3):115–137, 2003. doi:10.1016/S0024-3795(01)00480-3. 68

[114] SLEIJPEN, G. L. G., AND VAN DER VORST, H. A. A Jacobi-Davidson method for linear eigenvalue problems. *SIAM Journal on Matrix Analysis and Applications*, 17(2):401–425, 1996. Available from: http://www.math.uu.nl/people/sleijpen/Reprints/SIREV4200.ps.gz [cited 19 October 2008], doi:10.1137/S0036144599363084. 6, 64, 68, 77, 78, 79, 82, 84, 111

[115] ———. The Jacobi-Davidson method for eigenvalue problems and its relation with accelerated inexact Newton schemes. In MARGENOV, S. D., AND VASSILEVSKI, P. S., Editors, *Iterative methods in Linear Algebra, II.*, volume 3 of *IMACS Series in Computational and Applied Mathematics*, pages 377–389, New Brunswick, NJ, U.S.A., 1996. IMACS. Proceedings of the Second IMACS International Symposium on Iterative Methods in Linear Algebra, June 17-20, 1995, Blagoevgrad. Available from: http://www.math.uu.nl/people/sleijpen/JDaNEWTON.ps.gz [cited 18 October 2008]. 93

[116] SLEIJPEN, G. L. G., VAN DER VORST, H. A., AND MEIJERINK, E. Efficient expansion of subspaces in the Jacobi-Davidson method for standard and generalized eigenproblems. *Electronic Transactions on Numerical Analysis*, 7:75–89, 1998. Available from: http://etna.mcs.kent.edu/vol.7.1998/pp75-89.dir/pp75-89.html [cited 13. October 2008]. 214

[117] SORENSEN, D. C. Implicit application of polynomial filters in a k-step Arnoldi method. *SIAM Journal on Matrix Analysis and Applications*, 13(1):357–385, 1992. doi:10.1137/0613025. 72, 74

[118] STATHOPOULOS, A. Nearly Optimal Preconditioned Methods for Hermitian Eigenproblems under Limited Memory. Part I: Seeking One Eigenvalue. *SIAM Journal on Scientific Computing*, 29(2):481–514, 2007. doi:10.1137/050631574. 115

[119] STATHOPOULOS, A., AND MCCOMBS, J. R. PRIMME [online]. Available from: http://www.cs.wm.edu/~andreas/software/ [cited 09 October 2008]. 115

[120] _____ . Nearly Optimal Preconditioned Methods for Hermitian Eigenproblems under Limited Memory. Part II: Seeking many Eigenvalues. *SIAM Journal on Scientific Computing*, 29(5):2162–2188, 2007. doi:10.1137/060661910. 115

[121] STEWART, G. W. *Basic Decompositions*, volume I of *Matrix Algorithms*. SIAM, Philadelphia, 1998, ISBN 978-0898714142. 3

[122] _____ . *Eigensystems*, volume II of *Matrix Algorithms*. SIAM, Philadelphia, 2001, ISBN 978-0898715033. 3, 42, 46, 50, 59, 64, 68, 70, 94

[123] VAN DEN ESHOF, J. The convergence of Jacobi-Davidson iterations for hermitian eigenproblems. *Numerical Linear Algebra With Applications*, 9(2):163–179, 2002. doi:10.1002/nla.266. 93

[124] VAN DER SLUIS, A., AND VAN DER VORST, H. A. The Convergence Behaviour of Ritz Values in the Presence of Close Eigenvalues. *Linear Algebra and Its Applications*, 88/89:651–694, 1987. doi:10.1016/0024-3795(87)90129-7. 64

[125] VAN DER VORST, H. A. Computational methods for large eigenvalue problems. In CIARLET, P., AND LIONS, J., Editors, *Solution of Equations in $\mathbb{R}^n$ (Part 4), Techniques of Scientific Computing (Part 4), Numerical Methods for Fluids (Part 2)*, volume VIII of *Handbook of Numerical Analysis*, pages 3–179, Amsterdam, 2002. North-Holland (Elsevier), ISBN 978-0444509062. Available from: http://www.math.uu.nl/people/vorst/lecture.html [cited 09 October 2008]. 4, 58

[126] VAN LOAN, C. F. *Computational Frameworks for the Fast Fourier Transform*, volume 10 of *Frontiers in applied mathematics*. SIAM, Philadelphia, 1992, ISBN 978-0898712858. 36

[127] _____ . The Ubiquitous Kronecker Product. *Journal of Computational and Applied Mathematics*, 123:85–100, 2000. doi:10.1016/S0377-0427(00)00393-9. 253

[128] VON NEUMANN, J. *Mathematical Foundations of Quantum Mechanics*. Princeton University Press, Reprint edition, 1996, ISBN 978-0691028934. 124

[129] WANG, X.-G., AND CARRINGTON, JR., T. New ideas for using contracted basis functions with a Lanczos eigensolver for computing vibrational spectra of molecules with four or more atoms. *Journal of Chemical Physics*, 117(15):6923–6934, 2002. doi:10.1063/1.1506911. 6, 151, 156

[130] _____ . A contracted basis-lanczos calculation of vibrational levels of methane: Solving the Schrödinger equation in nine dimensions. *Journal of Chemical Physics*, 119(1):101–117, 2003. doi:10.1063/1.1574016. 6, 151, 156

[131] WERNER, J. *Numerische Mathematik 2*. Vieweg, Wiesbaden, 1st edition, 1991, ISBN 978-3528072339. 14

[132] WILKINSON, J. H. *The Algebraic Eigenvalue Problem.* Monographs on Numerical Analysis. Oxford University Press, new edition, 1988, ISBN 978-0198534181. 48

[133] ZARE, R. N. *Angular Momentum.* Understanding spatial aspects in chemistry and physics. John Wiley & Sons, 1988, ISBN 978-0471858928. 152, 177

List of Tables

List of Figures

List of Algorithms

List of Listings

List of Symbols

Linear Algebra and Numerical Analysis

$\mathrm{diag}(a_1, \ldots, a_p)$ $\mathrm{diag}(a_1, \ldots, a_p) = \begin{bmatrix} a_1 & 0 & \ldots & 0 \\ 0 & a_2 & \ddots & \vdots \\ \vdots & \ddots & \ddots & \vdots \\ 0 & \ldots & 0 & a_p \end{bmatrix}$ (diagonal matrix)

a_{ij} entry of matrix A in the i-th row and the j-th column

$A(i, j)$ alternative notation for a_{ij}

$A(i, :), A(:, j)$ i-th row, resp. j-th column of A

$A([i_1, i_2, i_3], :)$ submatrix consisting of rows i_1, i_2, i_3 of A

$A(:, [j_1, j_2, j_3])$ submatrix consisting of columns j_1, j_2, j_3 of A

$\mathbf{A}^{>\gamma}$ $\mathbf{A}^{>\gamma} = \left\{ \begin{array}{ll} a_{ij} &, \text{ if } |a_{ij}| > \gamma \\ 0 &, \text{ otherwise} \end{array} \right\}$

Q orthogonal matrix

U unitary matrix

v^* complex conjugate transpose of $v \in \mathbb{C}^n$

V^* complex conjugate transpose of $V \in \mathbb{C}^{n \times n}$

$\langle u, v \rangle$ $\langle u, v \rangle = u^* v$, Euclidean scalar product of $u, v \in \mathbb{C}^n$

$\|v\|_2$ $\|v\|_2 = \sqrt{v^* v}$, Euclidean norm of $v \in \mathbb{R}^n$

λ eigenvalue

x eigenvector

(λ, x) exact eigenpair

θ Ritz value

u Ritz vector

(θ, u)	Ritz pair
σ	singular value
$\mathcal{O}(f)$	Big Oh notation, algorithm has order of f
$G(i, k, \theta)$	Givens rotation associated with angle θ and index pair (i, k)
$J(p, q, \theta)$	Jacobi rotation associated with angle θ and index pair (p, q)
$\mathcal{K}, \mathcal{L}$	subspaces of $\mathbb{R}^n$

Quantum Chemistry and Double Renner Effect

$\widehat{H}$	Hamilton operator, Hamiltonian
$\widehat{T}$	Kinetic Energy Operator (KEO)
E	exact eigenvalue of $\widehat{H}$
ψ	exact eigenfunction of $\widehat{H}$
(E, ψ)	exact eigenpair of $\widehat{H}$
$\mathbf{H}$	Finite Basis Representation (FBR) of $\widehat{H}$
$\widetilde{E}$	eigenvalue of $\mathbf{H}$
$\widetilde{c}$	eigenvector of $\mathbf{H}$
$(\widetilde{E}, \widetilde{c})$	eigenpair of $\mathbf{H}$
$\widetilde{\mathbf{H}}$	contracted Hamiltonian matrix $\widetilde{\mathbf{H}} = \mathbf{V^*HV}$
$\widetilde{\widetilde{E}}$	eigenvalue of $\widetilde{\mathbf{H}}$
$\widetilde{\widetilde{c}}$	eigenvector of $\widetilde{\mathbf{H}}$
$(\widetilde{\widetilde{E}}, \widetilde{\widetilde{c}})$	eigenpair of $\widetilde{\mathbf{H}}$
S	spin multiplicity
J	rotational quantum number
N	quantum number related to angular momentum
K	quantum number related to angular momentum
N_R	vibrational quantum number, stretching along the R-bond
N_r	vibrational quantum number, stretching along the r-bond

$v_2^a,\ v_2^b$	vibrational quantum numbers, bending
Γ_{rve}	molecular symmetry
Γ_{vib}	vibrational symmetry
(N', N'')	address of an N-block
ΔK	$\Delta K = |K' - K''|$ (difference between K quantum numbers)
ΔN	$\Delta N = |N' - N''|$ (difference between N quantum numbers)
$(N', K'), (N'', K'')$	address of a K-block
$\mathbf{H}^{(J,S,\Gamma_{rve})}$	J-block associated with the triple (J, S, Γ_{rve})
$\mathbf{H}^{(J,S,\Gamma_{rve})}_{(N',K'),(N'',K'')}$	K-block determined by the address $(N', K'), (N'', K'')$

Functional Analysis

$\mathcal{H}$	Hilbert space
$\mathscr{F}, \mathscr{F}^{-1}$	Fourier transform resp. inverse Fourier transform
$L^2(\mathbb{R}^n)$	Hilbert space of square-integrable functions
$L^\infty(\mathbb{R}^n)$	Banach space of essentially bounded functions
$C^\infty(\mathbb{R}^n)$	space of infinetely often differentiable functions
$\mathcal{S}(\mathbb{R}^n)$	Schwartz space, space of rapidly decreasing functions
$H^2(\mathbb{R}^n)$	Sobolev space of order two
Δ	$\Delta = \sum\limits_{i=1}^{n} \frac{\partial^2}{\partial x_i^2}$ (Laplacian)
A^*	Adjoint operator of A
$D(A)$	Domain of operator A
$E(\lambda)_{\lambda \in \mathbb{R}}$	spectral family
$\sigma(A)$	spectrum of operator A
$\sigma_d(A)$	discrete spectrum of operator A
$\sigma_{\mathrm{ess}}(A)$	essential spectrum of operator A

Index

**A B C D E F G H I J K
L M N O P Q R S T U V
W X Y Z**

T